ESSENTIALS OF MECHANICS
A Unified First Course

ESSENTIALS OF MECHANICS
A Unified First Course

by
DONALD F. YOUNG
WILLIAM F. RILEY
KENNETH G. McCONNELL
THOMAS R. ROGGE

The Iowa State University Press, Ames, Iowa, U.S.A.

Library of Congress Cataloging in Publication Data

Main entry under title:

Essentials of mechanics.

    An expanded version of Introduction to applied mechanics, by D. F. Young.
    1. Mechanics, Applied. I. Young, Donald F.
TA350.E87   620.1   73-22115
ISBN 0-8138-1110-4

© 1974 The Iowa State University Press
Ames, Iowa 50010. All rights reserved.

Printed in the U.S.A.

First edition, 1974

Contents

PART V ENGINEERING MATERIALS AND APPLICATIONS

Preface

Although essentially all engineering curricula require some course work in mechanics, the needs of the various engineering disciplines vary widely. This book is intended as a text for a first course in mechanics for those students who require only a one- or two-semester course. It is the belief of the authors that the needs of these students can best be met by a broad, unified treatment of mechanics rather than an in-depth study of one or two specialized areas. The material contained in this book is an expanded version of Young's *Introduction to Applied Mechanics*. Young's book was written primarily for nonengineering students, whereas the present text is designed primarily for engineering students.

In the selection of specific subject matter, emphasis has been placed on topics which engineers with limited training in mechanics are most likely to find of value. Although the coverage is broad, an in-depth, rigorous treatment of the selected topics has been attempted and the material contained in the book is not intended as simply a "survey of mechanics."

The text is organized into five major parts, each containing one or more chapters, as follows:

Part   I—Review
Part   II—Equilibrium
Part   III—Motion and Deformation
Part   IV—Rigid-Body Dynamics and Vibrations
Part   V—Engineering Materials and Applications

Part I contains material normally found in calculus and physics courses which are presumed as prerequisites for a first course in mechanics. However, it has been the authors' experience that ideas associated with forces and moments and topics such as elementary particle dynamics, centroids, moments of inertia, units, etc. need to be reviewed to reinforce the students' understanding of these elementary but important facets of mechanics.

Part II emphasizes the fundamental nature of equilibrium by considering not only rigid bodies but equilibrium concepts applied to a continuum through the introduction of the concept of stress. Equilibrium of fluids is also treated in this section.

Kinematics of particles, rigid bodies in plane motion, and deformable bodies are included in Part III. The concept of strain is introduced, and the pertinent equations relating to strain transformations and the measurement of strain are included. This section is followed in Part IV with a discussion of rigid body dynamics, with emphasis on plane motion, and an introduction to the vibratory motion of single degree of freedom systems.

In the first four parts of the book material properties play a minimal role. Only the ideas of a rigid body and a fluid at rest are introduced. Thus the student is able to develop general concepts and equations which are applicable to broad classes of problems. In Part V important characteristics of materials are discussed and related to applications in solid and fluid mechanics. The final chapter provides an introduction to dimensional analysis and model theory.

Although the English gravitational system of units (pound-slug-second-foot) is used in the majority of examples the International System of Units (newton-kilogram-second-metre), commonly called SI, is also used in a number of examples. With the increasing usage of SI in engineering disciplines the need for engineers to become familiar with this metric system of units is apparent.

Students are encouraged to carefully study all examples, since the solutions frequently contain important points associated with practical aspects of problem-solving techniques. All data given in examples and problems are assumed to be accurate to at least three significant figures, consistent with the use of the slide rule for obtaining numerical answers. Over 350 problems are included with answers supplied to odd-numbered problems.

This text is organized so that it may be used in a number of ways. It is particularly well suited for a three-quarter (9-hour) or two-semester (6-hour) sequence, and essentially the entire text can be covered in this period of time. In two quarters (6 hours) an integrated course can be developed from the first ten chapters to include the topics of rigid body equilibrium, stress, fluid statics, kinematics, rigid body dynamics, vibrations, strain, materials, and applications in solid mechanics. The material in these chapters has been arranged so that only portions of a given chapter need be covered without loss of continuity with succeeding topics.

The authors gratefully acknowledge the help of many colleagues and students in the development of this book. We especially appreciate the support and encouragement of Dr. H. J. Weiss, Head, Department of Engineering Science and Mechanics at Iowa State University.

Donald F. Young
William F. Riley
Kenneth G. McConnell
Thomas R. Rogge

part I

REVIEW

Introduction to Mechanics

1.1 Scope of Mechanics

When a collection of matter is acted on by a system of forces, this action will in general induce resisting forces, internal stresses, deformation, and motion. The science of mechanics consists of the study and analysis of these factors and their interrelationships.

The scope and range of topics normally considered within the province of mechanics is indeed broad. The following list of topics and descriptive phrases is suggestive, but not all-inclusive, of the numerous classical subdivisions of mechanics:

1. Statics—rigid bodies in equilibrium.
2. Dynamics—rigid bodies in motion.
3. Mechanics of materials—stresses and deformations in solids.
4. Fluid mechanics—behavior of liquids and gases at rest and in motion.
5. Vibrations—periodic and transient motion of machines, structures, and systems.
6. Elasticity—mathematical analysis of stress and deformation in elastic systems.
7. Rheology—flow and deformation of materials.

Applications for each of these topics can be found not only in the fields of engineering and physics but also in agriculture, biology, geology, medicine, oceanography, etc.

Although each of the topics noted above represents a highly developed specialized area in mechanics, principles and concepts common to all provide the necessary framework for the development of the specialty. In this book we will focus on these common ingredients and consider a broad range of topics, each topic logically building on the preceding one. This approach will provide an integrated, overall view of the science of mechanics, with a detailed working knowledge of the aspects that commonly arise in many engineering and interdisciplinary applications.

1.2 Continuum Model

In the analysis of a given problem, regardless of the particular application, certain idealizations are made so that a tractable model can be established. It is hoped that the model used will be a satisfactory representation of the physical system of interest. The type and number of idealizations required depend to a certain extent on what information is desired. An important concept, which may be considered as an idealization of a real system, is the *continuum model.* Although we recognize that matter consists of a collection of discrete particles at the molecular level, in most instances we are concerned with a system whose dimensions are very large in comparison with molecular dimensions. Thus for all practical purposes we may assume that matter is continuously distributed over the region of interest. When such an assumption is made, we are utilizing a continuum model, a basic assumption made throughout this text. However, it should be recognized that as the size of the system of interest approaches molecular dimensions, as may be the case in certain biological problems, the continuum model must be discarded. Alternatively, as the distance between molecules becomes large in comparison with the dimensions of the system, the continuum model is no longer applicable. This situation may be encountered in the high-altitude flight of missiles and spacecraft.

A special type of continuum model is the *rigid body.* As the name implies, a rigid body is one that does not deform under the action of external forces. Since all material will deform under loading to some extent, the rigid body concept is an approximation. However, this idealization is frequently adequate for the analysis of certain types of problems. A *particle* is defined as a small quantity of matter whose dimensions are large compared with molecular dimensions, but negligibly small when compared with the dimensions of the overall system of interest.

1.3 Forces

If an object such as a block (Fig. 1.1a) is placed on a plane surface and we push on it, the block may remain at rest or start to slide along the surface. We say that we are applying a "force" to the block. In addition, the block is exerting a "force" on the supporting surface. Since forces always occur in equal and opposite pairs, the surface also exerts a resultant force on the block (Fig. 1.1b). Thus a *force* can be defined as the action of one body on another. In this

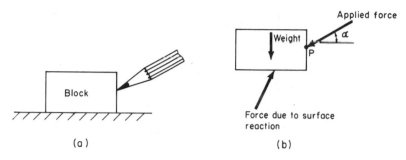

Fig. 1.1

particular example the complete system of interest consists of four "bodies"—the pencil, the block, the object (such as a table) on which the block rests, and the earth that develops the pull of gravity (weight).

The important characteristics of a force are (1) magnitude, (2) direction, and (3) location of line of action. It can also be demonstrated experimentally that the applied force of Fig. 1.1b can be broken down into components in accordance with the parallelogram law (Fig. 1.2). If these components pass through the application point P or any point along the line of action of the force, they would have the same external effect on the body as the original force. The *external effect* of a force either accelerates the body or develops reactions on the body opposing the motion.

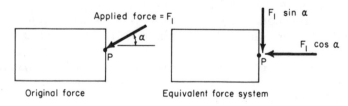

Fig. 1.2

By definition a *vector* is a quantity that has both direction and magnitude and conforms to the parallelogram law of addition. Since these are precisely the characteristics of a force, we arrive at the important conclusion that a force is a vector quantity and must be treated accordingly.

A common requirement in the analysis of elementary mechanics problems is the *resolution* of a force into its components. The force A in Fig. 1.3 can first be resolved in accordance with the parallelogram law into a component along the x axis and a component in the

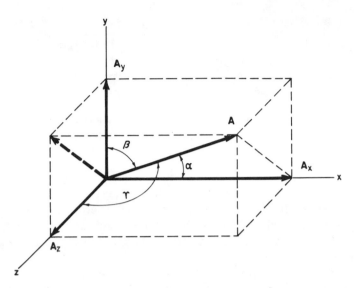

Fig. 1.3

yz plane.[1]   This latter component can then be resolved into components along the y axis and the z axis. By this process the original
force **A** has been resolved into three orthogonal components commonly called *rectangular components*. The magnitudes of the three
rectangular components are:

$$A_x = A \cos \alpha, \quad A_y = A \cos \beta, \quad A_z = A \cos \gamma \qquad (1.1)$$

The use of vector algebra provides a very efficient method for
determining the components of a force. Consider vectors of unit
magnitude that are directed along the positive x, y, and z axes (Fig.
1.4) and defined as the *unit vectors*, **i**, **j**, and **k**. Consider a force **F** in
the xy plane making an angle α with the x axis (Fig. 1.5). We can
immediately obtain the two rectangular components $\mathbf{F}_x$ and $\mathbf{F}_y$ of
the force **F** from the parallelogram law; i.e.,

$$\mathbf{F}_x = F \cos \alpha \, \mathbf{i}, \quad \mathbf{F}_y = F \sin \alpha \, \mathbf{j} \qquad (1.2)$$

where F is the magnitude of the force **F**. The *dot product* of two
vectors **A** and **B** is given by the expression

$$\mathbf{A} \cdot \mathbf{B} = AB \cos \beta \qquad (1.3)$$

where β is the angle between the two vectors (Fig. 1.6). Note that
the dot product is a *scalar*. From Fig. 1.6 it is observed that $\mathbf{A} \cdot \mathbf{B}$
can be interpreted as the magnitude of the component of **B** in the

---

1. Vectors will be indicated by boldface symbols.

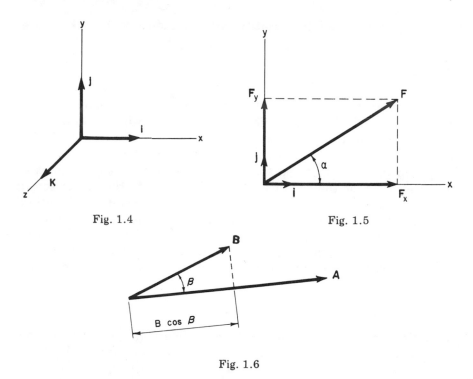

Fig. 1.4

Fig. 1.5

Fig. 1.6

direction of **A** times the magnitude of **A**, or the magnitude of the component of **A** in the direction of **B** times the magnitude of **B**. We see from this that the dot product of a given vector and a unit vector gives the magnitude of the rectangular component of the vector in the direction of the unit vector. Thus for the example in Fig. 1.5 we can write

$$F_x = \mathbf{F} \cdot \mathbf{i} = F \cos \alpha, \qquad F_y = \mathbf{F} \cdot \mathbf{j} = F \cos (90 - \alpha) = F \sin \alpha \qquad (1.4)$$

which checks with the results previously obtained.

In certain cases we may be given the components and wish to obtain the resultant. This can be done by the application of the parallelogram law and is known as the process of *composition*.

On a diagram or figure a vector quantity can be designated in various ways; e.g., in Fig. 1.5 the vector component F_x could be indicated as shown in this figure or as $\rightarrow F_x$ or $\rightarrow F_x$ **i**, where F_x is understood to represent the magnitude of the vector and the arrow indicates that the quantity is a vector. These ways of designating a vector quantity are used interchangeably in this text. However, it is not convenient to use arrows to distinguish between scalars and vectors in equations, and it should be clearly understood that in any given equation a vector quantity will be indicated by a boldface symbol.

1.4   Moments

The *moment* of a force with respect to an *axis* perpendicular to a plane containing the force is defined as a vector whose magnitude is the product of the magnitude of the force and the perpendicular distance between the line of action of the force and the axis. The direction of the vector representing the moment is along the axis, and the sense of the vector is obtained from the "right-hand screw rule"; i.e., we let our fingers "curl" around the axis in the direction of the force, and the thumb points in the direction of the required vector. A simple example is shown in Fig. 1.7. The force **F** lies in the xy plane, and the moment with respect to an axis through 0 perpendicular to the xy plane is required. From the definition of the moment we obtain

$$\mathbf{M_0} = Fd\ (-\mathbf{k}) \tag{1.5}$$

where d is the perpendicular distance between the point 0 and the extension of the force **F**. The direction of the vector is in the negative z direction, and this is indicated by $-\mathbf{k}$. We could also write $\mathbf{M_0} = Fd\ \curvearrowright$, where the curved arrow is used to specify the direction.

An alternate method for determining the moment of **F** in Fig. 1.7 is to resolve **F** into two components and determine the vector

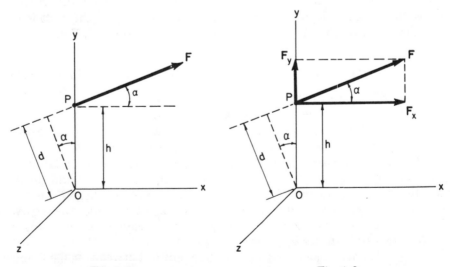

Fig. 1.7   Fig. 1.8

sum of the moments of the components. We resolve **F** into rectangular components at the point P (Fig. 1.8). Since the line of action of the force $\mathbf{F}_y$ passes through the z axis, its moment is zero with respect to this axis. Therefore $\mathbf{M}_0 = -F_x h\,\mathbf{k} + (0)\,\mathbf{k}$. However, $F_x = F\cos\alpha$ and $d = h\cos\alpha$ so that $\mathbf{M}_0 = -Fd\mathbf{k}$, which checks with the previous result. Regardless of the location of **F** along its line of action the same result is obtained. We have thus demonstrated two important principles that can be proved more generally.

1. *Principle of transmissibility.* The moment of a force is independent of the location of the force along its line of action.
2. *Principle of moments.* The moment of the resultant of a system of forces is equal to the vector sum of the moments of the forces of the system.

The *resultant* of a force system is defined as the simplest force system that can replace the original system and have the same external effect. In the example just considered, $\mathbf{F}_x$ and $\mathbf{F}_y$ represent the original force system and **F** the resultant.

EXAMPLE 1.1

A rigid bracket is loaded with a 100-lb force **F** (Fig. E1.1). Determine the moment of this force with respect to the z axis. The force lies in the xy plane.

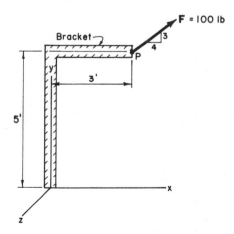

Fig. E1.1

SOLUTION

The x component of **F** is

$$\mathbf{F}_x = (4/5)(100)\mathbf{i} = 80\mathbf{i} = 80\text{ lb} \rightarrow \text{through } P$$

and the y component is

$$\mathbf{F}_y = (3/5)(100)\mathbf{j} = 60\mathbf{j} = 60 \text{ lb} \uparrow \text{ through } P$$

We now use the definition of the moment with respect to an axis and the principle of moments to obtain the moment of $\mathbf{F}$ with respect to the z axis; i.e.,

$$\mathbf{M}_z = -(5)(80)\mathbf{k} + (3)(60)\mathbf{k} = -220\mathbf{k} = 220 \text{ ft-lb} \nearrow \text{along } z \text{ axis} \quad Ans.$$

The answer could also be written in the form

$$\mathbf{M}_z = 220 \text{ ft-lb} \, \text{)} \qquad\qquad Ans.$$

We may also define the *moment* of a force with respect to a *point* as a vector whose magnitude is the product of the magnitude of the force and the perpendicular distance between the line of action of the force and the point. The direction of the moment is perpendicular to the plane containing the force and point, and its sense is obtained from the right-hand screw rule. For this more general case it is frequently convenient to use vector algebra to determine the moment.

The *position vector* $\mathbf{r}$ of a point P is defined as the directed line segment between some selected reference point, usually the origin of our coordinate system, and the point P (Fig. 1.9). If the coordinates

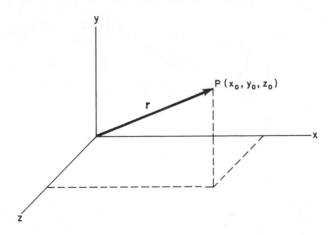

Fig. 1.9

of the point P are (x_0, y_0, z_0), we can write

$$\mathbf{r} = x_0\mathbf{i} + y_0\mathbf{j} + z_0\mathbf{k} \qquad\qquad (1.6)$$

where vector addition is implied. The *cross product* of two vectors

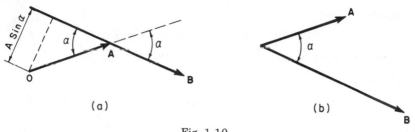

Fig. 1.10

A and B (Fig. 1.10a) is by definition

$$\mathbf{A} \times \mathbf{B} = AB \sin \alpha \, \mathbf{n} \qquad (1.7)$$

where $\mathbf{n}$ is a unit vector normal to the plane containing the two vectors. The sense of the vector $\mathbf{n}$ is obtained by placing the two vectors "tail" to "tail" (Fig. 1.10b) and applying the right-hand screw rule; i.e., let our fingers curl in the direction from the first vector, $\mathbf{A}$, toward the second, $\mathbf{B}$, and the direction in which the thumb points is the proper direction for $\mathbf{n}$. Note that $\mathbf{A} \times \mathbf{B} \neq \mathbf{B} \times \mathbf{A}$; i.e., the cross product is not *commutative*. From the definition we see that the cross product is a vector; it is commonly referred to as the *vector product*. The magnitude of $\mathbf{A} \times \mathbf{B}$ can be seen to be equal to the product of the perpendicular distance between the point 0 and the vector $\mathbf{B}$ and the magnitude of $\mathbf{B}$. This immediately suggests that moments and cross products can be related.

We again consider the simple example in Fig. 1.7. For this special case the moment with respect to an axis through 0 perpendicular to the plane containing the force and with respect to the point 0 are the same. Let $\mathbf{r}$ be the position vector from the origin 0 to any point P on the line of action of the force $\mathbf{F}$ (Fig. 1.11a). The force and position vectors can be written in terms of their rectangular com-

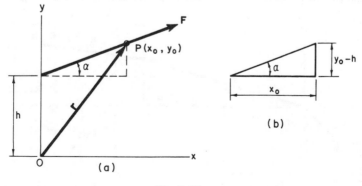

Fig. 1.11

ponents as

$$F = F_x i + F_y j, \qquad r = x_0 i + y_0 j$$

The cross product of the position vector and the force is

$$r \times F = (x_0 i + y_0 j) \times (F_x i + F_y j)$$
$$= x_0 F_x (i \times i) + x_0 F_y (i \times j) + y_0 F_x (j \times i) + y_0 F_y (j \times j)$$

However, from the basic definition of the cross product

$$i \times i = j \times j = 0, \qquad i \times j = k, \qquad j \times i = -k \qquad (1.8)$$

so that $r \times F = (x_0 F_y - y_0 F_x)k$. Also

$$F_x = F \cos \alpha, \qquad F_y = F \sin \alpha, \qquad (y_0 - h)/x_0 = \sin \alpha / \cos \alpha$$

(see Fig. 1.11b) so that

$$r \times F = F(x_0 \sin \alpha - y_0 \cos \alpha)k$$
$$= F\{x_0 [(y_0 - h)/x_0] \cos \alpha - y_0 \cos \alpha\}k$$
$$= -Fh \cos \alpha \, k = -Fdk = M_0$$

which checks with the previously determined expression for the moment. This relationship between the cross product and moment can be shown to be true for the more general case, i.e., when the force has any orientation in space. Therefore we may consider the relationship

$$r \times F = M_0 \qquad (1.9)$$

as the general expression for the moment of the force F, with respect to a point 0, which is the reference point for the position vector r.

EXAMPLE 1.2

Determine the moment of the force F of Fig. E1.2 with respect to (a) the y axis and (b) the point 0.

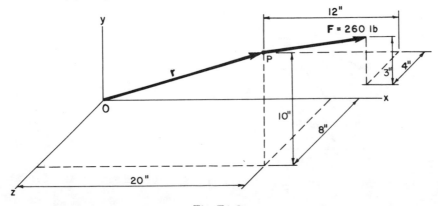

Fig. E1.2

SOLUTION

The rectangular components of **F** through P are

$$F_x = \frac{12}{\sqrt{(3)^2 + (4)^2 + (12)^2}} (260)i = 240i \text{ lb}$$

$$F_y = (3/13)(260)j = 60j \text{ lb}$$

$$F_z = (4/13)(260)k = 80k \text{ lb}$$

(a) The moment with respect to the y axis is

$$M_y = (8)(240)j - (20)(80)j = 320j \text{ in.-lb} \qquad \textit{Ans.}$$

Note that F_y is parallel to the y axis and does not contribute to the moment with respect to this axis.

(b) To determine the moment with respect to the point 0 it is convenient to use the cross product $M_0 = r \times F$, where

$$r = 20i + 10j + 8k, \qquad F = 240i + 60j + 80k$$

The moment about 0 is

$$M_0 = r \times F = \begin{vmatrix} i & j & k \\ 20 & 10 & 8 \\ 240 & 60 & 80 \end{vmatrix} = (800 - 480)i - (1600 - 1920)j + (1200 - 2400)k$$

$$M_0 = (320i + 320j - 1200k) \text{ in.-lb} \qquad \textit{Ans.}$$

Note that the component of M_0 in the x direction is obtained by subtracting the two terms $(10)(80) - (8)(60)$. An inspection of Fig. E1.2 reveals this is the magnitude of the moment of **F** about the x axis. Thus if we take the dot product $i \cdot M_0$, we obtain the magnitude of the moment about the x axis. This result can be generalized so that the moment of a force with respect to a line can be obtained by calculating the moment with respect to some point on the line and taking the dot product of this vector with a unit vector **n** along the axis of interest. Thus $M_{\text{axis along } n} = (n \cdot M_0)n$. Part (a) of this example could be solved on this basis; i.e.

$$M_y = [j \cdot (320i + 320j - 1200k)]j = 320j \text{ in.-lb}$$

which checks the result previously obtained.

1.5 Couples

A *couple* is defined as a pair of parallel forces that have equal magnitudes but opposite senses. A typical couple formed by the forces F_1 and F_2 is shown in Fig. 1.12. A couple will have a "twisting" effect on a body, but since the forces are equal and opposite, there will be no tendency for translation of the body. The moment of the couple of Fig. 1.12 with respect to the origin is $M_0 = r_1 \times F_1 + r_2 \times F_2$.

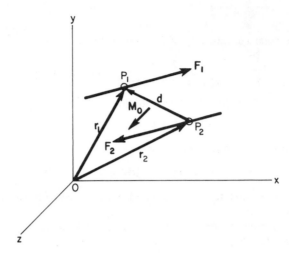

Fig. 1.12

By definition $F_2 = -F_1$, so

$$M_0 = r_1 \times (F_1) + r_2 \times (-F_1) = (r_1 - r_2) \times F_1$$

However, as noted from Fig. 1.12, $r_1 - r_2 = d$, so we may write $M_0 = d \times F_1$. Since d is a vector in the plane of the forces, the direction of the moment is perpendicular to this plane. The two points P_1 and P_2 are arbitrary so that we can choose them in such a manner that d is perpendicular to the forces. Thus the moment of a couple can be easily determined as

$$M_0 = Fd\mathbf{n} \qquad\qquad (1.10)$$

where F is the magnitude of one of the forces, d the magnitude of the perpendicular distance between the two forces, and $\mathbf{n}$ a unit vector normal to the plane of the forces. The sense of $\mathbf{n}$ is obtained from the right-hand screw rule.

From the derivation of the expression for the moment of the couple we observe that the moment is independent of the choice of the reference point 0; i.e., the vector d is the pertinent quantity. In effect this means that the moment of the couple is unchanged if (1) it is moved to a parallel position in its plane; (2) it is moved to a parallel plane; (3) it is rotated in its plane; or (4) the magnitude of the forces, or the perpendicular distance between the forces, is changed as long as the product of the force and distance remains unchanged. Since the only effect a couple has on a system is due to its moment, we may conclude that all these four operations may be performed without changing the external effect of a couple on a given system.

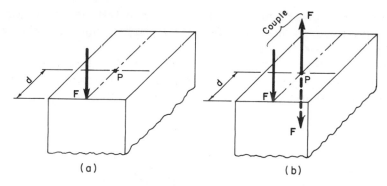

Fig. 1.13

In certain problems it is convenient to replace a force by a force and a couple. This can be accomplished in the following manner. Consider the single force **F** of Fig. 1.13a. We wish to replace this force by a parallel force acting through the point P and a couple. To do this, we add two equal and opposite forces at P (Fig. 1.13b). This is permissible since their net effect is zero. However, we now can consider this new system of forces as a couple, in a plane containing the original force and the point P, and a downward force **F** at P, so that the original force has been replaced by a force and a couple that will have the same external effect on the system.

1.6 Newton's Laws of Motion

The exact relationship between force and motion was not understood from the time of the Greek philosophers (who were among the first to consider such ideas) to the time of Galileo (1564-1642), even though it was thought that force and motion were somehow related. Galileo was the first to disprove some of the earlier false ideas through his experiments on falling balls and swinging pendulums. Near the beginning of the seventeenth century Johann Kepler (1571-1630) published his well-known laws on planetary motion, which were based on experimental observations. In his study to explain Kepler's laws of planetary motion, Sir Isaac Newton (1642-1727)[2] developed the law of universal attraction and was the first to state what today are called *Newton's laws of motion*.[3] Great scientific and engineering progress has been made since Newton's time,

---

2. Note that Newton was born in the same year that Galileo died.

3. These laws are a special case of the more recent relativistic laws governing the motion of particles moving near the speed of light. Sometimes Newton's laws of motion are called the "classical laws of motion" as opposed to the "modern relativistic laws of motion."

but his three laws of motion and law of gravitational attraction are still recognized as the basis for engineering dynamics where relativistic effects are not significant.

Before stating Newton's laws of motion, we should recall some basic concepts of particle kinematics. Consider a particle P moving along a straight line (Fig. 1.14a). The position of P with respect to 0

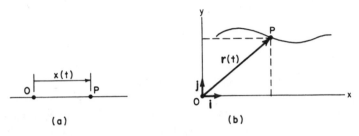

Fig. 1.14

is given by the single coordinate $x(t)$. Recall that the velocity of P is given by the time rate of change of x; i.e.,

$$v = \frac{dx}{dt}$$

while the acceleration of P is the time rate of change of the velocity; i.e.,

$$a = \frac{dv}{dt} = \frac{d^2 x}{dt^2}$$

When particle P moves along a curved path in a plane (Fig. 1.14b), we need the two coordinates $x(t)$ and $y(t)$ to locate P with respect to 0. For this situation, it is convenient to use the position vector $r(t)$ to locate P. The vector $r(t)$ can be written as

$$r(t) = x(t)i + y(t)j$$

where i and j are unit vectors and $x(t)$ and $y(t)$ are the scalar components of $r(t)$. Point P has two velocity components, $v_x = dx/dt$ and $v_y = dy/dt$, so that the resultant velocity becomes

$$v = \frac{dr}{dt} = v_x i + v_y j = \frac{dx}{dt} i + \frac{dy}{dt} j$$

which is a vector quantity with magnitude and direction. Similarly, the acceleration has two components,

$$a_x = dv_x/dt = d^2 x/dt^2, \qquad a_y = dv_y/dt = d^2 y/dt^2$$

so that the resultant acceleration becomes

$$\mathbf{a} = \frac{d\mathbf{v}}{dt} = \frac{d^2\mathbf{r}}{dt^2} = a_x\mathbf{i} + a_y\mathbf{j} = \frac{dv_x}{dt}\mathbf{i} + \frac{dv_y}{dt}\mathbf{j} = \frac{d^2x}{dt^2}\mathbf{i} + \frac{d^2y}{dt^2}\mathbf{j}$$

which is also a vector quantity with magnitude and direction. Hence we have that the position vector, the velocity vector, and the acceleration vector are related by

$$\mathbf{v} = \frac{d\mathbf{r}}{dt} \tag{1.11}$$

and

$$\mathbf{a} = \frac{d\mathbf{v}}{dt} = \frac{d^2\mathbf{r}}{dt^2} \tag{1.12}$$

The units of velocity are usually expressed in feet per second (fps, ft/s), inches per second (ips, in./s), metres per second (m/s), or centimetres per second (cm/s); while acceleration is usually expressed in feet per second per second (fps^2, ft/s^2), metres per second per second (m/s^2), etc.

We can state Newton's laws of motion[4] for a single particle as follows:

1. *Newton's first law* (*inertia*). A particle will remain at rest or will continue with uniform motion along a straight line as long as the resultant force acting on the particle is zero. This resistance to change in the state of motion (or rest) of a particle is a property of matter called *inertia*. The *mass* of a particle is a measure of its inertia, i.e., its resistance to changes in motion.
2. *Newton's second law* (*motion*). The acceleration of a particle is proportional to and in the direction of the resultant force acting on the particle. The constant of proportionality is equal to or proportional to the mass of the particle, depending on the system of units used.
3. *Newton's third law* (*action and reaction*). The forces of mutual action between any two particles must be equal in magnitude, opposite in sense, and directed along the same straight line; i.e., to every force action there exists an equal and opposite force reaction.

---

4. Newton's laws of motion were first published in his *Principia* in 1687 and are restated here in modern terminology. These results were published nearly seventy years after Kepler published his first two laws in 1609 and his third law in 1619.

In particle dynamics, Newton's second law can be written as

$$\mathbf{F} = m\mathbf{a} = m\frac{d\mathbf{v}}{dt} \qquad (1.13)$$

where $\mathbf{F}$ is the resultant force acting on the particle, m the inertial mass of the particle, $\mathbf{v}$ the velocity of the particle, and $\mathbf{a}$ the acceleration of the particle.

Newton's law of universal gravitation was published in 1686 and states that the mutual force of attraction F between two point masses m_1 and m_2, which are separated by a distance r, has the magnitude

$$F = G\,(m_1 m_2 /r^2) \qquad (1.14)$$

and is directed along the line joining the mass centers. The proportionality factor G is the *universal gravitational constant*.

When a particle of mass m is resting on or near the earth's surface (Fig. 1.15), we call the force of attraction the body weight W.[5] If

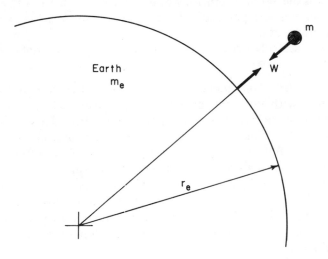

Fig. 1.15

we let $m_1 = m_e$, $r = r_e$, and $m_2 = m$ in Eq. (1.14), then

$$W = (Gm_e /r_e^2)m \qquad (1.15)$$

Substitution of Eq. (1.15) into Newton's second law, Eq. (1.13),

---

5. An elevation of 2 mi above sea level will cause approximately 1% reduction in the body weight W of mass m. Hence we can consider the weight of the body to be invariant for most engineering calculations as long as we are near sea level.

gives

$$(Gm_e/r_e^2)m = ma$$

or

$$a = Gm_e/r_e^2 = g \tag{1.16}$$

where g is the *local free-fall acceleration* of the particle in a vacuum at the earth's surface.[6]   Finally, substitution of Eq. (1.16) into Eq. (1.15) gives

$$W = mg \tag{1.17}$$

which relates the measured weight W, the mass of the body m, and the local value of the acceleration due to gravity g. Of these variables, only the mass is an invariant; i.e., the mass m is a property of the particle matter that remains constant whether the particle is on the earth, the moon, or in free space. The weight of the particle (force of gravitational attraction), however, is dependent on the location of the particle with respect to other bodies such as the earth or moon.

The local weight of a body (force of attraction between the body and the earth) is one force that must always be considered to act on a body in establishing the force system that will produce motion according to Newton's second law. If the weight is the only force acting on a body, we have the free-fall situation described by Eq. (1.16). We shall use Eq. (1.17) throughout this text to convert either from mass units to weight units or from weight units to mass units when needed.

1.7 Dimensions and Units

When dealing with quantities in physical problems, an important characteristic is the qualitative nature of each quantity; i.e., What type of physical characteristic does it describe? Common examples of such characteristics are length, time, force, and velocity. We associate with each quantity one, or more, *basic dimensions* that serves to describe the quantity qualitatively. Basic dimensions of importance in mechanics include: length L, time T, force F and mass M. Other quantities of interest can be described in terms of these; e.g., velocity $\sim L/T$, area $\sim L^2$, volume $\sim L^3$, and density $\sim M/L^3$.

In addition to this qualitative description it is also generally

---

6. Due to the fact that the earth is not a perfect sphere and slight differences in local mass density exist, small variations in the value of the local acceleration due to gravity occur.

necessary to have a quantitative measure of a given quantity. For example, if we measure the width of this page in the book and say that it is 10 units wide, the statement has no meaning until the basic unit of length is defined. However, if we indicate that the basic unit is a centimetre and specify a standard length representing a centimetre, a *system of units* has been established in which various lengths can be quantitatively compared. In addition to the length unit we must also establish a system of units for force, mass, and time.

From Newton's second law, expressed in the form of Eq. (1.13), we see that the basic dimensions of force, mass, and acceleration must be related for this equation to be *dimensionally homogeneous*. A dimensionally homogeneous equation is one in which all terms have the same basic dimensions, and we take as a fundamental axiom that all equations describing physical phenomena must be dimensionally homogeneous. Thus we can arbitrarily establish a unit system for length, time, and force, but the corresponding mass unit is then fixed; or we can establish a unit system for length, time, and mass with the force unit then being fixed. This latter procedure is used in the establishment of the *absolute mks metric system* in which the unit of length is the metre (m), the unit of time the second (s), and the basic unit of mass is taken as the standard kilogram (kg), which is a cylinder of platinum iridium alloy kept by the International Bureau of Weights and Measures in Sevres, France. The basic unit for force, called a newton (N), is then defined as the force required to accelerate a 1-kg mass 1 m/s^2; i.e.,

$$1 \text{ unit force (N)} = (1 \text{ kg})(1 \text{ m/s}^2) \qquad (1.18)$$

The mass unit in the *absolute cgs metric system* is the gram (gm), which is 1/1000 kg, and the basic unit of force is the *dyne*, where

$$1 \text{ dyne} = (1 \text{ gm})(1 \text{ cm/s}^2) \qquad (1.19)$$

The standard *pound mass* (lbm) is defined as a body of mass 0.45359237 kg, and the standard pound force (lb) is the gravitational force of attraction between a pound mass and the earth at some specified point on the earth's surface. The point is taken at a location for which the *acceleration of gravity* (g) is 32.1740 ft/s^2 (commonly approximated as 32.2 ft/s^2).

For the *absolute English system* the unit of mass is the pound mass, the unit of length is the foot (ft), and the unit of time the second. The unit of force is the *poundal* (pdl), defined through the equation

$$1 \text{ pdl} = (1 \text{ lbm})(1 \text{ ft/s}^2) \qquad (1.20)$$

For these three so-called absolute unit systems basic units for length, time, and mass are established; and the basic unit of force is defined from Newton's second law.

For the *gravitational English system* the basic unit of length is the foot, of time the second, and the force unit is the standard pound force. The basic mass unit, called a *slug*, is defined from the relationship

$$1 \text{ lb} = (1 \text{ slug})(1 \text{ ft/s}^2) \tag{1.21}$$

Thus by definition a 1-lb force acting on a mass of 1 slug will give this mass an acceleration of 1 ft/s². For all four of these systems of units the weight of a certain quantity of material has a different numerical value than its mass. For example, in the gravitational English system a quantity of matter having a 1-slug mass will have a weight of 32.2 lb at a location on the earth's surface where the acceleration of gravity is 32.2 ft/s².

For two other systems of units the basic units of length, time, force, *and* mass are defined independently; thus special care must be exercised when using these systems in conjunction with Newton's second law. In the *English engineering system* the basic unit of mass is taken as the pound mass, the unit of force the pound force, the unit of length the foot, and the unit of time the second. To make the equation expressing Newton's second law dimensionally homogeneous, we state that the force acting on a mass is *proportional* to the mass times acceleration; i.e., $F = kma$ and thus

$$1 \text{ lb} = k(1 \text{ lbm})a(\text{ft/s}^2) \tag{1.22}$$

By definition a pound force acting on a pound mass at a specified location where $g = g_0$ ft/s² will accelerate the mass with an acceleration of g_0 ft/s². Thus for Eq. (1.22) to be both numerically and dimensionally correct,

$$1 \text{ lb} = k(1 \text{ lbm})(g_0 \text{ ft/s}^2)$$

and

$$k = \frac{1 \text{ lb}}{(1 \text{ lbm})(g_0 \text{ ft/s}^2)}$$

Thus in using the English engineering system, we must write Newton's second law in the form

$$F = ma/g_c \tag{1.23}$$

where

$$g_c = [(1 \text{ lbm})(g_0 \text{ ft/s}^2)]/1 \text{ lb} = 1/k$$

and numerically g_c is usually taken as 32.2, corresponding to $g_0 = 32.2 \text{ ft/s}^2$.

In a similar manner the *gravitational metric system* is established such that a mass of 1 gm is the basic unit of mass and the force unit is the gravitational attraction of the gram mass, i.e., the gram weight or force (gmf). Thus 1 gmf = $K(1 \text{ gm})(g_0 \text{ cm/s}^2)$ with

$$K = 1 \text{ gmf}/[(1 \text{ gm})(g_0 \text{ cm/s}^2)]$$

Newton's law must be written as

$$F = ma/G_c \qquad (1.24)$$

where

$$G_c = [(1 \text{ gm})(g_0 \text{ cm/s}^2)]/1 \text{ gmf} = 1/K$$

The numerical value for G_c is 981, which is the magnitude of g_0 in cm/s^2. For both the English engineering system and the gravitational metric system the weight of a body and its mass are numerically equal at the point on the earth's surface where $g = g_0$.

In Table 1.1 the units for the various systems are tabulated. For all systems the basic unit of time is the second. In 1960 the Eleventh General Conference on Weights and Measures, the international organization responsible for maintaining precise uniform standards of measurements, formally adopted the *International System of Units* as the international standard. In this system (commonly called SI) there are six basic units (Table 1.2). The derived unit for force is the

Table 1.1. Systems of units

System of units	Force	Mass	Length
Absolute mks metric (SI)	newton (N)	kilogram (kg)	metre (m)
Absolute cgs metric	dyne	gram (gm)	centimetre (cm)
Gravitational English	pound (lb)	slug	foot (ft)
Absolute English	poundal (pdl)	pound (lbm)	foot (ft)
English engineering	pound (lb)	pound (lbm)	foot (ft)
Gravitational metric	gram (gmf)	gram (gm)	centimetre (cm)

Table 1.2. Basic units for SI

Physical quantity	Unit	Symbol
Length	metre	m
Mass	kilogram	kg
Time	second	s
Electric current	ampere	A
Temperature	degree Kelvin	$^\circ$K
Luminous intensity	candela	cd

newton as defined by Eq. (1.18). Examples of other derived SI units of special interest in mechanics are given in Table 1.3.

The SI units for length, time, mass, and force are the same as those for the absolute mks metric system. Prefixes for forming multiples and fractions of SI units are given in Table 1.4. As the use of SI becomes more commonplace in the United States and throughout the world, engineers will be required to be familiar with both SI and the English unit systems commonly used by engineers in the United

Table 1.3. Derived SI units

Physical quantity	SI units	Symbol
Area	square metre	m^2
Volume	cubic metre	m^3
Frequency	hertz	Hz
Density	kilogram per cubic metre	kg/m^3
Velocity	metre per second	m/s
Angular velocity	radian per second	rad/s
Acceleration	metre per second squared	m/s^2
Force	newton	N
Pressure	newton per square metre	N/m^2
Kinematic viscosity	square metre per second	m^2/s
Dynamic viscosity	newton-second per square metre	$N\text{-}s/m^2$
Work, energy, quantity of heat*	joule	J
Power†	watt	W

*The *joule* is the work done when the point of application of a force of 1 N is displaced through a distance of 1 m in the direction of the force.

†The *watt* is the unit of power and is equal to 1 J/s.

Table 1.4. Prefixes

Factor by which unit is multiplied	Prefix	Symbol
10^{12}	tera	T
10^9	giga	G
10^6	mega	M
10^3	kilo	k
10^2	hecto	h
10	deka	da
10^{-1}	deci	d
10^{-2}	centi	c
10^{-3}	milli	m
10^{-6}	micro	μ
10^{-9}	nano	n
10^{-12}	pico	p
10^{-15}	femto	f
10^{-18}	atto	a

States.   Conversion tables suitable for many engineering computations are given in Appendix A.

In this text the gravitational English system (lb, slug, ft, s) will
be used primarily as the basic system of units. However, some examples and problems will be specified in terms of the metric SI system. Some basic conversions include:

$$1 \text{ ft} = 0.3048^* \text{ m}, \quad 1 \text{ lb} = 4.4482216152605^* \text{ N}$$

$$1 \text{ slug} = 14.5939029 \text{ kg}$$

*Exact, by definition.

1.8 Centroids of Volumes and Areas

Volume integrals of the form

$$\int_V x\, dV, \qquad \int_V y\, dV, \qquad \int_V z\, dV$$

are frequently encountered in mechanics and are called the *first
moments of the volume V* with respect to the yz, xz, and xy planes
respectively.   A point in space (x_c, y_c, z_c) (Fig. 1.16) is called the
centroid of the volume V when defined such that

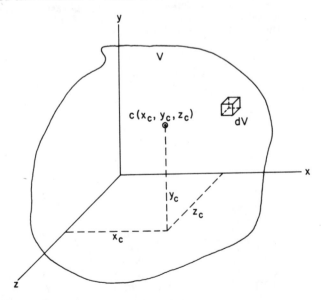

Fig. 1.16

$$x_c V = \int_V x \, dV \qquad (1.25a)$$

$$y_c V = \int_V y \, dV \qquad (1.25b)$$

$$z_c V = \int_V z \, dV \qquad (1.25c)$$

where

$$V = \int_V dV$$

The centroid is a geometrical property of the volume and depends only on the shape of the volume and not on any of its physical properties. It can be shown that if a volume has a plane of symmetry, the centroid must lie in this plane.

EXAMPLE 1.3

Find the centroid of a solid hemisphere of radius a.

SOLUTION

We choose a rectangular Cartesian coordinate system with the origin at the center of the base of the hemisphere (Fig. E1.3). Since the xz and yz planes are

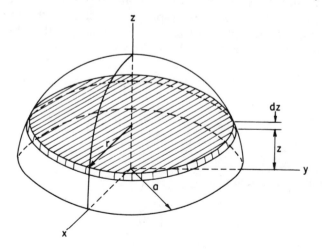

Fig. E1.3

planes of symmetry, the centroid of this volume must lie on their intersection. This immediately leads to $x_c = y_c = 0$. To find z_c, we use Eq. (1.25c). The volume dV of the infinitesimal slice in Fig. E1.3 depends on the elevation z as indicated by the expression $dV = \pi[r(z)]^2\,dz$. From the equation of the hemisphere $x^2 + y^2 + z^2 = a^2$; and from the fact that $[r(z)]^2 = x^2 + y^2$, it follows that $dV = \pi(a^2 - z^2)\,dz$. Integration gives the volume

$$V = \int_0^a \pi(a^2 - z^2)\,dz = \pi\left(a^2 z - \frac{z^3}{3}\right)\Big|_0^a = \frac{2\pi a^3}{3}$$

With the above volume element we then have

$$\int_V z\,dV = \int_0^a \pi(a^2 - z^2)z\,dz = \pi\left(a^2\frac{z^2}{2} - \frac{z^4}{4}\right)\Big|_0^a = \frac{\pi a^4}{4}$$

and

$$z_c = \frac{\displaystyle\int_V z\,dV}{V} = \frac{\pi a^4}{4}\cdot\frac{3}{2\pi a^3} = \frac{3a}{8}$$

The centroid thus is located at

$$x_c = 0, \qquad y_c = 0, \qquad z_c = 3a/8 \qquad\qquad Ans.$$

The concept of the centroid of a volume is easily carried over to thin plates and areas. Let the xy plane be taken as the middle surface of a plate or the plane of an area so that $z_c = 0$. The x and y coordinates of the centroid are then given by

$$x_c A = \int_A x\,dA \qquad\qquad (1.26a)$$

$$y_c A = \int_A y\,dA \qquad\qquad (1.26b)$$

where the area A is given by

$$A = \int_A dA$$

The integrals in Eqs. (1.26) are called the first moments of the area with respect to the y and x axes respectively. If an area has an axis of symmetry, the centroid must lie on this axis.

EXAMPLE 1.4

Find the centroid of the thin plate having the shape of a semicircle in Fig. E1.4.

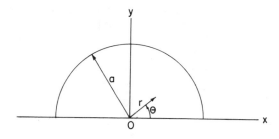

Fig. E1.4

SOLUTION

An appropriate coordinate system to use here is a plane polar coordinate system with the pole at point 0. Then $x = r \cos \theta$, $y = r \sin \theta$, and the increment of area is $dA = r\, dr\, d\theta$.

To find the centroid of the area, we first note that the y axis is a line of symmetry; therefore the centroid must be on the line $x = 0$. The area is one-half the area of a circle, $A = \pi a^2 / 2$; and using Eq. (1.26b), we find

$$\frac{\pi a^2}{2} y_c = \int_A y\, dA = \int_0^a r^2\, dr \int_0^\pi \sin \theta\, d\theta = \frac{2a^3}{3}$$

Thus the coordinates of the centroid are

$$x_c = 0, \qquad y_c = 4a/3\pi, \qquad z_c = 0 \qquad\qquad Ans.$$

EXAMPLE 1.5

Find the centroid of the shaded composite area in Fig. E1.5.

SOLUTION

We note that the composite area can be divided into the semi-circular area A, the rectangular area B, and the rectangle C (to be re-moved). Since the y axis is an axis of symmetry, the x coordinate of the centroid is zero. To find the y coordinate, we sum moments with respect to the x axis. Thus

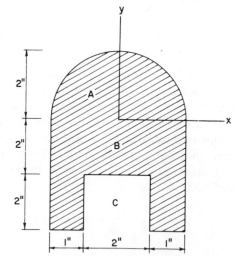

Fig. E1.5

$$y_c A = \int_A y \, dA = \int_{\text{area A}} y \, dA + \int_{\text{area B}} y \, dA - \int_{\text{area C}} y \, dA$$

$$= (y_c)_A (A)_A + (y_c)_B (A)_B - (y_c)_C (A)_C$$

The negative sign is required for the last integral since the area is to be subtracted. The pertinent quantities are tabulated below:

Area designation	Area (in.^{2})	y_c (in.)
A	2π	$8/3\pi$
B	16	-2
C	-4	-3
Total	18.28	

The y coordinate can now be computed as

$$y_c 18.28 = (8/3\pi) \; (2\pi) - (2)(16) + (3)(4) = -14.67$$

and

$$y_c = -14.67/18.28 = -0.802$$

Thus the centroid of the area is located at

$$x_c = 0, \qquad y_c = -0.802 \text{ in.} \hspace{2cm} Ans.$$

1.9 Moment of Inertia of Areas

In the study of dynamics and strength of materials an integral of the form

$$\int_A r^2 \, dA$$

frequently appears. The quantity r is the perpendicular distance from some line BC to the incremental area dA, and the integration is carried out over some prescribed area A (Fig. 1.17). This integral is the *second moment of the area* about some axis and in mechanics is more commonly called the *moment of inertia* of the area. If one refers an area to a rectangular Cartesian coordinate system, the following moments of inertia with respect to the coordinate axis can be defined:

$$I_x = \int_A y^2 \, dA, \qquad I_y = \int_A x^2 \, dA \hspace{2cm} (1.27)$$

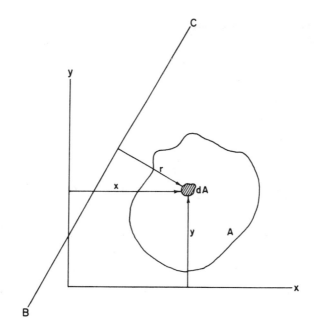

Fig. 1.17

Another expression that will arise in future study is the integral

$$I_{xy} = \int_A xy \, dA \tag{1.28}$$

This quantity is known as the *product of inertia* of the area with respect to the x and y axes. Here x and y are the distances from the y and x axes respectively to the element of area dA. If either the x or y axes are lines of symmetry for the area A, the product of inertia I_{xy} is zero.

The second moments of frequently used areas have been tabulated and are available in handbooks. Table 1.5 gives the area, centroid, and moments of inertia of a few commonly encountered areas.

A useful theorem that relates the moment of inertia about any line to the moment of inertia about a parallel line through the centroid is known as the *parallel axis theorem*, which states that the moment of inertia I_d of an area A about any axis is equal to the moment of inertia I_c of the area about a parallel axis through the centroid of the area plus the area times the distance squared between the two lines. In equation form the theorem can be written as

$$I_d = I_c + Ad^2 \tag{1.29}$$

Table 1.5.   Some geometrical properties of common areas

Rectangular Plate

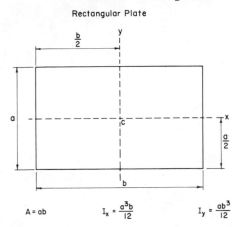

$$A = ab \qquad I_x = \frac{a^3b}{12} \qquad I_y = \frac{ab^3}{12}$$

Circular Plate

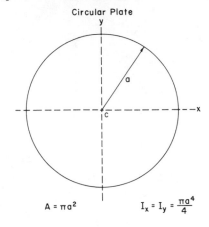

$$A = \pi a^2 \qquad I_x = I_y = \frac{\pi a^4}{4}$$

Triangular Plate

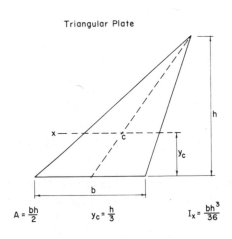

$$A = \frac{bh}{2} \qquad y_c = \frac{h}{3} \qquad I_x = \frac{bh^3}{36}$$

Circular Annulus

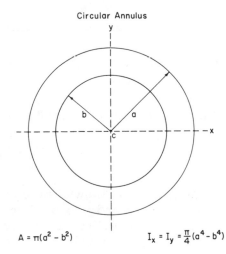

$$A = \pi(a^2 - b^2) \qquad I_x = I_y = \frac{\pi}{4}(a^4 - b^4)$$

Semicircular Plate

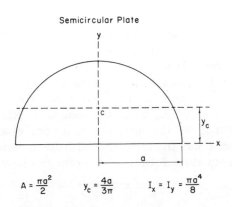

$$A = \frac{\pi a^2}{2} \qquad y_c = \frac{4a}{3\pi} \qquad I_x = I_y = \frac{\pi a^4}{8}$$

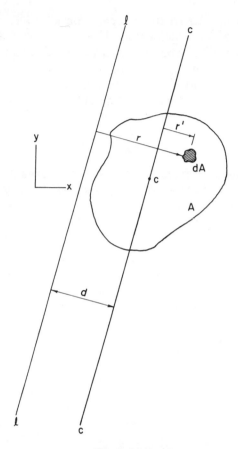

Fig. 1.18

where cc is the line through the centroid, $\ell\ell$ is the line parallel to cc, and d is the distance between cc and $\ell\ell$ (Fig. 1.18).

To prove the parallel axis theorem, consider the moment of inertia of the incremental area dA about the line $\ell\ell$ (Fig. 1.18). This gives $dI_\varrho = r^2\,dA$. But $r = r' + d$, where r' is the distance from line cc to the incremental area. Substituting for r and integrating over the area gives

$$I_\varrho = \int_A (r' + d)^2\,dA = \int_A (r')^2\,dA + 2d \int_A r'\,dA + d^2 \int_A dA$$

The last integral is the area A, while the second integral is the first moment of the area about a centroidal axis and hence must vanish. The first integral is the second moment of the area about the centroidal line cc. We can therefore express I_ϱ as $I_\varrho = I_c + Ad^2$.

In a manner similar to the above, a parallel axis theorem can be proved for the product of inertia. If we let the x' and y' axes pass through the centroid of the area and be parallel to the x and y axes, we have $I_{xy} = I_{x'y'} + d_x d_y A$ where d_x and d_y are the distances between the x', x axes and y', y axes respectively.

The *polar moment of inertia* of an area is defined as

$$J = \int_A \rho^2 \, dA \tag{1.30}$$

where ρ is the distance from the increment of area A to a point in the plane. If the point in the plane is chosen as the origin of a Cartesian coordinate system, we have

$$J = I_x + I_y \tag{1.31}$$

EXAMPLE 1.6

Find the moments of inertia and product of inertia of the rectangular area about the x and y axes (Fig. E1.6).

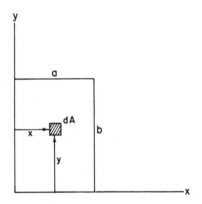

Fig. E1.6

SOLUTION

The increment of area $dA = dx \, dy$. The moment of inertia about the x axis is

$$I_x = \int_A y^2 \, dA = \int_{y=0}^{b} \int_{x=0}^{a} y^2 \, dx \, dy = a \left[\frac{y^3}{3} \Big|_0^b \right] = \frac{ab^3}{3} \qquad Ans.$$

and about the y axis

$$I_y = \int_A x^2 \, dA = \int_{y=0}^{b} \int_{x=0}^{a} x^2 \, dx \, dy = b \left[\frac{x^3}{3} \Big|_0^a \right] = \frac{ba^3}{3} \qquad Ans.$$

The product of inertia is

$$I_{xy} = \int_A xy \, dA = \int_{y=0}^{b} \int_{x=0}^{a} xy \, dx \, dy = \frac{x^2}{2} \Big|_0^a \frac{y^2}{2} \Big|_0^b = \frac{a^2 b^2}{4} \qquad Ans.$$

EXAMPLE 1.7

Find the moment of inertia of the triangle ABC about the line $x = a$ (Fig. E1.7).

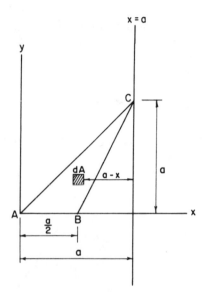

Fig. E1.7

SOLUTION

The integral defining the moment of inertia of the triangular area about $x = a$ is

$$I_a = \int_A (a - x)^2 \, dA$$

To determine the limits of integration, we first write the equations of the lines comprising the sides of the triangle. Moving from left to right, we have

$$y = x \quad \text{or} \quad x = y, \qquad y = 2x - a \quad \text{or} \quad x = (y + a)/2$$

Integrating first in the x direction and then in the y direction, I_a is

$$I_a = \int_{y=0}^{y=a} dy \int_{y}^{(y+a)/2} (a - x)^2 \, dx = \int_{y=0}^{a} \left[\frac{-(a-x)^3}{3} \, \Big|_{y}^{(y+a)/2} \right] dy$$

$$I_a = \frac{7}{24} \int_{y=0}^{a} (a-y)^3 \, dy = \frac{-7}{24} \frac{(a-y)^4}{4} \Big|_0^a$$

$$I_a = 7a^4/96 \hspace{7cm} Ans.$$

EXAMPLE 1.8

For the rectangular area in Fig. E1.6 find the moment of inertia about a line parallel to the y axis passing through the centroid.

SOLUTION

The centroid of the rectangular area is at $(a/2, b/2)$. The parallel axis theorem gives $I_d = I_c + Ad^2$. From Example 1.6 $I_y = ba^3/3$, and $d = a/2$. The moment of inertia desired is

$$I_c = I_y - Ad^2 = (ba^3/3) - (ab)(a^2/4) = a^3b/12 \hspace{2cm} Ans.$$

EXAMPLE 1.9

Find the centroid, the moments of inertia I_x, I_y, and the product of inertia I_{xy} for the shaded area in Fig. E1.9a.

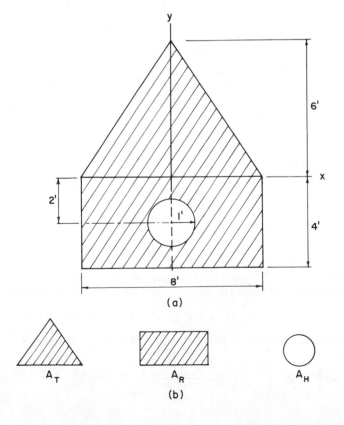

(a)

(b)

Fig. E1.9

SOLUTION

The area can be divided into a set of simpler areas (Fig. E1.9b). The centroid of each section can be computed, and the centroid of the total area with respect to the coordinate axes in Fig. E1.9a is found from

$$x_c A = x_T A_T + x_R A_R - x_H A_H, \qquad y_c A = y_T A_T + y_R A_R - y_H A_H$$

Here A_R is the area of the rectangle without the hole, and the last term in each of the above equations accounts for the hole in the rectangle. From symmetry we see that

$$x_c = 0 \qquad\qquad Ans.$$

The area of the shaded figure is $A = (1/2)(6)(8) + (4)(8) - \pi = 52.9 \text{ ft}^2$, and thus

$$y_c A = [(6)(8)/2](6/3) - (4)(8)(4/2) + (4/2)\pi = -9.72 \text{ ft}^3$$
$$y_c = -0.1840 \text{ ft} \qquad\qquad Ans.$$

The moment of inertia about the x axis is computed using

$$I_x = (I_x)_T + (I_x)_R - (I_x)_c$$

With the aid of Table 1.5 and the parallel axis theorem we have

$$I_x = \left[\frac{(8)(6)^3}{36} + \frac{(6)(8)}{2}\left(\frac{6}{3}\right)^2\right] + \left[\frac{(4)^3(8)}{12} + (4)(8)\left(\frac{4}{2}\right)^2\right] - \left[\frac{\pi}{4} + \pi\left(\frac{4}{2}\right)^2\right]$$
$$= 302 \text{ ft}^4 \qquad\qquad Ans.$$

In a similar manner we obtain the moment of inertia about the y axis,

$$I_y = 2\left[\frac{(6)}{36}\left(\frac{8}{2}\right)^3 + \frac{(6)(8)}{4}\left(\frac{8}{6}\right)^2\right] + \frac{(4)(8)^3}{12} - \frac{\pi}{4} = 234 \text{ ft}^4 \qquad\qquad Ans.$$

Since the region is symmetric about the y axis, the product of inertia I_{xy} vanishes; i.e.,

$$I_{xy} = 0 \qquad\qquad Ans.$$

PROBLEMS

1.1. Resolve the 325-lb force in Fig. P1.1 into two components, one having a line of action along AB and the other parallel to CD.

Ans. $\mathbf{F}_{AB} = 170 \text{ lb}$ ⟋⁴₃, $\mathbf{F}_{CD} = 322 \text{ lb}$ ⟍⁵₁₂,   both through 0

1.2. Two forces, $\mathbf{A} = 24\mathbf{i} + 30\mathbf{j} - 32\mathbf{k}$ lb and $\mathbf{B} = 3\mathbf{i} + 4\mathbf{k}$ lb, act through a common point Q. Determine the component of $\mathbf{A}$ acting in the direction of $\mathbf{B}$.

1.3. The 130-lb force in Fig. P1.3 is the resultant of two forces, one of which

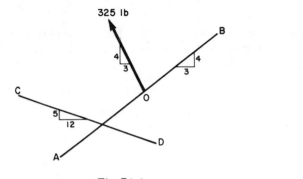

Fig. P1.1   Fig. P1.3

is *P* as shown.   Determine the other force, which has a vertical line of action.

Ans.   82.5 lb ↑

1.4.   Determine the magnitude of the moment of the force in Fig. P1.4 with respect to (a) the point 0 and (b) the *x* axis.

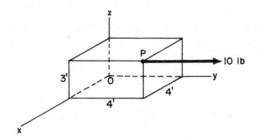

Fig. P1.4

1.5.   Determine the moment of the force in Fig. P1.5 with respect to the *aa* axis.

Ans.   **M = 20 in.-lb**

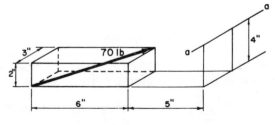

Fig. P1.5

1.6.   A 180-lb force acts along the line from *D* to *F* (Fig. P1.6).   Determine the

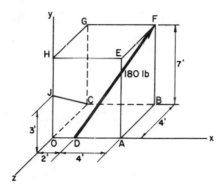

Fig. P1.6

moment of the 180-lb force with respect to line *JC*. Line *JC* is in the *yz* plane.

1.7. A 90-lb force acts along a diagonal line from *B* to *N* (Fig. P1.7). Determine the moment of the 90-lb force with respect to line *EP*.

Ans. **M** = 432(0.8**j** + 0.6**k**) in.-lb

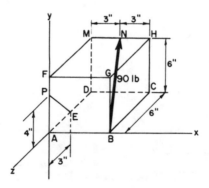

Fig. P1.7

1.8. Determine the moment of the 54-N force of Fig. P1.8 (a) with respect to point 0, and (b) with respect to the *AA* axis (which is parallel to the *z* axis), using the basic definition of the moment with respect to an axis.

1.9. The 24-lb forces in Fig. P1.9 are applied at the corners *A* and *B* of the parallelepiped and act along *AE* and *BF* respectively. Show that the given couple may be replaced by a set of vertical forces at points *C* and *D*. Specify the magnitude as well as the direction of these forces.

Ans. **F** = 16 lb, ↑ at *C* and ↓ at *D*

1.10. Using the transformations of a couple, replace the force system in Fig. P1.10 by a single force. Show steps used in your solution.

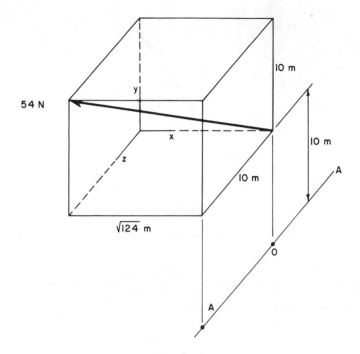

Fig. P1.8

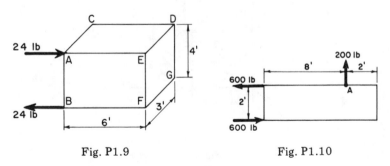

Fig. P1.9   Fig. P1.10

1.11. Replace the 260-lb force and the 800 ft-lb couple in Fig. P1.11 with two vertical forces, one at A and the other at B, without changing the external effects on the body.

Ans.   **B** = 567 lb ↓,   **A** = 307 lb ↑

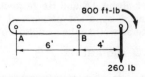

Fig. P1.11

1.12. A particle in rectilinear motion is located by $s = Bt^3 - 6t$, where s is in feet when t is in seconds. When $t = 2$ s, the velocity is 18 fps to the right. Determine (a) the position and acceleration when $t = 4.0$ s and (b) the total distance traveled between $t = 0$ and $t = 4.0$ s.

1.13. A particle P moves along the curve $y = 3x^2$ in Fig. P1.13, where $x = 4t - 1$ and x and y are in feet when t is in seconds. Determine (a) the position vector $\mathbf{r}$ and (b) the velocity and the acceleration of P when $t = 2.0$ s.

Ans. (a) $\mathbf{r} = (4t - 1)\mathbf{i} + (48t^2 - 24t + 3)\mathbf{j}$
(b) $\mathbf{v} = 4\mathbf{i} + 168\mathbf{j}$ fps, $\qquad \mathbf{a} = 0\mathbf{i} + 96\mathbf{j}$ fps^2

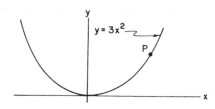

Fig. P1.13

1.14. The position vector locating a particle in the xy plane is given by

$$\mathbf{r} = (t^3 - 11.5t^2 + 54t)\mathbf{i} + (1.667t^3 + 1.5t^2 - 11t + 2)\mathbf{j}$$

where $\mathbf{r}$ is in metres when t is in seconds. Determine, at $t = 2.0$ s, (a) the rectangular components of the velocity and acceleration and (b) the speed of the particle.

1.15. Verify the following:

(a) 1 in.$^2 = 6.45 \times 10^{-4}$ m^2 (d) 1 ft-lb $= 1.356$ N-m (joule)
(b) 1 ft/s$^2 = 3.05 \times 10^{-1}$ m/s^2 (e) 1 lb-s/ft$^2 = 47.88$ N-s/m^2
(c) 1 psi $= 6.89 \times 10^3$ N/m^2

1.16. Verify the following:

(a) 1 m$^3 = 35.3$ ft^3 (c) 1 N/m$^3 = 6.37 \times 10^{-3}$ lb/ft^3
(b) 1 m/s $= 3.28$ ft/s   (d) 1 N/m $= 6.85 \times 10^{-2}$ lb/ft

1.17. A certain body weighs 10 lb on the earth's surface where the acceleration of gravity is 32.2 ft/s^2. Determine its mass in kilograms.

Ans. 4.53 kg

1.18. Find the coordinates of the centroid for the area bounded above by the x axis, on the left by the y axis, and below by the curve $y = x^2 - 4$.

1.19. Locate the centroid of the shaded area under the curve in Fig. P1.19 by integration. Check your answer by considering this problem as a composite area.

Ans. $x_c = 3$, $\quad y_c = 1$

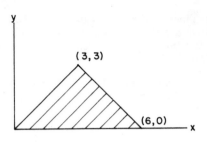

Fig. P1.19

1.20. Locate the centroid of the shaded composite area in Fig. P1.20.

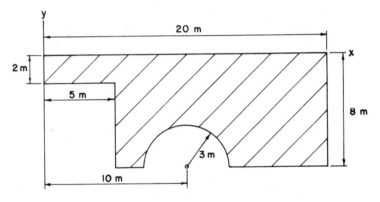

Fig. P1.20

1.21. For the area between the curves in Fig. P1.21 determine (a) I_x and (b) I_{xy} by integration.

Ans. (a) $I_x = 0.571$,   (b) $I_{xy} = 0.667$

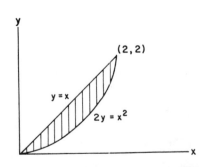

Fig. P1.21

1.22. Find I_x, I_y, and I_{xy} for the shaded area in Fig. P1.22.

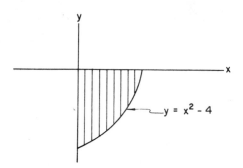

Fig. P1.22

1.23. Determine the moment of inertia of a circular area and a semicircular area about axis y', which is parallel to the y axis and tangent to the circular rim. Use the parallel axis theorem (Fig. P1.23).

Ans. $I_{y'} = 3.927\pi r^4$ (circle) $I_{y'} = 0.393r^4$ (semicircle)

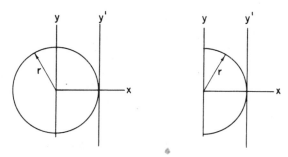

Fig. P1.23

1.24. For the area in Fig. P1.24, find I_x and I_y.

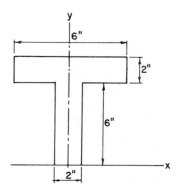

Fig. P1.24

part II

EQUILIBRIUM

Equilibrium of Rigid Bodies

2.1 Introduction

Newton's second law indicates that a particle will either be at rest or moving with a uniform velocity if the resultant force acting on it is zero. A particle under the action of a zero resultant force is said to be in equilibrium. The concept of equilibrium can be applied to systems of particles, which can be thought of as comprising finite bodies. A rigid body acted on by external forces is said to be in equilibrium if the resultant of these forces is zero, and the analysis of such a force system is termed a problem in *statics*. An important part of the analysis of a statics problem is the determination of the resultant of a force system acting on the body. The resultant of a given force system acting on a rigid body is defined to be the simplest force system that can replace the original system without changing the external effect on the body. A resultant is not necessarily a single force but may be a force, a couple, or a force and a couple.

2.2 Resultants of Force Systems

Force systems in general can be classified as *concurrent; coplanar, nonconcurrent;* and *noncoplanar, nonconcurrent* or *general.*

A *concurrent* force system is one in which all lines of action of the forces of the system pass through a common point (Fig. 2.1). We can combine various pairs of forces of this system by the parallelogram law, e.g., reduce F_1 and F_2 to a single force and repeat the process until a single force is obtained. In this manner the concurrent force system can always be reduced to a single force passing through the original point of concurrence, and this force is the resultant for this force system. In general, the simplest method for determining this resultant is to resolve each force into its rectangular components and determine the rectangular components of the resultant from the

45

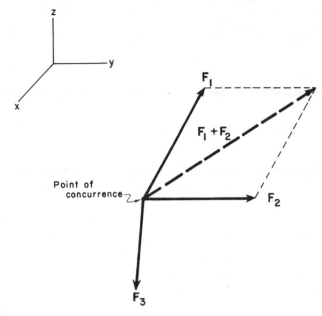

Fig. 2.1

equations

$$R_x = \sum_{i=1}^{n} (F_x)_i, \qquad R_y = \sum_{i=1}^{n} (F_y)_i, \qquad R_z = \sum_{i=1}^{n} (F_z)_i \qquad (2.1)$$

Here $(F_x)_i$ represents the x component of the ith force, and n is the number of forces. In vector form the resultant can be written as

$$\mathbf{R} = R_x \mathbf{i} + R_y \mathbf{j} + R_z \mathbf{k} \qquad (2.2)$$

where $\mathbf{i}$, $\mathbf{j}$, $\mathbf{k}$ are the unit vectors in the positive x, y, z directions respectively.

The *coplanar, nonconcurrent* force system is composed of forces which all lie in the same plane (coplanar) but do not all intersect at a common point (nonconcurrent) (Fig. 2.2).

The resultant of a nonparallel, coplanar force system can be obtained by combining the various pairs of forces by the parallelogram law, making use of the fact that the forces are transmissible along their lines of action, until a single force is obtained. This force, which is the resultant, passes through a particular point in the plane of the forces, and the location of this point depends on the characteristics of the various forces of the system. A special situation arises when the vector sum of all the forces except one results in a force that is

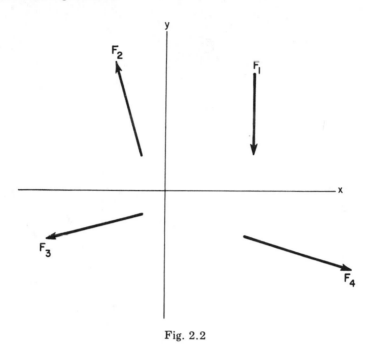

Fig. 2.2

parallel to, equal in magnitude to, and has an opposite sense to the remaining force (Fig. 2.3).

Another conceptual approach to finding the resultant of a coplanar, nonconcurrent force system (including a parallel force system) is to replace each force by a force-couple system at an arbitrary point 0. In Fig. 2.4a the arbitrary point is chosen as the origin of the

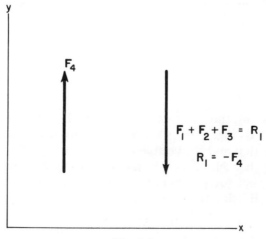

Fig. 2.3

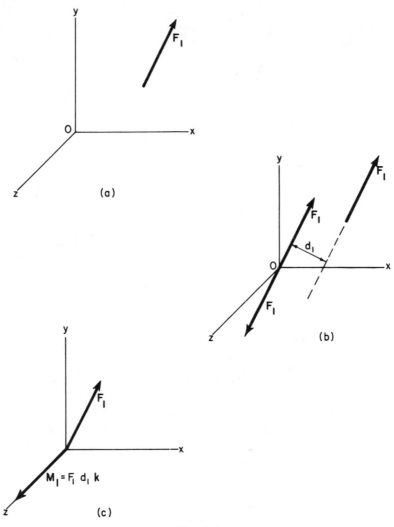

Fig. 2.4

coordinate system, and two forces parallel to $\mathbf{F}_1$, opposite in sense, are drawn through 0 (Fig. 2.4b). These three forces are equivalent to the force $\mathbf{F}_1$ through 0 and the couple $\mathbf{M}_1 = F_1 d_1 \mathbf{k}$ (Fig. 2.4c). The above is repeated with each force of the original system so that at 0 we have a concurrent force system that can be combined by the parallelogram law of addition to yield

$$\mathbf{R} = R_x \mathbf{i} + R_y \mathbf{j} \tag{2.3}$$

The couples of each of these forces at 0 all act to produce a vector moment along the z direction. Adding these couples results in a moment $M_0 = M_0 \mathbf{k}$. Hence the original force system can be replaced by a single force $\mathbf{R}$ at the point 0 and a single moment $\mathbf{M_0}$.

If $\mathbf{R} = \mathbf{0}$, the resultant is the couple $\mathbf{M_0}$; while if $\mathbf{R} \neq \mathbf{0}$, the force $\mathbf{R}$ and couple $\mathbf{M_0}$ are equivalent to a single force as demonstrated in Chapter 1. To find the line of action of this single force, use the principle of moments which states that the sum of the moments due to the individual forces about point 0 is equal to the moment of the resultant force $\mathbf{R}$ about point 0. This leads to

$$M_0 = \Sigma \ M_i = Rd \qquad (2.4)$$

where d is the distance from point 0 to the line of action of the force $\mathbf{R}$ (Fig. 2.5).

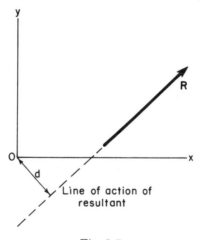

Fig. 2.5

We conclude that the resultant is either a force or a couple. It will not be a coplanar force *and* a couple since this combination can always be reduced to a single force.

EXAMPLE 2.1

Find the resultant force and the resultant moment about the origin and locate the line of action of the resultant for the coplanar forces acting in Fig. E2.1a.

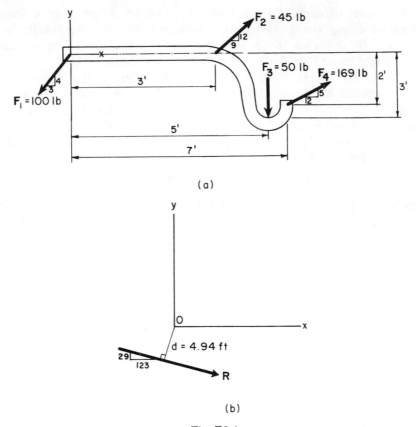

(a)

(b)

Fig. E2.1

SOLUTION

Determine the rectangular components of the forces starting at the left end:

$$\mathbf{F}_{1x} = -100\,(3/5)\mathbf{i} = 60\ \text{lb} \leftarrow \qquad \mathbf{F}_{3x} = 0$$

$$\mathbf{F}_{1y} = -100\,(4/5)\mathbf{j} = 80\ \text{lb} \downarrow \qquad \mathbf{F}_{3y} = 50\ \text{lb} \downarrow$$

$$\mathbf{F}_{2x} = 45\,(9/15)\mathbf{i} = 27\ \text{lb} \rightarrow \qquad \mathbf{F}_{4x} = 169\,(12/13)\mathbf{i} = 156\ \text{lb} \rightarrow$$

$$\mathbf{F}_{2y} = 45\,(12/15)\mathbf{j} = 36\ \text{lb} \uparrow \qquad \mathbf{F}_{4y} = 169\,(5/13)\mathbf{j} = 65\ \text{lb} \uparrow$$

Find R_x and R_y by summing forces (positive to the right and positive upward) so that

$$\mathbf{R}_x = (-60 + 27 + 156)\mathbf{i} = 123\ \text{lb} \rightarrow$$

$$\mathbf{R}_y = (-80 + 36 - 50 + 65)\mathbf{j} = 29\ \text{lb} \downarrow \qquad\qquad Ans.$$

or in vector form

$$\mathbf{R} = R_x\mathbf{i} + R_y\mathbf{j} = 123\mathbf{i} - 29\mathbf{j} \text{ lb} \qquad\qquad Ans.$$

The moment about the origin of $\mathbf{F}_1$ is

$$\text{(+)}\, M_1 = 60(0) + 80(0) = 0$$

where (+) is arbitrarily chosen as the direction of positive moment. Similarly,

$$\text{(+)}\, M_2 = (27)(0) + 36(3) = 108 \text{ ft-lb}$$

$$\text{(+)}\, M_3 = -50(5) = -250 \text{ ft-lb}$$

$$\text{(+)}\, M_4 = +(156)(2) + 65(7) = 767 \text{ ft-lb}$$

The negative sign on the moment M_3 indicates that the moment due to $\mathbf{F}_3$ actually acts in a direction opposite to that chosen as positive moment.

Since each moment tends to produce a rotation about the same axis through the origin, algebraic addition can be used to find the total moment

$$\mathbf{M_0} = 625 \text{ ft-lb} \;\curvearrowleft \qquad\qquad Ans.$$

or

$$\mathbf{M_0} = 625\mathbf{k} \text{ ft-lb} \qquad\qquad Ans.$$

Here the resultant force $\mathbf{R}$ is not zero; hence the resultant will be a single force. The location of this force relative to the origin 0 is found from Eq. (2.4), which gives

$$d = 625/\sqrt{(123)^2 + (29)^2} = 4.94 \text{ ft}$$

The resultant force is shown in Fig. E2.1b.

The *noncoplanar, nonconcurrent* or *general* force system is a system of forces which are not all in the same plane and whose lines of action do not all intersect at a common point. To determine the resultant of a force system of this nature, proceed as in the case of the coplanar, nonconcurrent force system. Here we replace each force by a parallel force and a couple at an arbitrary point P in space. The set of forces at this point constitutes a concurrent set of forces that can be combined by the parallelogram law to yield

$$\mathbf{R} = \sum_{i=1}^{n} \mathbf{F}_i \qquad\qquad (2.5)$$

The couple needed for each force transferred from its original position to the point P is $\mathbf{M} = \mathbf{r} \times \mathbf{F}$, where $\mathbf{r}$ is a vector from P to any point on the line of action of $\mathbf{F}$. For each force of the system that is transferred to point P we obtain a couple, and the vector addition of these couples produces a resultant couple $\mathbf{M}_P$.

$$M_P = \sum_{i=1}^{n} (\mathbf{r}_i \times \mathbf{F}_i) \tag{2.6}$$

In Fig. 2.6a we consider a general system of forces. Each of these

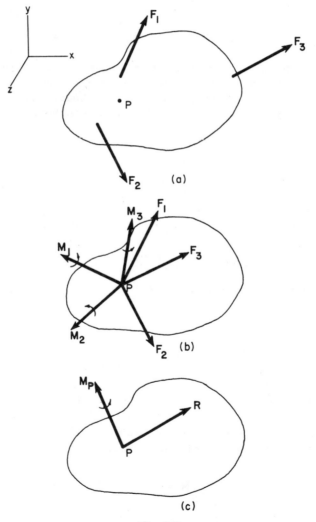

Fig. 2.6

forces is moved parallel to itself to act through point P, and they are shown in Fig. 2.6b with their accompanying couples. The forces and couples at P are added vectorially to produce a resultant force $\mathbf{R}$ and a resultant couple $\mathbf{M}_P$ (Fig. 2.6c).

If we choose the origin of a Cartesian coordinate system at P, Eqs. (2.5) and (2.6) can be written in component form as

$$R_x = \sum_{i=1}^{n} (F_x)_i \qquad M_x = \sum_{i}^{n} [y_i (F_z)_i - z_i (F_y)_i]$$

$$R_y = \sum_{i=1}^{n} (F_y)_i \qquad M_y = \sum_{i=1}^{n} [z_i (F_x)_i - x_i (F_z)_i] \qquad (2.7)$$

$$R_z = \sum_{i=1}^{n} (F_z)_i \qquad M_z = \sum_{i=1}^{n} [x_i (F_y)_i - y_i (F_x)_i]$$

Here R_x, R_y, R_z are the components of the resultant force in the x, y, z directions respectively, and M_x, M_y, M_z are the resultant moments about the x, y, z axes respectively.

In general, the resultant for a general force system consists of a force and a moment, but a special case arises if the resultant force $\mathbf{R}$ and the resultant couple $\mathbf{M}_P$ are orthogonal to one another; i.e., $\mathbf{R} \cdot \mathbf{M}_P = 0$. If the two vectors $\mathbf{R}$ and $\mathbf{M}_P$ are orthogonal and $\mathbf{R} \neq \mathbf{0}$, the resultant of the force system can be reduced to a single force. To see this, we note that the force $\mathbf{R}$ can be replaced by an equivalent force and a couple in the plane perpendicular to the moment. A couple can be found that is equal in magnitude but directed opposite to $\mathbf{M}_P$, and the two couples then combine to give a zero moment. We are thus left with the force $\mathbf{R}$ acting at a distance d from P in a plane perpendicular to $\mathbf{M}_P$. However, if $\mathbf{R} = \mathbf{0}$, the resultant will be a couple.

EXAMPLE 2.2

Reduce the system of forces acting on the rectangular block in Fig. E2.2 to a force-couple system.

SOLUTION

Write each of the forces as vectors, with $\mathbf{i}$, $\mathbf{j}$, $\mathbf{k}$ being the unit vectors in the x, y, z directions respectively. Thus

$\mathbf{F}_1 = 26[(12/13)\mathbf{j} + (5/13)\mathbf{k}] = 24\mathbf{j} + 10\mathbf{k}$ N $\qquad \mathbf{F}_4 = -10\mathbf{j}$ N

$\mathbf{F}_2 = 10\mathbf{k}$ N $\qquad\qquad\qquad\qquad\qquad\qquad \mathbf{F}_5 = -15\mathbf{i}$ N

$\mathbf{F}_3 = -5\mathbf{k}$ N

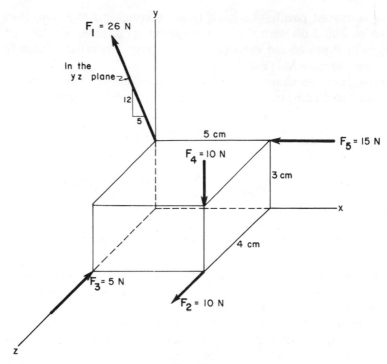

Fig. E2.2

Summation of forces gives $R = -15i + 14j + 15k$ N. Now compute the moments of each force about the origin. To do this, write the position vector from 0 to the point of application of each force:

$$r_1 = 3j, \quad r_2 = 5i + 4k, \quad r_3 = 4k$$
$$r_4 = 5i + 3j + 4k, \quad\quad\quad r_5 = 5i + 3j$$

The moment is then

$$M_0 = \sum_{i=1}^{5} (r_i \times F_i)$$

$$= (3j) \times (24j + 10k) + (5i + 4k) \times (10k) + 4k \times (-5k)$$
$$+ (5i + 3j + 4k) \times (-10j) + (5i + 3j) \times (-15i) \text{ N-cm}$$
$$= (70i - 50j - 5k) \text{ N-cm}$$

To see whether any further reduction is feasible, compute the dot product of R and M_0 to obtain

$$R \cdot M_0 = (-15)(70) + (14)(-50) + (15)(-5) \neq 0$$

Here **R** and M_0 are not perpendicular to one another, and the resultant of the force system is a force through 0 given by

$$R = (-15i + 14j + 15k) \text{ N}$$

and a couple

$$C = (70i - 50j - 5k) \text{ N-cm} \qquad\qquad Ans.$$

2.3   Center of Gravity and Distributed Loads

In the preceding sections we have worked with "concentrated forces." This type of load is generally a mathematical concept and is rarely found in physical reality, although in many physical problems a concentrated force can be used as a good approximation for the actual load. A common type of load found in physical situations is the distributed load, i.e., a load which may vary from point to point over an area or in a volume. Examples of distributed loads are (1) your weight distributed by your feet on the floor, (2) the impact a fullback in football feels when tackled by a defensive lineman, or (3) a number of automobiles on a bridge.

The most commonly encountered distributed load is the weight of a body, which is usually given only in magnitude; e.g., we say that a steel block weighs 100 lb. It is understood that the direction of the force is for all practical purposes directed toward the center of the earth. In actuality when specifying the weight of a body in this manner, we are really giving the resultant force of the infinitesimal forces due to the earth's gravitational field of attraction acting on each mass particle making up the total body. To replace these infinitesimal forces by a resultant force, we must know the magnitude of the resultant force and its point of application.

Consider a body of material filling a volume V with *specific weight* γ, defined as the weight per unit volume. Here γ may be a function of the position in the body; i.e., $\gamma = \gamma(x, y, z)$. Partition the volume V into infinitesimal volumes dV (Fig. 2.7). The magnitude of the weight of an infinitesimal volume is $dW = \gamma\, dV$, where γ is an average specific weight for the volume element dV. Here we have assumed that the gravitational attraction of each particle is in the direction of the negative z axis, and for all practical purposes the forces on each element are parallel. The assumption of a parallel force system, even though not exactly correct, is for our purposes sufficiently accurate if the size of the body is small compared to the size of the earth. This type is termed a noncoplanar parallel force system. We sum all the infinitesimal forces dW, with the positive direction of the

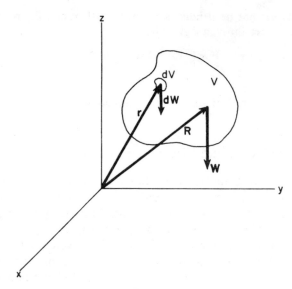

Fig. 2.7

coordinate axis taken as the direction of positive force. The result is

$$W = -Wk = -k \int_V dW = -k \int_V \gamma(x, y, z)\, dV$$

where W is the actual weight of the body.

To find the point of application of the resultant force W, equate the moment of the force W about some arbitrarily chosen point to the sum of the moments about the same point of the infinitesimal forces dW. For convenience we choose the origin of our coordinate system as the point about which moments are computed. Let r be the position vector to the force dW and R be the position vector to the point of application of W. Equating the moments gives

$$-R \times Wk = -\int_V r \times k\, dW = -\int_V r \times k\, \gamma(x, y, z)\, dV$$

The above equation can be written as

$$\left[-RW + \int_V r\, \gamma(x, y, z)\, dV \right] \times k = 0$$

where in general the vector in the brackets is not parallel to k. This leads to

$$RW = \int_V r\gamma(x, y, z)\, dV$$

The resultant weight W is thus equivalent to the distributed weight if

$$W = \int_V \gamma(x, y, z)\, dV \qquad (2.8)$$

and

$$RW = \int_V r\gamma\, dV \qquad (2.9)$$

Let $R = x_g i + y_g j + z_g k$ and $r = xi + yj + zk$ so that Eq. (2.9) can be written as

$$x_g W = \int_V x\gamma(x, y, z)\, dV$$

$$y_g W = \int_V y\gamma(x, y, z)\, dV \qquad (2.10)$$

$$z_g W = \int_V z\gamma(x, y, z)\, dV$$

The integrals on the right-hand side of Eqs. (2.10) represent the sum of the moments of the incremental weight dW about the yz, xz, and xy planes, respectively.   The terms on the left-hand sides of Eq. (2.10) are the moments of W about the yz, xz, and xy planes respectively.

The point whose coordinates are (x_g, y_g, z_g) is called the *center of gravity* of the body and physically represents the point at which the body in any position can be supported by a single force and have no tendency to rotate. If the specific weight γ is constant, the center of gravity and the centroid of the volume coincide; i.e., $x_g = x_c$, $y_g = y_c$, and $z_g = z_c$.

EXAMPLE 2.3

Find the center of gravity of the block in Fig. E2.3 if the specific weight of the material is given by the equation $\gamma = kx$.

SOLUTION

For this geometry a rectangular coordinate system is the most convenient, and the differential element of volume is $dV = dx\, dy\, dz$. The weight is then

$$W = \int_V kx\, dx\, dy\, dz = \int_{x=0}^{x=2a} kx\, dx \int_{y=0}^{y=2b} dy \int_{z=0}^{z=2c} dz = 8\, ka^2 bc$$

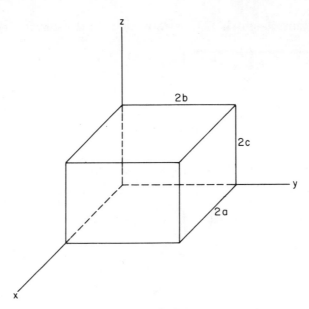

Fig. E2.3

To find x_g, y_g, z_g, integrate Eqs. (2.10). Thus

$$\int_V x\gamma \, dV = \int_0^{2a} kx^2 \, dx \int_0^{2b} dy \int_0^{2c} dz = \frac{32ka^3bc}{3}$$

$$\int_V y\gamma \, dV = \int_0^{2a} kx \, dx \int_0^{2b} y \, dy \int_0^{2c} dz = 8ka^2b^2c$$

$$\int_V z\gamma \, dV = \int_0^{2a} kx \, dx \int_0^{2b} dy \int_0^{2c} z \, dz = 8 \, ka^2bc^2$$

and

$$x_g = (4/3) \, a, \qquad y_g = b, \qquad z_g = c \qquad\qquad Ans.$$

Note that the block as a geometric figure has three planes of symmetry given by $x = a$, $y = b$, $z = c$; since the centroid of the volume must be on all these planes of symmetry, it must then be located at $x_c = a$, $y_c = b$, $z_c = c$. Thus the center of gravity and the centroid of the volume do not coincide in this case.

The concept of the center of gravity of volumes is easily carried over to thin plates. A plate of constant thickness is considered thin when its thickness is small compared to its cross-sectional dimensions.

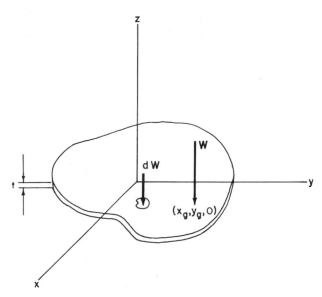

Fig. 2.8

If we consider a plate oriented as in Fig. 2.8 and assume that $z = 0$ is the middle surface of the plate and if the plate is thin, we can assume that the specific weight does not change significantly as a function of z. We can then approximate the specific weight as $\gamma = \gamma(x, y)$.

For a thin plate of constant thickness t, oriented as in Fig. 2.8, the weight as given by Eq. (2.8) is

$$W = \int_A t\gamma(x, y)\, dA \tag{2.11}$$

where dA is the increment of area and the integral is taken over the area of the plate. From Eqs. (2.10) we then have for the x and y coordinates of the center of gravity

$$x_g W = \int_A tx\gamma(x, y)\, dA, \quad y_g W = \int_A ty\gamma(x, y)\, dA \tag{2.12}$$

The z coordinate of the center of gravity is $z_g = 0$ since $z = 0$ is a plane of symmetry of the plate and the specific weight is independent of z. If γ is constant, the center of gravity of the thin plate coincides with the centroid of the planar area.

EXAMPLE 2.4

Find the center of gravity of the thin plate having the shape of a semicircle as in Fig. E2.4. The specific weight is given by the expression $\gamma = k(x^2 + y^2)^{1/2}$.

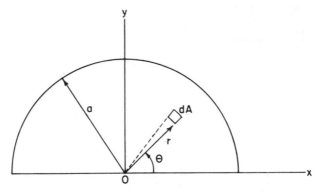

Fig. E2.4

SOLUTION

An appropriate coordinate system to use here is a plane polar coordinate system with the pole at point 0. Then $x = r \cos \theta$, $y = r \sin \theta$, and the specific weight becomes $\gamma = kr$. The increment of area is $dA = r\,dr\,d\theta$ and the plate is assumed to be of constant thickness t. The weight is

$$W = \int_A tkr^2\,dr\,d\theta = \int_{r=0}^a tkr^2\,dr \int_0^\pi d\theta = \frac{\pi ka^3 t}{3}$$

To find the center of gravity, compute the two integrals in Eqs. (2.12):

$$\int_A tx\gamma\,dA = \int_A tkr^3 \cos\theta\,dr\,d\theta = \int_0^a tkr^3\,dr \int_0^\pi \cos\theta\,d\theta = 0$$

$$\int_A ty\gamma\,dA = \int_A tkr^2 \sin\theta\,dr\,d\theta = \int_0^a tkr^3\,dr \int_0^\pi \sin\theta\,d\theta = \frac{ka^4 t}{2}$$

Thus

$$x_g = 0, \quad y_g = 3a/2\pi, \quad z_g = 0 \qquad\qquad Ans.$$

Another type of distributed load frequently encountered is surface tractions, i.e., a load distributed over the surface of a body. The distributed load on a surface can be thought of as a vector function $\mathbf{f}(x, y, z)$ that has units of force per unit area. The resultant force due to this distributed load is given by

$$\mathbf{R} = \int_A \mathbf{f}(x, y, z)\,dA \qquad\qquad (2.13)$$

where the integration is over the surface area on which the load is applied. The moment about some arbitrary chosen point 0 is

$$M_0 = \int_A r \times f \, dA$$

where r is a position vector from point 0 to a point on the loaded surface.

In general, a resultant for the distributed load could be a force **R** acting at some arbitrary point 0 and a moment M_0 about the same point. Any further reduction of the resultant depends on the conditions discussed in Section 2.2.

In following chapters we will be concerned with distributed loads on beams. A *beam* can be defined as a general cylinder whose length is large compared to the maximum cross-sectional dimensions. Common examples are long circular rods, I beams, or long beams of rectangular cross sections. The distributed loads most often encountered in beam analysis are coplanar and parallel (Fig. 2.9). Since in many

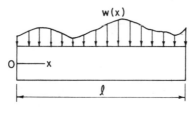

Fig. 2.9

cases these loads are uniform across the thickness of the beam, they are sometimes called *line loads* and expressed as a force per unit length. The resultant of a coplanar-parallel force system, all of whose forces are in the same direction, will be a force whose direction is the same as the parallel forces. The resultant of the distributed line load $w(x)$ from Eq. (2.13) is

$$R = \int_0^\ell w(x) \, dx \tag{2.14}$$

where the direction of the force is directed downward. The moment of the distributed load about 0 is found to be

$$M_0 = \int_0^\ell x w(x) \, dx \tag{2.15}$$

The vector representing the moment is directed into the plane of the paper.

The point of application relative to 0 is given by the condition that the moment due to R must be equal to M_0. Thus

$$x_R R = \int_0^\ell xw(x)\, dx \qquad (2.16)$$

where x_R is the point of application of the resultant R. Note that the integral in Eq. (2.16) is the first moment of the area bounded by $w(x)$ about the y axis, and x_R is the x coordinate of the centroid of the area bounded by $w(x)$. The resultant force R in Eq. (2.14) is the area under the distributed load curve. These two facts sometimes allow easy computation of the resultant force and its point of application for a distributed load, as illustrated in Example 2.5.

EXAMPLE 2.5

Find the resultant of the distributed load in Fig. E2.5.

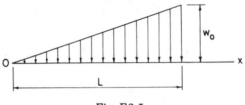

Fig. E2.5

SOLUTION

The magnitude of the resultant force is the area under the distributed load, which for a right triangle is $R = w_0 L/2$. To locate the point of application, find the first moment of this area about the vertical line through point 0. Here $w(x) = w_0 x/L$ and

$$\int_0^L xw(x)\, dx = \int_0^L \frac{w_0}{L} x^2\, dx = \frac{w_0 L^3}{3L} = \frac{w_0 L^2}{3}$$

$$x_R = \left[\int_0^L xw(x)\, dx \right] \bigg/ R = \frac{2w_0 L^2}{3w_0 L} = \frac{2L}{3}$$

Note that x_R is the x coordinate of the centroid of the triangle and could have been obtained from Table 1.5. The resultant is then

$$R = w_0L/2 \downarrow$$

acting at a distance of 2L/3 to the right of point 0.   .   *Ans.*

2.4 Free-Body Diagram

In previous sections consideration has been given primarily to a system of forces without regard to any physical problem or to the

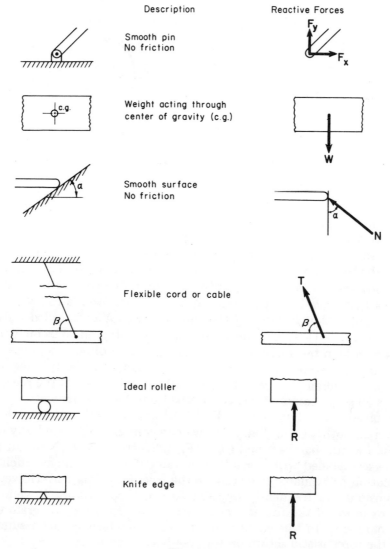

Fig. 2.10

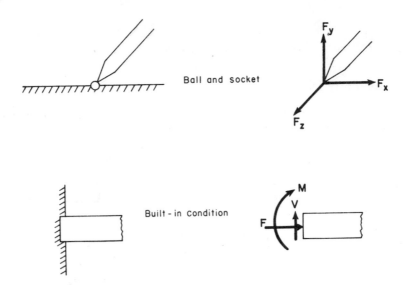

Fig. 2.10 (*continued*)

origin of the forces. However, the various forces acting on a system must first be determined before the resultant can be obtained or any conclusion drawn with respect to equilibrium conditions. To determine the forces, we isolate the body of interest from its surroundings and show the forces and couples that act on it due to the surrounding bodies that have been removed. A body so isolated is called a *free body*, and a sketch showing the forces and couples is called a *free-body diagram* (FBD). Free-body diagrams play an extremely important role in the analysis of many mechanics problems, and the importance of constructing complete, correct, free-body diagrams cannot be overemphasized. Figure 2.10 is a collection of the reactive forces on some commonly encountered bodies in mechanics.

Consider the two rods in Fig. 2.11a, where all surfaces are assumed smooth. Using Fig. 2.10, we can construct the free-body diagrams for the bars AB and CD. (Figs. 2.11b, 2.11c). Note all the forces are labeled with a symbol, or value if known, and as much information as available is placed on the free-body diagram. However, dimensions are frequently omitted to avoid confusion. In general, one or more of the forces that appear on the free-body diagram will be unknown. In this case the unknown characteristics are assumed, and the force is still shown on the free-body diagram.

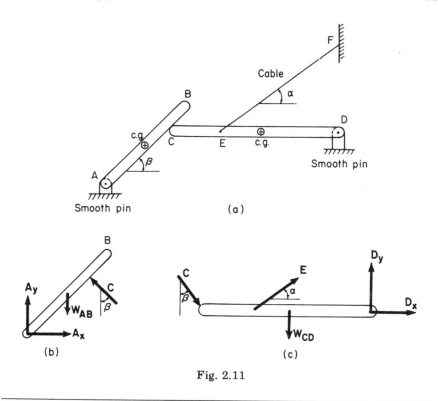

Fig. 2.11

EXAMPLE 2.6

Draw the free-body diagrams for the weight hanging from the cable in Fig. E2.6a, the block resting on the smooth plane in Fig. E2.6b, and the mechanism loaded and supported as in Fig. E2.6c.

(a) If we cut the rope and effectively remove the rigid support from the rope and weight, we can ask what forces must be exerted on the rope-weight combination by the removed portion. Gravity exerts a force downward if we assume that downward is in the direction of the center of the earth, and the rope exerts an upward force on the removed portion. The free-body diagram is shown in Fig. E2.6d.

(b) We remove the block from the inclined plane and are led to a free-body diagram as in Fig. E2.6e. The forces in Fig. E2.6e are the weight of the block W, the force N of the plane in a normal direction on the block, and the externally applied force P.

(c) The mechanism is isolated from the pin and roller to give the free-body diagram in Fig. E2.6f. Here F_1 is the reaction of the rollers on the mechanism, and F_2 and F_3 are the vertical and horizontal reactions of the pin on the mechanism. The external applied load is P. No internal members need be considered for this particular free-body diagram.

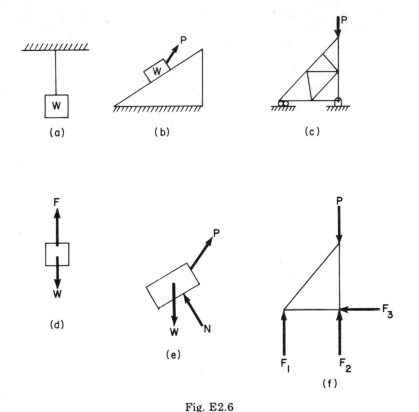

Fig. E2.6

2.5 Equilibrium

Any general force system can always be replaced by a force and a couple. From Newton's second law we also know that if the resultant force on a particle is zero, the particle will be in equilibrium; i.e., it will either be at rest or translating with a constant velocity. Extending this concept of equilibrium to a rigid body requires the resultant of the force system to vanish. Therefore to ensure equilibrium we write

$$R = 0, \qquad M_p = 0 \tag{2.17}$$

where R and M_p are the force and couple respectively that make up the resultant of the general force system. That these are actually the necessary and sufficient conditions for equilibrium can be proved from the principles of dynamics. Equations (2.17) can be written as

$$\sum_{i=1}^{n} \mathbf{F}_i = 0, \qquad \sum_{i=1}^{n} (\mathbf{r}_i \times \mathbf{F}_i) = 0 \qquad (2.18)$$

where $\mathbf{F}_i$ represents any force of the general force system.

From Eqs. (2.7) the equilibrium conditions stated in Eqs. (2.18) can be written in component form as

$$\sum_{i=1}^{n} (F_x)_i = 0 \qquad \sum_{i=1}^{n} [y_i(F_z)_i - z_i(F_y)_i] = 0$$

$$\sum_{i=1}^{n} (F_y)_i = 0 \qquad \sum_{i=1}^{n} [z_i(F_x)_i - x_i(F_z)_i] = 0 \qquad (2.19)$$

$$\sum_{i=1}^{n} (F_z)_i = 0 \qquad \sum_{i=1}^{n} [x_i(F_y)_i - y_i(F_x)_i] = 0$$

where the equations in the second column represent the moments of the forces about the x, y, z axes respectively.

The x, y, z axes have been arbitrarily chosen as the direction for summing forces and the axes about which moments are computed. If the system of forces is in equilibrium, the sum of the components of the forces in any direction must vanish and the moment of the forces about any line in space must also equal zero. This fact allows the freedom to choose a direction in space, then to sum forces in this direction, and/or to compute the moment of the forces about a line along that direction. These ideas will be demonstrated in the following examples.

Though not explicitly shown in the moment equations of Eqs. (2.19), the general force system can include couples. These couples are included in Eqs. (2.19) as oppositely directed, equal, parallel forces. In many problems it is convenient at times to merely add the couples to the moment equations.

It is clear from Eqs. (2.19) that for any given general force system we have *six* equations and therefore can solve for a maximum of *six* unknowns. However, as can be seen from a consideration of the following special cases, we are frequently not able to solve for as many as six unknowns.

Concurrent force system. The resultant of this force system is a force; therefore the equations necessary to ensure equilibrium are

$$\Sigma F_x = 0, \qquad \Sigma F_y = 0, \qquad \Sigma F_z = 0 \qquad (2.20)$$

where the subscript on the summation is understood. There are only *three* equations in this case. These results also follow from Eqs. (2.19) since if we select the origin of the coordinate system as the point of concurrence of the forces, the moment equations are identically satisfied. Although it may appear that if we selected the origin at some other point six equations would be available, further considerations would reveal that only three of the equations are independent.

Coplanar, nonconcurrent force system. For this special case the resultant is either a force or a couple. Thus to ensure equilibrium we write

$$\Sigma F_x = 0, \qquad \Sigma F_y = 0 \qquad\qquad (2.21)$$

which eliminates the possibility of a resultant force, and

$$\Sigma M_z = \Sigma\,[x(F_y) - y(F_x)] = 0 \qquad\qquad (2.22)$$

which eliminates the couple. The xy plane is arbitrarily selected as the plane of the forces. For this force system *three* equations are required to ensure equilibrium. This result also follows from Eqs. (2.19) since three of the equations are satisfied identically for the coplanar force system.

EXAMPLE 2.7

The homogeneous prism of Fig. E2.7a weighs 500 N and has a 100 N·m couple acting on it in the ADE plane. Determine the tension in the three supporting wires at A, B, C.

SOLUTION

A free-body diagram of the prism (Fig. E2.7b) indicates we have a special case of a general force system that is a parallel force system. Since all the forces act in the z direction, the two equations $\Sigma F_x = 0$ and $\Sigma F_y = 0$ are satisfied identically. The only moments produced by the forces are about the x and y axes; hence $\Sigma M_z = 0$ is identically satisfied. This leaves only three equilibrium equations to be considered. The unknown forces here are labeled **A**, **B**, and **C**; and if we take moments about a line through A and B, the only unknown involved is the force **C**.

$$+\negmedspace\nwarrow \Sigma M_{AB} = 0, \qquad (1)(500) + 100 - 3C = 0$$

or

$$\mathbf{C} = 200 \text{ N} \uparrow \qquad\qquad\qquad\qquad Ans.$$

If we next take moments about a line ED, we obtain

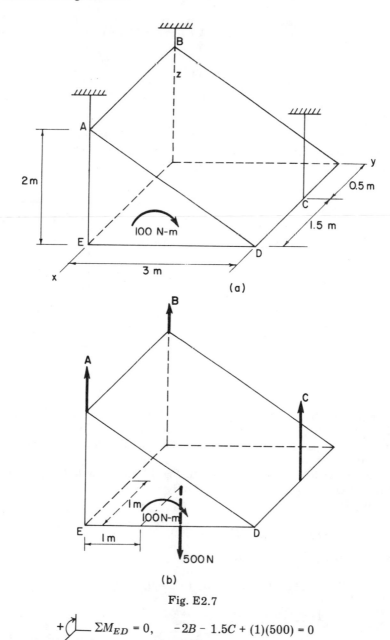

(a)

(b)

Fig. E2.7

$$+\,\circlearrowleft\quad \Sigma M_{ED} = 0, \qquad -2B - 1.5C + (1)(500) = 0$$

Substituting for **C** and solving for **B** gives

$$B = 100 \text{ N} \uparrow \qquad\qquad Ans.$$

Summing forces in the vertical direction yields

$$\textcircled{+}\, \Sigma F_z = 0, \quad A + B + C - 500 = 0$$

or

$$\mathbf{A} = 200 \text{ N} \uparrow \qquad\qquad Ans.$$

EXAMPLE 2.8

Find the reactions on the ladder (Fig. E2.8) which rests on a frictionless floor and is pinned with a frictionless pin at the top, when a 200-lb man is half-way up the ladder. The weight of the ladder is to be neglected.

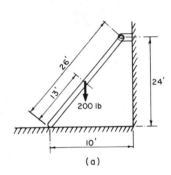 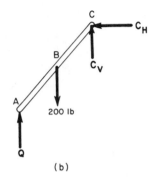

(a)   (b)

Fig. E2.8

SOLUTION

Draw a free-body diagram of the ladder (Fig. E2.8b), noting that since the floor is frictionless, it exerts a force normal to itself on the ladder. The reaction of the pin is displayed as two components in the horizontal and vertical directions.

Examining the free-body diagram in Fig. E2.8b, we note three unknowns, and if we take moments about point C only the unknown Q is involved in the equation. Thus

$$\textcircled{+}\, \Sigma M_c = 0, \quad 10Q - 5(200) = 0, \quad \mathbf{Q} = 100 \text{ lb} \uparrow \qquad Ans.$$

The two force equations then give

$$\textcircled{+}\, \Sigma F_y = 0, \quad Q - 200 + C_v = 0, \quad \mathbf{C_v} = 100 \text{ lb} \uparrow \qquad Ans.$$

$$\textcircled{+}\, \Sigma F_H = 0, \quad \mathbf{C_H} = \mathbf{0} \qquad\qquad Ans.$$

EXAMPLE 2.9

Determine the horizontal and vertical components of the pin reaction at B on member $ABCD$ for the structure in Fig. E2.9a. The 400-ft-lb couple C is in the plane of the paper and is applied to member BE.

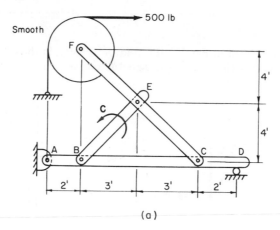

(a)

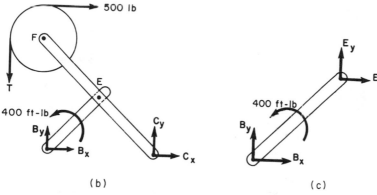

(b)   (c)

Fig. E2.9

SOLUTION

Draw a free-body diagram containing the unknowns; in this case it is convenient to use the diagram in Fig. E2.9b. A free-body diagram of the pulley F reveals that the tension T in the cable is 500 lb. Summation of moments around C gives

$$\underset{\curvearrowleft}{(+)} \; 500(8) - B_y(6) + 400 - 500(10) = 0, \quad B_y = -100$$

Thus on member $ABCD$ the pin reaction is $B_y = 100$ lb ↑. To find B_x, use the free-body diagram of Fig. E2.9c and sum moments about E. Thus

$$\underset{\curvearrowleft}{(+)} \; 400 - (-100)(3) + B_x(4) = 0, \quad B_x = -175$$

On member $ABCD$ $B_x = 175$ lb →; therefore the pin reactions at B on $ABCD$ are

$$B_x = 175 \text{ lb} \rightarrow, \quad B_y = 100 \text{ lb} \uparrow \qquad\qquad Ans.$$

EXAMPLE 2.10

For the mechanism in Fig. E2.10a find the tension in the cable AG, the pin reactions on member CD, and the horizontal and vertical pin reaction at E on member $BGDE$.

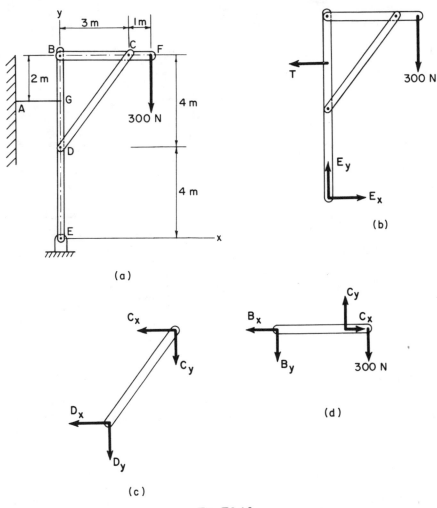

Fig. E2.10

SOLUTION

Consider a free-body diagram of the total mechanism in Fig. E2.10b. From equilibrium considerations

$$\text{(+↺)} \; E_y = 300 \text{ N} \uparrow, \qquad \text{(+→)} \; E_x = T$$

$$\text{(+↻)} \Sigma M_E = 0, \qquad -6T + 4(300) = 0 \qquad\qquad Ans.$$

Hence we have

$$\mathbf{T} = 200 \text{ N} \leftarrow, \qquad \mathbf{E}_x = 200 \text{ N} \rightarrow \qquad\qquad Ans.$$

To find the pin reactions on member CD, draw the free-body diagram of the member (Fig. E2.10c). Equilibrium of the member CD leads to

$$\text{(+→)} D_x + C_x = 0, \qquad \text{(+↺)} D_y + C_y = 0$$

$$\text{(+↻)} \Sigma M_D = 0, \qquad 3C_y - 4C_x = 0 \qquad\qquad \text{(a)}$$

Equation (a) gives $C_y/C_x = 4/3$; and since the tangent of the angle that the resultant force of C_x and C_y makes with the x axis is C_y/C_x, the line of action of the resultant $\mathbf{C}$ is along the member. Similarly, the resultant $\mathbf{D}$ has its line of action along the member but is oppositely directed to the resultant $\mathbf{C}$. This is an example of a *two-force member*; i.e., a two-force member will be in equilibrium only if the two forces have the same line of action and equal magnitude but opposite sense. We can then write

$$C_x = (3/5)C, \qquad C_y = (4/5)C, \qquad D_x = (3/5)D, \qquad D_y = (4/5)D$$

The information obtained from the free-body diagram of member CD is not sufficient to completely solve the problem. To obtain more information, consider the free-body diagram of member BF (Fig. E2.10d).

If we sum moments about a line perpendicular to the plane of the paper through B, the only unknown in the equation will be C_y, so that

$$\text{(+↻)} \Sigma M_B = 0, \qquad -3C_y + 4(300) = 0$$

Hence

$$\mathbf{C} = 500 \text{ N} \; \underset{3}{\diagup}^4 \text{ on } CD, \qquad \mathbf{D} = 500 \text{ N} \; \underset{3}{\diagup}^4 \text{ on } CD \qquad Ans.$$

EXAMPLE 2.11

A plane truss is loaded as in Fig. E2.11a. Find the forces in members DE, CD, DF.

SOLUTION

A plane truss is normally considered to be a structure comprised of beam elements held together at their ends to form a rigid body such that the loads, reactions, and members are all in the same plane. The forces we wish to find in the appropriate members are really *internal forces*. The determination of the internal forces in each member of a truss is based on the following assump-

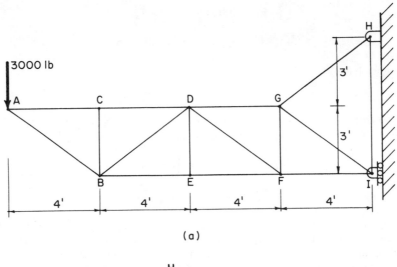

(a)

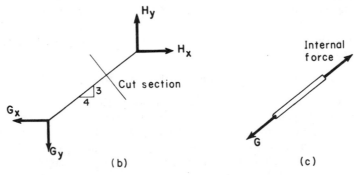

(b)   (c)

Fig. E2.11 (a), (b), (c)

tions: (1) the weight of the beam elements are neglected, (2) the ends of the elements are fastened together with smooth pins, and (3) the loads and reactions act only at the point of connection of the ends of the elements. These points of connection are called *joints*. Under these assumptions the action of the pins at the two ends of the member hold it in equilibrium.

If we consider a free-body diagram of a single member, such as *GH* in Fig. E2.11b, the member is clearly a two-force member. This can be shown to be true of all members of a truss, provided the above assumptions are valid. If we cut a two-force member between two joints (Fig. E2.11c), the internal force developed at the cut section acts along the member and has a magnitude equal to the load carried by the member at the joint.

There are two approaches to analyzing a truss, the *method of joints* and the *method of sections*. For this example we will demonstrate both.

Method of joints. The idea in this method is to isolate each individual joint of the structure. For each joint we have a free-body diagram and a set of equilibrium equations. This set of equations can then be solved for the forces in the required members.

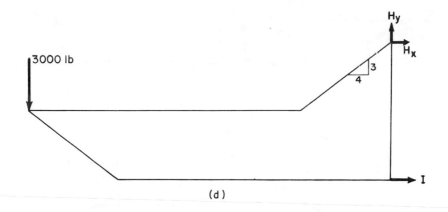

(d)

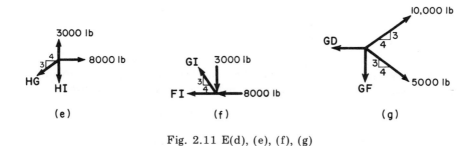

(e)   (f)   (g)

Fig. 2.11 E(d), (e), (f), (g)

The reactions at H and I are found by considering a free-body diagram of the entire structure (Fig. E2.11d). From equilibrium we have

$$I = -8000 \text{ lb}, \qquad H_y = 3000 \text{ lb}, \qquad H_x = 8000 \text{ lb}$$

or

$$H_x = 8000 \text{ lb} \rightarrow, \qquad H_y = 3000 \text{ lb} \uparrow, \qquad I = 8000 \text{ lb} \leftarrow$$

The joint at H is isolated, and the free-body diagram is shown in Fig. E2.11e. The equilibrium of forces in the horizontal direction gives

$$HG = 10,000 \text{ lb}$$

while equilibrium in the vertical direction leads to $HI = -3000$ lb. Thus at joint H we have

$$HG = 10,000 \text{ lb} \diagdown^3_4, \qquad HI = 3000 \uparrow$$

The next joint we isolate is at I, and the free-body diagram is shown in Fig. E2.11f. Equilibrium leads to

$$GI = 5000 \text{ lb} \diagup^3_4, \qquad FI = 12,000 \text{ lb} \rightarrow$$

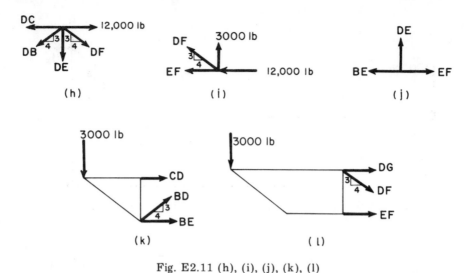

Fig. E2.11 (h), (i), (j), (k), (l)

Next we isolate the joint at G and draw the free-body diagram of this joint (Fig. E2.11g). The equilibrium of forces in the horizontal direction gives

$$GD = (4/5)(10,000) + (4/5)(5000) = 12,000 \text{ lb}$$

while equilibrium of forces in the vertical direction leads to

$$GF = (3/5)(10,000) - (3/5)(5000) = 3000 \text{ lb}$$

Thus we have at joint G

$$GD = 12,000 \text{ lb} \leftarrow, \qquad GF = 3000 \text{ lb} \downarrow$$

We next examine joint D, which has a free-body diagram as in Fig. E2.11h. From equilibrium we have

$$\overset{+}{\longrightarrow} \quad -DC + 12,000 + (4/5)\,DF - (4/5)\,DB = 0 \qquad \text{(a)}$$

$$\overset{+}{\uparrow} \quad DE + (3/5)\,DF + (3/5)\,DB = 0 \qquad \text{(b)}$$

We have four unknowns and two equations, which means we must have two more independent equations involving the unknowns to get a solution. We consider joint F since we know the force in member FI and GF. The free-body diagram of this joint is shown in Fig. E2.11i. Equilibrium in the vertical direction gives

$$\overset{+}{\uparrow} \quad 3000 + (3/5)\,DF = 0$$

or

DF = 5000 lb $^3\diagdown_4$ at joint *F*   *Ans.*

The equilibrium equation in the horizontal direction involves none of the unknowns in Eqs. (a) and (b) and hence is not used here. We next examine joint *E*
as seen in Fig. E2.11j. Equilibrium in the vertical direction immediately gives

$$DE = 0 \qquad\qquad Ans.$$

Substituting the values of *DE* and *DF* in Eqs. (a) and (b) leads to

DB = 5000 lb $_4\diagup^3$ at joint *D*

and

DC = 4000 lb ← at joint *D*   *Ans.*

   Method of sections. The method of sections considers the equilibrium of
sections of the structure rather than equilibrium of the individual joints. The
section to be examined is chosen to include the unknown force we want to find.
As in the analysis by joints, one section to choose is about joint *E*, (Fig. E2.11j).
We then have, as before,

$$DE = 0 \qquad\qquad Ans.$$

   To find the force in member *CD*, consider the section in Fig. E2.11k. Here
we have made an imaginary cut of the structure with a plane between the joints
C, *D* and *B*, *E*. The forces in the members are drawn as before, keeping in mind
that each member is a two-force member. Notice that if we sum the moments of
the forces about a line perpendicular to the plane of the paper through joint *B*,
only the unknown *CD* is involved in the equation.

$$\overset{+}{\curvearrowright}\Sigma\ M_B = 0, \qquad 3CD - 4(3000) = 0$$

or

CD = 4000 → at joint *C*   *Ans.*

The remaining force in member *DF* can be found by considering the section in
Fig. E2.11l. The equilibrium equation in the vertical direction involves only the
unknown *DF* and leads to

DF = 5000 lb $^3\diagdown_4$ at joint *D*   *Ans.*

2.6 Shear and Moment Diagrams

   The study of deformable bodies concerns itself with the change
of shape a body undergoes due to external forces. In general, a body
resists this change of shape, and this resistance gives rise to internal

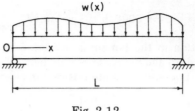

Fig. 2.12

forces.   In this case when the body is in static equilibrium, these internal forces can be related to the externally applied forces. We consider here the internal forces in beams under the action of co-planar, parallel loads. As a particular example, consider a beam on simple supports loaded with a distributed load $w(x)$ (Fig. 2.12). The free-body diagram of the total beam is shown in Fig. 2.13, and from

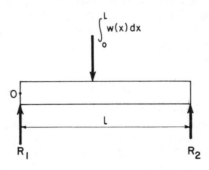

Fig. 2.13

equilibrium considerations we have

$$\overset{+}{\uparrow}\!\!\!\bigcirc \Sigma F = 0, \quad R_1 + R_2 = \int_0^L w(x)\, dx$$

and

$$\overset{+}{\frown}\!\!\!\bigcirc \Sigma M_0 = 0, \quad \int_0^L xw(x)\, dx - R_2 L = 0$$

Thus R_1 and R_2 can be expressed as

$$R_1 = \int_0^L \left(1 - \frac{x}{L}\right) w(x)\, dx \tag{2.23}$$

$$R_2 = \frac{1}{L} \int_0^L xw(x)\, dx \tag{2.24}$$

To relate the internal forces to the applied loads, we mentally cut and remove a portion of the beam and place the reactions of the removed portion on the retained section (Fig. 2.14). We then invoke the condition that if a body is in overall static equilibrium, every part of the body must also be in static equilibrium. In general, we would expect the portion removed to exert a moment $M(x)$ and a force $V(x)$ on the free body (Fig. 2.14). Since all forces on the body are

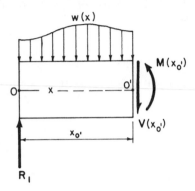

Fig. 2.14

coplanar and parallel, the reactive force $V(x)$ is also in the same plane and parallel to the applied force. The moment $M(x)$ is such that its vector points in a direction normal to the plane of the paper.

The summation of forces gives

$$\text{\textcircled{+}}\Sigma F = 0, \qquad R_1 - \int_0^{x_{0'}} w(x)\, dx - V(x_{0'}) = 0$$

or

$$V(x_{0'}) = R_1 - \int_0^{x_{0'}} w(x)\, dx \qquad (2.25)$$

Moment equilibrium leads to

$$\text{\textcircled{+}}\Sigma M_{0'} = 0, \qquad -M(x_{0'}) - \int_0^{x_{0'}} (x_{0'} - x) w(x)\, dx + R_1 x_{0'} = 0$$

or

$$M(x_{0'}) = R_1 x_{0'} - \int_0^{x_{0'}} (x_{0'} - x) w(x)\, dx \qquad (2.26)$$

Here $x_{0'}$ was an arbitrarily chosen distance from 0, and as such Eqs.

(2.25) and (2.26) give the force $V(x)$ and the moment $M(x)$ as a function of position along the length of the beam. The force $V(x)$ is termed a *shear force*, and it tends to slide two adjacent planes of the beam relative to each other. Equation (2.23) substituted into Eqs. (2.25) and (2.26) gives equations in terms of $w(x)$ for $V(x)$ and $M(x)$. A graphical representation of $V(x)$ and $M(x)$ as functions of position along the beam can thus be constructed. These diagrams are called *shear* and *moment diagrams.*

EXAMPLE 2.12

Draw the shear and moment diagrams for the beam loaded as in Fig. E2.12a.

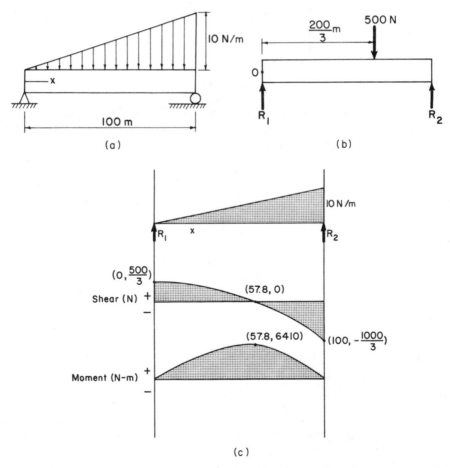

(a)

(b)

(c)

Fig. E2.12

SOLUTION

Draw a free-body diagram of the entire beam (Fig. E2.12b). Equilibrium of the beam requires

$$\circlearrowright{+}\ \Sigma F = 0, \quad R_1 + R_2 = 500$$

$$\circlearrowleft{+}\ \Sigma M_0 = 0, \quad (500)\,[(200)/3] - 100\,R_2 = 0$$

or

$$R_2 = (1000/3)\ \text{N} \uparrow, \quad R_1 = (500/3)\ \text{N} \uparrow$$

Now apply Eqs. (2.25) and (2.26) to obtain the shear force and the moment. With the origin of our coordinate system chosen as in Fig. E2.12a we have

$$w(x) = \frac{x}{10}, \quad V(x) = \frac{500}{3} - \int_0^x \frac{u\,du}{10} = \frac{500}{3} - \frac{x^2}{20}$$

$$M(x) = \frac{500}{3}\,x - \int_0^x (x - u)\,\frac{u}{10}\,du = \frac{500x}{3} - \frac{x^3}{60}$$

We now sketch the curves for the shear and moment (Fig. E2.12c).

Returning to Fig. 2.12, we cut an infinitesimal segment from the body (Fig. 2.15). Here we note that the moment and shear force

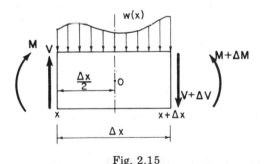

Fig. 2.15

change as we go from a section at x to a section at $x + \Delta x$. Since the overall body is in equilibrium, the removed portion must also be in equilibrium. The resultant of the distributed force can be approximated by $w(\xi)\,\Delta x$ where $x < \xi < x + \Delta x$ and is thought to act at $x + \Delta x/2$. We then have

$$\circlearrowright{+}\ \Sigma F = 0, \quad V - (V + \Delta V) - w(\xi)\,\Delta x = 0$$

which simplifies to

$$\frac{\Delta V}{\Delta x} = -w(\xi)$$

If we take the limit as $\Delta x \to 0$ and assume that $w(x)$ is a continuous function, we then obtain

$$\frac{dV}{dx} = -w(x) \tag{2.27}$$

Summing moments about 0, we get

$$\circlearrowright \Sigma M_0 = 0, \quad M + V\frac{\Delta x}{2} + (V + \Delta V)\frac{\Delta x}{2} - (M + \Delta M) = 0$$

This equation simplifies to

$$-\frac{\Delta M}{\Delta x} + V + \frac{\Delta V}{2} = 0$$

Taking the limit as $\Delta x \to 0$ and assuming $V(x)$ is continuous, we have

$$\frac{dM}{dx} = V(x) \tag{2.28}$$

Equations (2.27) and (2.28) relate the distributed load to the shear force and the shear force to the moment. These equations can be used to find the shear force and moment distribution. When using these equations, a sign convention for positive shear and moment is needed. For this book the shear forces and moments are positive if they act as shown in Fig. 2.15.

EXAMPLE 2.13

Draw the shear and moment diagram for a beam loaded as in Fig. E2.13 and find the maximum value of the moment.

SOLUTION

From the free-body diagram (Fig. E2.13b) we get the reactions

$$R_1 = R_2 = qL/2$$

From Eq. (2.27) we see that dV/dx must be a constant and hence $V(x)$ is a linear function of x. From Fig. E2.13b we see that $V(x)$ will be positive at the left end and equal to R_1. The linear function of shear force plots as a straight line with negative slope and must be such that it equals $-R_2$ at the right end of the beam. A sketch of this line can easily be made (Fig. E2.13c).

Since $V(x)$ is a linear function, $M(x)$ will be a quadratic function; and since

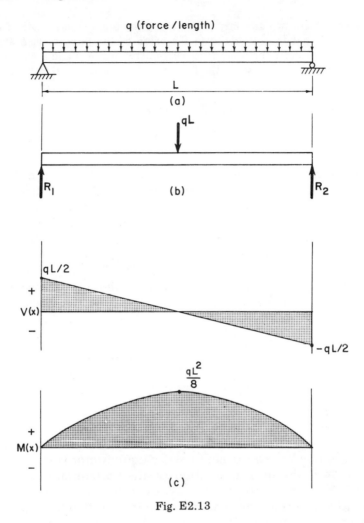

Fig. E2.13

there are no moments on the ends of the beam, $M(x)$ must vanish at the left and right support. We further note that the slope of the moment is merely the value of the shear force $V(x)$. Hence as we move from left to right, the slope of the moment decreases until we have a zero slope, or horizontal tangent, at the point where $V(x)$ is zero. From the shear diagram we see that the slope of the moment diagram is positive, passes through zero, and then is negative.

A sketch of the moment diagram is shown in Fig. E2.13c. From Eq. (2.28) we have

$$M_B - M_A = \int_{x=A}^{x=B} V(u)\, du$$

where A and B are two arbitrary points along the beam, and $B > A$. This last equation states that the difference in the value of the moment at B and A is equal to the area under the shear force curve between $x = A$ and $x = B$. For the problem at hand choose A to be the left end of the beam and B to be the point of maximum moment, then

$$M_{\max} = (qL/2)(L/2)(1/2) = qL^2/8 \qquad\qquad Ans.$$

2.7 Friction

In the previous sections we have assumed that if two bodies were in contact, the surface of contact between the two bodies was smooth. The reactive force between a body and a smooth surface is always normal to the smooth surface. The following demonstration shows that in general a surface cannot be considered smooth.

A block is placed on a flat surface, and a force parallel to that surface is applied to the block (Fig. 2.16). The magnitude of the force is initially zero and then increased. As the force increases from zero, no motion of the block is initially noted. A free-body diagram

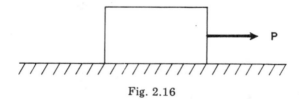

Fig. 2.16

of the block indicates that a force parallel to the surface must develop if the block is to remain in static equilibrium (Fig. 2.17). In Fig. 2.17, N is the normal reaction of the horizontal plane on the block and F is the *friction force*. We see that the friction force is tangent to the surface of contact and acts in a direction to oppose the direction of impending motion of the block. Horizontal force

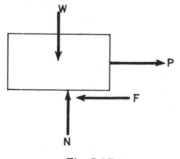

Fig. 2.17

equilibrium leads to $P = F$. If the block remains in equilibrium as P is increased, the friction force F must also increase. A value of P is eventually reached when the block moves, which indicates that the friction force is no longer large enough to maintain equilibrium. At this point of impending motion the friction force attains its maximum value, and experiments have shown that this maximum friction force is proportional to the normal force N. In equation form we can write this relation as

$$F_{max} = \mu N \tag{2.29}$$

where μ is the *coefficient of static friction*.

The coefficient of static friction between two materials is obtained experimentally. Since many factors are involved in determining μ, we can normally find a range of values of μ between any two surfaces. For a more detailed discussion of the phenomena of friction the interested reader can consult the many good texts on materials.

Once the block begins to move, the friction force is given by

$$F = \mu_k N \tag{2.30}$$

where μ_k is the *kinetic coefficient of friction*. This coefficient depends on the velocity of the block and in general is less than the coefficient of static friction. A representative plot of the friction force versus the applied load P and the relative velocity between contact surfaces is given in Fig. 2.18.

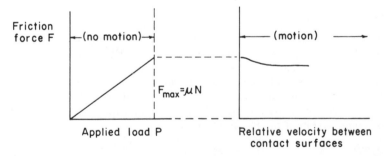

Fig. 2.18

The friction force and the normal force between a block and a plane surface can be combined into a single reactive force (Fig. 2.19). The magnitude of this single force is $R = \sqrt{N^2 + F^2}$, and its line of action can be described by an angle θ measured with respect to the normal force and is given by the expression $\tan \theta = F/N$. If the block has impending motion, then $\mu = F_{max}/N$, and for $F = F_{max}$ the angle

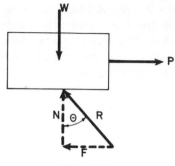

Fig. 2.19

θ takes on its maximum value θ_{max}. We then have

$$\tan \theta_{max} = \mu \qquad (2.31)$$

where θ_{max} is called the *angle of friction.*

In this section we will restrict our attention to static-friction problems of the three following types: (1) from the statement of the problem it is not known if the motion is impending; (2) on all contact surfaces on which there are frictional forces the motion is specified as impending; and (3) impending motion is indicated, but neither the surface on which motion impends is specified nor the type of impending motion (slipping or tipping) is specified.

The following examples demonstrate each of the above types.

EXAMPLE 2.14

A homogeneous block weighing 260 lb is to be moved up the plane by applying a force P through a rope to the block (Fig. E2.14a). The rope remains parallel to the plane and can at most support a maximum force of 125 lb. The coefficient of static friction between the block and the plane is 0.2. If the maximum force is applied to the rope, will the block move? If the block does not move under the allowable force applied to the rope, determine what force the rope must sustain if the block is to move.

SOLUTION

If the maximum force of 125 lb is applied to the rope, we do not know if the block remains in position, has impending motion, or actually is moving. We first assume that the block remains in position. A free-body diagram of the block is shown in Fig. E2.14b, where the friction force **F** is drawn in a direction opposite to possible motion. Force equilibrium in directions normal and tangent to the plane give

$$\text{(↖↗)} N = 240 \text{ lb}, \qquad \text{(↗)} P = 100 + F$$

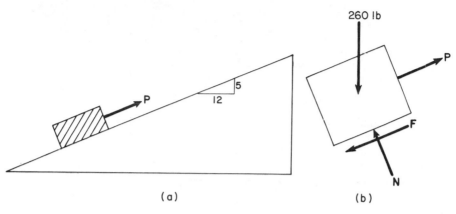

(a)   (b)

Fig. E2.14

For $P_{\text{max}} = 125$ lb the friction force developed is

$$F = 25 \text{ lb} \angle \begin{smallmatrix} 5 \\ 12 \end{smallmatrix}$$

If we assume the motion is impending, then $F = \mu N$, or

$$\mathbf{F}_{\text{max}} = 48 \text{ lb} \angle \begin{smallmatrix} 5 \\ 12 \end{smallmatrix}$$

We thus have that the maximum friction force is greater than the friction force developed by the maximum load the rope can sustain. Hence the block will not move.

If we assume that motion is impending, the friction force is F_{max} and the rope must be able to sustain the force. $P = 100 + F_{\text{max}}$ or

$$P = 148 \text{ lb} \qquad\qquad\qquad Ans.$$

EXAMPLE 2.15

A homogeneous block weighing 500 N is acted on by a horizontal force of 400 N and a vertical force P (Fig. E2.15a). If the coefficient of friction between the block and the plane is 0.25, what is the minimum value of P so that the block has impending motion?

SOLUTION

The statement of the problem indicates that the block is considered to have impending motion, and the friction force thus developed is $F = \mu N$.

A free-body diagram of the block is shown in Fig. E2.15b, and equilibrium of forces leads to

$$\oplus\uparrow N - 500 - P = 0, \qquad \oplus\rightarrow 400 - F = 0$$

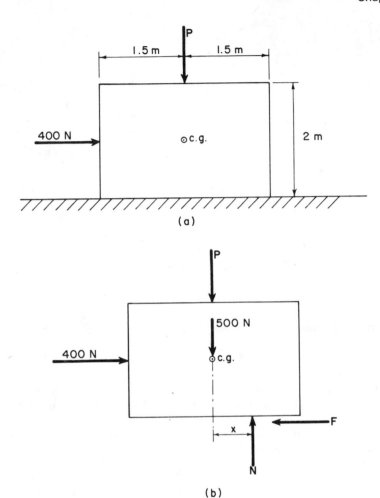

Fig. E2.15

Thus

$$\mathbf{F} = 400 \text{ N} \leftarrow, \quad \mathbf{N} = 1600 \text{ N} \uparrow, \quad \mathbf{P} = 1100 \text{ N} \downarrow \qquad Ans.$$

In the free-body diagram we note that the normal force **N** does not act through the center of gravity (c.g.) of the block. To locate the point of application of the normal force with respect to the center of gravity, the moment equilibrium condition is used; i.e.,

$$\overset{\curvearrowleft}{(+)}\Sigma M_{\text{c.g.}} = 0, \quad -xN + F = 0$$

or $x = 0.25$ m (to the right of the center of gravity). The position of the point of application of the normal force is used to determine if the block will tip before it slides. Thus if the distance x was greater than 1.5 m, this woulu imply that the block would tip before it would slide.

EXAMPLE 2.16

Given two homogeneous blocks arranged and loaded as in Fig. E2.16a, find the value of P such that motion will be impending. The weights of the blocks are $W_A = 200$ lb, $W_B = 100$ lb, and the coefficients of friction are $\mu_{AP} = 0.2$ and $\mu_{AB} = 0.3$.

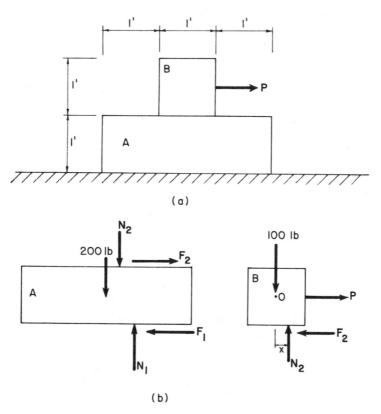

(a)

(b)

Fig. E2.16

SOLUTION

The problem statement indicates that motion is impending, but we do not know at which surface or surfaces. We thus assume that motion is impending between blocks A and B. The appropriate free-body diagrams under this assumption are shown in Fig. E2.16b. Equilibrium conditions applied to block B give

$$\left(\uparrow+\right) N_2 = W_A = 100 \text{ lb}, \qquad \left(\xrightarrow{+}\right) F_2 = P$$

But since motion is assumed to impend between A and B, $F_2 = \mu_{AB}N_2$ and $P = 30$ lb $\rightarrow$. We now check to see if B tips under this load.

$$\circlearrowleft \Sigma M_0 = 0, \quad -xN_2 + 0.5F_2 = 0$$

or $x = 0.15'$ to the right of 0. Hence the block does not tip.

A further check must now be made to ensure that for the value of P obtained, block A does not move on the plane. Equilibrium considerations of block A give

$$\text{(\uparrow+)} N_1 = 200 + N_2, \quad \text{(+)} F_1 = F_2$$

Hence

$$N_1 = 300 \text{ lb} \uparrow, \quad F_1 = 30 \text{ lb} \leftarrow$$

If block A has impending motion on the plane, the friction force is maximum and given by $F_{1 \text{ max}} = \mu_{AP}N_1 = 60$ lb. Since F_1 is not this large, block A does not have impending motion on the plane and our original assumption on where impending motion takes place is correct.

The value of P for impending motion is therefore

$$P = 30 \text{ lb} \rightarrow \qquad\qquad Ans.$$

PROBLEMS

2.1. For the force system in Fig. P2.1 find (a) the resultant force and (b) the moment of the force system with respect to the *aa* axis.

Ans: (a) $R = 136.7i + 31.6j + 27.8k$ lb through 0, (b) $M_{aa} = 136.7k$ ft-lb

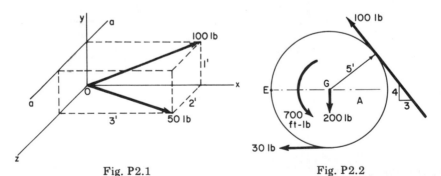

Fig. P2.1   Fig. P2.2

2.2. Determine the resultant of the force system acting on disk A in Fig. P2.2 and locate it with respect to point E.

2.3. Determine the resultant of the force system in Fig. P2.3 and locate it with respect to point A.

Ans:

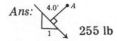

255 lb

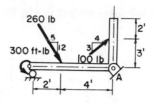

Fig. P2.3

2.4. The force system in Fig. P2.4 consists of the three forces on the figure, an unknown couple C, and an unknown force through point A. The resultant of the force system is the 260-lb force through B as shown. Determine the unknown force and couple.

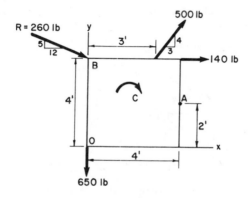

Fig. P2.4

2.5. In the parallel force system in Fig. P2.5, the 30-lb force is the resultant of three forces, two of which are shown. Determine the third force and locate it on a sketch.

Ans: $\mathbf{R} = 40$ lb $\downarrow$, $x_0 = -4.25$ ft, $z_0 = 2.00$ ft

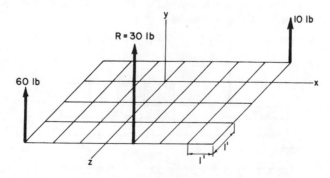

Fig. P2.5

2.6. Find the resultant of the force system in Fig. P2.6 and locate the resultant with respect to point A.

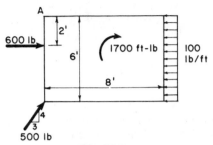

Fig. P2.6

2.7. Determine the resultant of the force system in Fig. P2.7 and locate it with respect to point A.

Ans: $\mathbf{R} = 100$ lb ↑ through a point 20 ft to the left of A

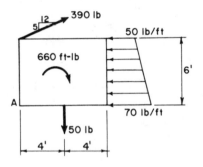

Fig. P2.7

2.8. Find the center of gravity of (a) a thin homogeneous plate covering the first quadrant of the circle $x^2 + y^2 = b^2$ and (b) a thin plate with density $\rho = k/r$ kg/m^3 covering the first quadrant of the circle $x^2 + y^2 = b^2$.

2.9. Find the reactions on the beam in Fig. P2.9 at A and B if the beam is in equilibrium under the given loads.

Ans: $\mathbf{A} = (3P/4) + (wL/8)$ ↑, $\mathbf{B} = (P/4) + (3wL/8)$ ↑

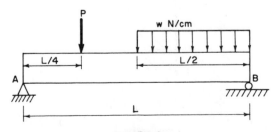

Fig. P2.9

2.10. For the system in Fig. P2.10 find the tension in the cord AB and the reaction of the plane on the block when $P = 100$ lb and $w = 500$ lb. Assume the surface of the plane and block are smooth.

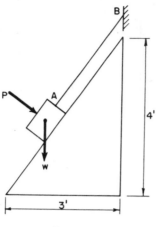

Fig. P2.10

2.11. Find the pin reaction at C on member CD for the mechanism in Fig. P2.11.

Ans: $\mathbf{C} = 240$ lb $\diagup^5_4$ through C

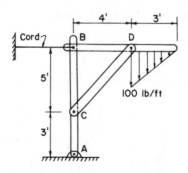

Fig. P2.11

2.12. Determine the horizontal and vertical components of the pin reaction at B (Fig. P2.12) on member ABC. Body E weighs 1000 lb, and the weights of the other members may be neglected.

2.13. The rigid frame in Fig. P2.13 carries a distributed load. Determine the pin reaction at B on member BD. All pins are smooth.

Ans: $B_x = 0,$ $B_y = 300$ lb $\downarrow$

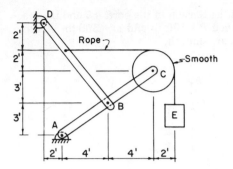

Fig. P2.12

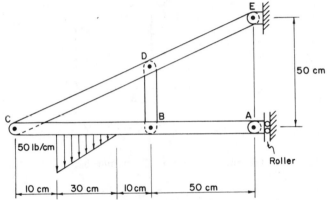

Fig. P2.13

2.14. For the mechanism in Fig. P2.14 find the force in the cord *CD* and the pin reaction at *E* on member *ADEF* in terms of *w*. Note all pins are smooth.

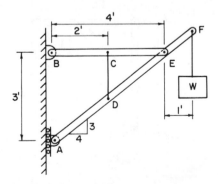

Fig. P2.14

2.15. Determine the components of the pin reaction at *B* on member *BCE* of the pin-connected structure in Fig. P2.15. The couple is applied to member *BCE*.

Ans: $B_x = 33.3$ lb $\rightarrow$, $B_y = 600$ lb $\downarrow$

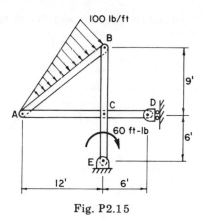

Fig. P2.15

2.16. The homogeneous horizontal plate in Fig. P2.16 weighs 120 lb. Determine the tensions in wires A, B, and C.

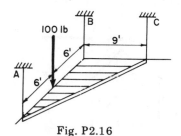

Fig. P2.16

2.17. The boom DE in Fig. P2.17 carries a load $F = (-300i - 500j)$ and is supported by a ball and socket at E and two wires. Determine the forces acting on ECD at E.

Ans: $E_x = 1300$ lb $\rightarrow$, $E_y = 125$ lb $\downarrow$, $E_z = 0$

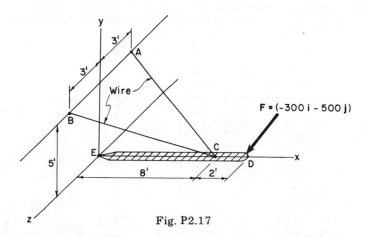

Fig. P2.17

2.18. Find the forces in members *CB*, *CG*, and *BG* of the plane truss in Fig. P2.18.

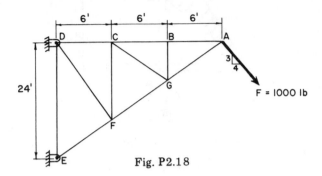

Fig. P2.18

2.19. Determine the forces in members *EF* and *DC* of the truss in Fig. P2.19.

Ans: *EF* = 141.4 lb C,   *DC* = 150 lb T

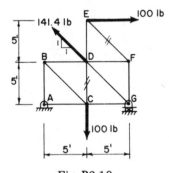

Fig. P2.19

2.20. The 2000-lb weight of Fig. P2.20 is being raised with a constant velocity by the winch at *C*. Determine the torque *T* and the forces in members *AB* and *FG*. The pulley at *A* is smooth.

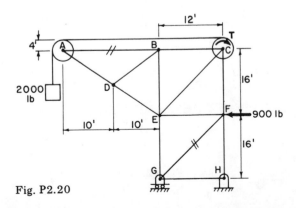

Fig. P2.20

2.21. For the truss in Fig. P2.21 find the pin reactions at F and E and find the force in the member BD.

Ans: $\mathbf{F}_x = 5.77 \text{ N} \leftarrow$, $F_y = 0$, $\mathbf{E} = 34.2 \text{ N} \leftarrow$, $\mathbf{BD} = 5.77 \text{ N} \leftarrow$

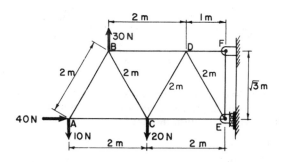

Fig. P2.21

2.22. Draw the shear and moment diagram for the beam loaded as in Fig. P2.22.

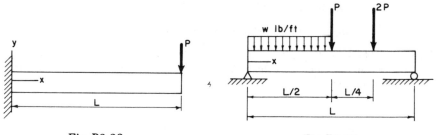

Fig. P2.22   Fig. P2.23

2.23. Draw the shear and moment diagram for the beam loaded as in Fig. P2.23. Here let $w = 100 \text{ lb/ft}$, $L = 8 \text{ ft}$, $P = 50 \text{ lb}$.

Ans: $V_{max} = 350 \text{ lb at } x = 0$, $V_{min} = -200 \text{ lb for } 6 < x < 8$
$M_{max} = 612.5 \text{ ft-lb at } x = 3.5 \text{ ft}$

2.24. The beam in Fig. P2.24 is loaded by a triangular load varying from zero on the left end to 100 lb/ft on the right end. (a) Find the reactions R_1 and R_2 and (b) determine the maximum moment and its location.

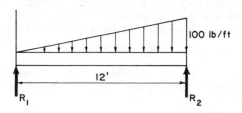

Fig. P2.24

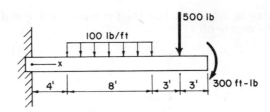

Fig. P2.25

2.25. For the beam loaded and supported as in Fig. P2.25 (a) write the shear and moment equations, (b) draw the shear and moment diagrams, and (c) determine the maximum moment.

Ans: (a) $0 < x < 4$ $4 < x < 12$

$V = 1300$ lb $V = 1700 - 100x$ lb

$M = 1300x - 14{,}200$ ft-lb $M = -50x^2 + 1700x - 15{,}000$ ft-lb

(b) $12 < x < 15$ $15 < x < 18$

$V = 500$ lb $V = 0$

$M = 500x - 7800$ ft-lb $M = -300$ ft-lb

(c) $\mathbf{M}_{max} = 14{,}200$ ft-lb $\curvearrowleft$ at fixed end

2.26. The block in Fig. P2.26 weighs 500 lb, and the coefficient of friction between the block and the plane is 0.25. Determine the frictional force acting on the box.

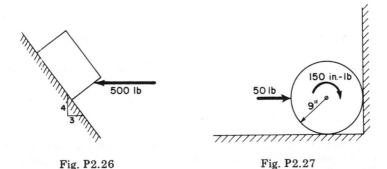

Fig. P2.26   Fig. P2.27

2.27. The homogeneous disk in Fig. P2.27 weighs 200 lb. The horizontal surface is smooth and the coefficient of friction between the vertical surface and the disk is 0.4. Determine the reactions of the surfaces on the disk.

Ans: $\mathbf{N}_H = 183.3$ lb ↑, $\mathbf{N}_V = 50$ lb ←, $\mathbf{F}_V = 16.7$ lb ↑

2.28. For the system in Fig. P2.28, find the force P that will cause B to have impending motion up the plane. Block A weighs 100 lb and block B weighs 150 lb.

2.29. Body B in Fig. P2.29 weighs 500 lb. The coefficients of friction are:

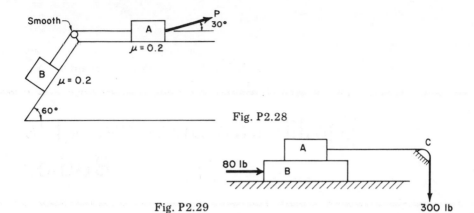

Fig. P2.28

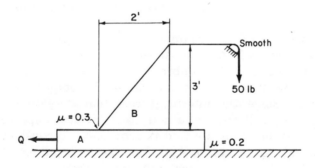

Fig. P2.29   300 lb

between A and B, 0.4; between B and the plane, 0.3; the fixed drum C is smooth. Determine the minimum weight that A must have to prevent motion.

Ans: $W_A = 767$ lb

2.30. Determine the force Q in Fig. P2.30 that will cause body A to have impending motion to the left. Body A weighs 300 lb and the homogeneous body B weighs 450 lb.

Fig. P2.30

chapter 3

Equilibrium of Deformable Bodies

3.1 Introduction

In Chapter 2 our attention was focused on rigid bodies at rest or in motion. Deformation of the body played no role in the analysis, and the distribution of forces in the interior of the body was not considered. In a wide variety of engineering problems involving design of structural elements and machine components, internal force distributions must be established to evaluate response of the components to service loads.

The problem of determining internal force distributions in deformable bodies will be approached through equilibrium consider-ations. If the body is in equilibrium under the action of a system of applied external forces, any small part of the body must also be in equilibrium. Since the internal forces that develop to maintain equilibrium of small parts of a body of arbitrary shape will vary in magnitude and direction, we find it convenient to consider intensity of force over a given area rather than the forces themselves. "In-tensity of force" or "force per unit area" leads to the concept of stress.

The discussions that follow cover two types of applied force; *surface forces* and *body forces*. Surface forces are applied at the boundary of the body, usually as the result of contact with another body. Body forces occur as the result of gravity, electric or magnetic fields, etc., and are applied to every element of volume of the body. Surface forces are distributed over an area, while body forces are distributed throughout a volume.

3.2 Concept of Stress

Consider a body of arbitrary shape that is in equilibrium under the action of a system of applied forces. Due to this action, internal

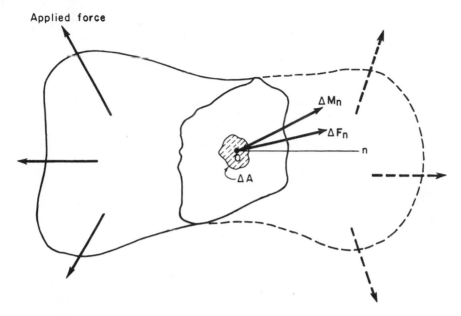

Fig. 3.1

forces develop at all points of the body. The nature of the internal force distribution at an arbitrary interior point 0 can be studied by exposing an interior plane through 0 with an imaginary cut (Fig. 3.1). Since the force distribution on the cut that is required to maintain equilibrium of the isolated part of the body will not generally be uniform, the distributed force acting on a small area ΔA surrounding the point 0 can be replaced (as shown in Chapter 2 for a general force system) by a statically equivalent resultant force $\Delta \mathbf{F}_n$ through 0 and a couple $\Delta \mathbf{M}_n$. The subscript n indicates that the force and couple are associated with a particular surface through 0, namely, the one having an outer normal in the n direction at 0. For any other surface through 0 the values of $\Delta \mathbf{F}$ and $\Delta \mathbf{M}$ could be different. Note also that the directions of $\Delta \mathbf{F}$ and $\Delta \mathbf{M}$ are not generally in the direction of n.

With $\Delta \mathbf{F}_n$ and $\Delta \mathbf{M}_n$ established, an average force per unit area $\mathbf{S}_{n\,\text{avg}}$ and an average couple per unit area $\mathbf{C}_{n\,\text{avg}}$ can be computed as

$$\mathbf{S}_{n\,\text{avg}} = \frac{\Delta \mathbf{F}_n}{\Delta A}, \qquad \mathbf{C}_{n\,\text{avg}} = \frac{\Delta \mathbf{M}_n}{\Delta A}$$

In the limit as ΔA is made smaller and smaller, the force $\Delta \mathbf{F}_n$ tends to become more and more uniform, and we obtain a quantity known as the *stress vector* $\mathbf{S}_n$. Thus

$$S_n = \lim_{\Delta A \to 0} \frac{\Delta \mathbf{F}_n}{\Delta A} \tag{3.1}$$

In a similar manner a quantity known as the *couple-stress vector* $\mathbf{C}_n$ is defined as

$$\mathbf{C}_n = \lim_{\Delta A \to 0} \frac{\Delta \mathbf{M}_n}{\Delta A} \tag{3.2}$$

Experimental evidence indicates that the couple-stress vector is extremely small in comparison to the stress vector. Thus no significant error is made by neglecting its effects as we will do in all future analyses of internal forces and deformations.

Different values for the stress vector $\mathbf{S}_n$ at 0 will be obtained by passing different planes through 0. An infinite number of stress vectors are possible, depending on the particular plane through 0 with which each is associated. This necessary association of the stress vector with a specific plane makes stress something more than an ordinary vector and is the reason for the subscript n. As long as we restrict our discussion to the stress at 0 on a particular plane, stress may be treated as an ordinary vector. In discussing the total concept of "stress at a point," however, we must recognize and handle stress as a quantity that is different from a vector quantity.

Later, we will relate material response to components of the stress vector rather than to the stress vector itself. In particular, the components normal and parallel to the plane will be useful. As shown in Fig. 3.2, the resultant force $\Delta \mathbf{F}_n$ can be resolved into the components ΔF_{nn} and ΔF_{nt}. A normal stress σ_n and a shear stress τ_n are then defined as

$$\sigma_n = \lim_{\Delta A \to 0} \frac{\Delta F_{nn}}{\Delta A} \tag{3.3}$$

and

$$\tau_n = \lim_{\Delta A \to 0} \frac{\Delta F_{nt}}{\Delta A} \tag{3.4}$$

The components σ_n and τ_n are related to the magnitude of the stress vector by the relation

$$S_n = \sqrt{\sigma_n^2 + \tau_n^2} \tag{3.5}$$

The intensity of force per unit area at a point within a body thus can be readily characterized by a normal stress σ_n and a shearing stress τ_n if the orientation of the area is specified.

For purposes of analysis we usually reference the area to some

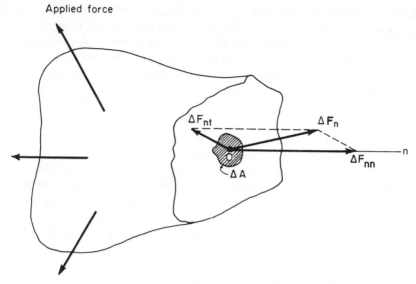

Fig. 3.2

coordinate system. For example, in a rectangular coordinate system we may choose to consider the stresses acting on planes having outer normals in the x, y, and z directions. Consider the plane having an outer normal in the x direction. In this case the normal and shear stresses on the plane will be σ_x and τ_x respectively. Since τ_x gen-

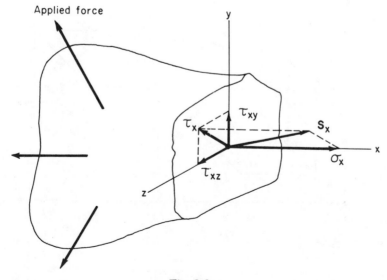

Fig. 3.3

erally will not coincide with the y or z axes, we can resolve τ_x into
the components τ_{xy} and τ_{xz} (Fig. 3.3). In this double subscript notation, the first indicates the direction of the normal to the plane on
which the stress acts and the second indicates the direction of the
stress. The relationships between the rectangular stress components
and the magnitude of the stress vector can be expressed as

$$\tau_x = \sqrt{\tau_{xy}^2 + \tau_{xz}^2}, \qquad S_x = \sqrt{\sigma_x^2 + \tau_{xy}^2 + \tau_{xz}^2} \qquad (3.6)$$

At this point we must establish a sign convention for the stresses.
For all the developments that follow, we will define the positive direction for stress as the positive coordinate direction if the outer
normal is in a positive direction. This is illustrated in Fig. 3.4a where
the outer normal to the area $ABCD$ is in the positive x direction.
The stress components σ_x, τ_{xy}, and τ_{xz} are positive as shown on area
$ABCD$. If the outer normal points in the negative coordinate direc-
tion, as in Fig. 3.4b for the area $EFGH$, then the stresses are positive
if directed in a negative coordinate direction. Thus the stresses in
Fig. 3.4b are positive when directed as shown. Note that positive
normal stresses are known as *tensile stresses;* i.e., they tend to stretch
the material. Negative normal stresses are known as *compressive
stresses.* Positive and negative shear stresses are not given special
names.

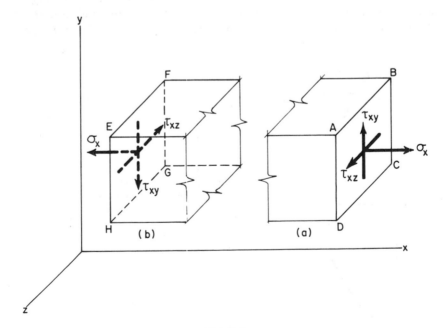

Fig. 3.4

3.3 Stress at a Point

The state of stress at a point in a material is not completely defined by the three components of a stress vector since the vector depends on the orientation of the plane with which it is associated. An infinite number of planes can be passed through the point, resulting in an infinite number of stress vectors being associated with that point. Fortunately, the specification of stresses acting on only three mutually perpendicular planes is sufficient to completely describe the state of stress at a point. Consider the stresses on a small element of volume containing the point P.

The characteristic element of volume associated with a Cartesian coordinate system is a rectangular parallelepiped with its faces oriented perpendicular to the coordinate axes. Assume that the element is extremely small so that variations in stress from point to point within the element can be neglected without introducing significant error. As a result the stresses can be assumed uniform over the faces and equal on parallel faces. Body forces can also be neglected since they are of higher order than the surface forces and therefore are of negligible magnitude for the element being considered. The rectangular components of stress vectors on planes having outer normals in the x, y, and z directions are shown on the appropriate faces of the element in Fig. 3.5; all stress components are positive. The three different sets of three Cartesian stress components can be written in the rectangular array

$$
\begin{bmatrix}
\sigma_x & \tau_{xy} & \tau_{xz} \\
\tau_{yx} & \sigma_y & \tau_{yz} \\
\tau_{zx} & \tau_{zy} & \sigma_z
\end{bmatrix}
$$

This array of nine stress components is the *stress tensor* and represents a complete description of the state of stress at the point.

At this stage it appears that nine stress components are needed to specify the state of stress at a point. Equilibrium considerations, however, reveal that all nine components are not independent. Three force equations and three moment equations must be satisfied for a body to be in equilibrium. A typical force equation for our small element yield

$$\Sigma F_x = \sigma_x (\Delta y)(\Delta z) - \sigma_x (\Delta y)(\Delta z) + \tau_{yx} (\Delta x)(\Delta z) - \tau_{yx} (\Delta x)(\Delta z)$$
$$+ \tau_{zx} (\Delta x)(\Delta y) - \tau_{zx} (\Delta x)(\Delta y) = 0$$

From this equation and two similar force equations for the y and z directions, we see that three of the conditions for equilibrium are

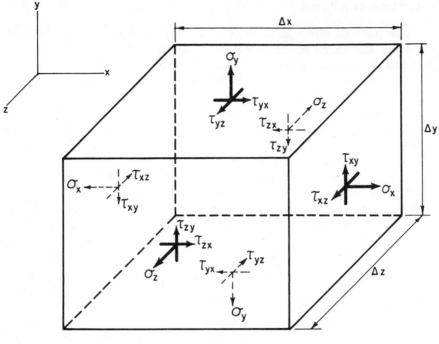

Fig. 3.5

identically satisfied by the stress distribution present on the element. Moment considerations about a line parallel to the z axis through the center of the element yield

$$\Sigma M_z = \tau_{xy}(\Delta y\ \Delta z)(\Delta x) - \tau_{yx}(\Delta z\ \Delta x)(\Delta y)$$

Since $\Sigma M_z = 0$ for the element to be in equilibrium, this equation and two similar equations for moments with respect to the x and y axes yield the following significant results:

$$\tau_{xy} = \tau_{yx}, \qquad \tau_{yz} = \tau_{zy}, \qquad \tau_{zx} = \tau_{xz} \tag{3.7}$$

Therefore, only six Cartesian stress components are necessary for a complete specification of the stress tensor. The components normally used are σ_x, σ_y, σ_z, τ_{xy}, τ_{yz}, τ_{zx}.

We have established one of the infinite number of stress tensors that could be used to specify the state of stress at the point of interest. Different tensors could be obtained by selecting different orientations for the reference parallelepiped. The planes of greatest interest from the design point of view are generally those on which the maximum normal stress or the maximum shear stress occurs. Since these are not normally the reference planes, ways must be found to locate them. Once expressions for the normal and shear stresses on the arbitrary plane are developed in terms of stresses on the reference planes, maximum normal and maximum shear stresses and their orientations can be evaluated quite easily in most practical cases.

3.4 Two-dimensional or Plane Stress

We can gain considerable insight into the nature of stress distributions by considering several special states of stress that frequently occur in practical design situations before we proceed to the general three-dimensional state of stress. A state of stress occurring on the unloaded surfaces of machine components or in thin plates when the loads are applied in the plane of the plate is referred to as *two-dimensional* or *plane*. For this case two parallel faces of the small element in Fig. 3.5 are assumed to be free of stress. For purposes of analysis let these faces be perpendicular to the reference z axis. Thus $\sigma_z = \tau_{zx} = \tau_{zy} = 0$. From Eqs. (3.7), however, this also implies that $\tau_{xz} = \tau_{yz} = 0$. Therefore, the components of the stress tensor of interest in this two-dimensional or plane stress analysis will be σ_x, σ_y, and $\tau_{xy} = \tau_{yx}$ (Fig. 3.6a). For convenience this state of stress

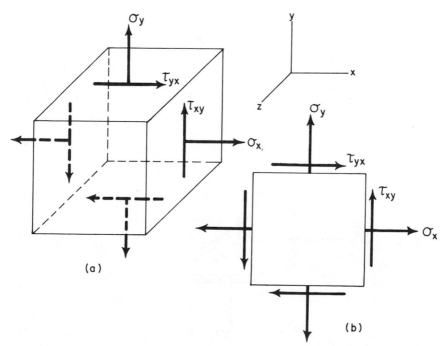

Fig. 3.6

is usually represented by a two-dimensional sketch (Fig. 3.6b). The three-dimensional element shown in Fig. 3.6a, however, should be kept in mind at all times.

The arbitrary planes on which we evaluate normal and shear stresses σ_n and τ_{nt} for the plane state of stress will have outer normals that lie in the reference xy plane. Thus a convenient element to use for equilibrium considerations is the wedge-shaped portion of the reference parallelepiped in Fig. 3.7a. A simplified two-dimensional form of representation of this element is shown in Fig. 3.7b.

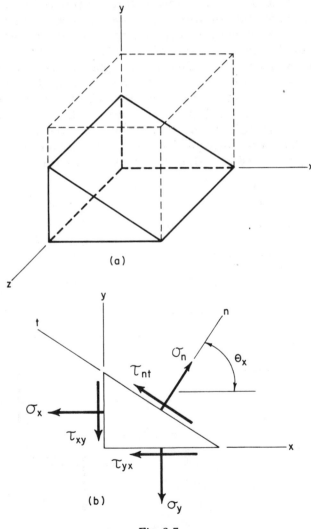

(a)

(b)

Fig. 3.7

Equations relating the normal and shear stresses σ_n and τ_{nt} (on an arbitrary plane oriented at an angle θ_x with respect to the reference x axis) and the stresses σ_x, σ_y, and $\tau_{xy} = \tau_{yx}$ (on the reference planes) can be developed by considering equilibrium of the wedge. For example, let the area of the inclined face of the wedge be ΔA. The areas of the vertical and horizontal faces having outer normals in the x and y directions respectively are then $\Delta A \cos \theta_x$ and $\Delta A \sin \theta_x$. Summation of forces in the n direction yields

$$\textcircled{+\nearrow}\, \Sigma F_n = 0$$

$$\sigma_n \Delta A - \tau_{yx}(\Delta A \sin \theta_x)\cos \theta_x - \tau_{xy}(\Delta A \cos \theta_x)\sin \theta_x$$
$$- \sigma_x(\Delta A \cos \theta_x)\cos \theta_x - \sigma_y(\Delta A \sin \theta_x)\sin \theta_x = 0$$

Since $\tau_{xy} = \tau_{yx}$,

$$\sigma_n = \sigma_x \cos^2 \theta_x + \sigma_y \sin^2 \theta_x + 2\tau_{xy} \sin \theta_x \cos \theta_x \qquad (3.8a)$$

In a similar manner, summation of forces in the t direction yields

$$\textcircled{+\nwarrow}\, \Sigma F_t = 0$$

$$\tau_{nt} \Delta A - \tau_{xy}(\Delta A \cos \theta_x)\cos \theta_x + \tau_{yx}(\Delta A \sin \theta_x)\sin \theta_x$$
$$+ \sigma_x(\Delta A \cos \theta_x)\sin \theta_x - \sigma_y(\Delta A \sin \theta_x)\cos \theta_x = 0$$

$$\tau_{nt} = -(\sigma_x - \sigma_y)\sin \theta_x \cos \theta_x + \tau_{xy}(\cos^2 \theta_x - \sin^2 \theta_x) \qquad (3.9a)$$

For convenience Eqs. (3.8a) and (3.9a) are frequently expressed in double-angle form as

$$\sigma_n = (1/2)(\sigma_x + \sigma_y) + (1/2)(\sigma_x - \sigma_y)\cos 2\theta_x + \tau_{xy} \sin 2\theta_x \qquad (3.8b)$$
$$\tau_{nt} = -(1/2)(\sigma_x - \sigma_y)\sin 2\theta_x + \tau_{xy} \cos 2\theta_x \qquad (3.9b)$$

Equations (3.8) and (3.9) indicate that normal and shear stresses σ_n and τ_{nt} can be determined on any plane oriented at an arbitrary angle θ_x with respect to the reference x axis if the stresses on the reference planes are known. Thus for the two-dimensional or plane stress case these equations provide a means for determining quantities of interest such as (1) planes that are free of shear stress; (2) planes on which extreme (maximum and minimum) values of normal stress occur, and the magnitudes of these stresses; and (3) planes on which extreme values of shear stress occur, and the magnitudes of these stresses.

Consider planes that are free of shear stress, $\tau_{nt} = 0$. If such planes exist, Eq. (3.9b) can be used to locate them. Thus

$$- (1/2)(\sigma_x - \sigma_y)\sin 2\theta_x + \tau_{xy} \cos 2\theta_x = 0$$

Solving for the angular orientation of the plane that is free of shear stress yields

$$\tan 2\theta_{xp} = 2\tau_{xy}/(\sigma_x - \sigma_y) \qquad (3.10a)$$

Equation (3.10a) defines two mutually perpendicular planes having outer normals in the reference xy planes. These planes, which are free of shear stress, are known as *principal planes*. In our original specification of the plane stress state we indicated that the plane having outer normal in the z direction was completely stress-free; therefore, this plane is also a principal plane. A more general analysis for an arbitrary three-dimensional state of stress, which is beyond the scope of this book, indicates that three mutually perpendicular principal planes always exist.

Planes on which the maximum and minimum normal stress occur for the two-dimensional or plane stress state are obtained from Eq. (3.8b). Differentiating with respect to θ_x and setting the resulting expression equal to zero, we obtain

$$\frac{d\sigma_n}{d\theta_x} = -(\sigma_x - \sigma_y)\sin 2\theta_x + 2\tau_{xy}\cos 2\theta_x = 0$$

from which

$$\tan 2\theta_{x\sigma} = 2\tau_{xy}/(\sigma_x - \sigma_y) \qquad (3.10b)$$

This expression is identical to Eq. (3.10a) which was developed to locate principal planes. Thus we can conclude that maximum and minimum normal stresses for the plane stress state occur on principal planes. These stresses are known as *principal* stresses, and their magnitudes are obtained by substituting the values of $\theta_{x\sigma}$ obtained from Eq. (3.10b) into Eq. (3.8b). Thus

$$\begin{aligned}
\sigma_{P1} &= \frac{\sigma_x + \sigma_y}{2} + \sqrt{\left(\frac{\sigma_x - \sigma_y}{2}\right)^2 + \tau_{xy}^2} \\
\sigma_{P2} &= \frac{\sigma_x + \sigma_y}{2} - \sqrt{\left(\frac{\sigma_x - \sigma_y}{2}\right)^2 + \tau_{xy}^2}
\end{aligned} \qquad (3.11)$$

Equations (3.11) give the maximum and minimum values of normal stress that can occur on planes having outer normals in the reference xy plane. As a result, they are frequently referred to as the *in-plane* principal stresses. These two stresses together with the *out-of-plane* principal stress $\sigma_z = 0$ represent the principal stresses associated with the plane or two-dimensional stress state. A general three-dimensional analysis indicates that three principal stresses will always exist on three mutually perpendicular planes through the point. One principal stress always represents the maximum value of normal stress on any plane through the point, while another always represents the minimum value of normal stress on any plane through the point. The third principal stress represents some intermediate value of normal stress and has no particular significance.

When Eq. (3.10a) is used to establish the orientation of the principal planes with respect to the reference axes, one value of the angle will always lie in the range $-45° \leqslant \theta_{xp} \leqslant 45°$. Here a positive angle indicates counterclockwise rotation and a negative angle indicates clockwise rotation about the positive z axis. For the case of $\sigma_x > \sigma_y$ the acute angle θ_{xp} denotes the position of σ_{P1} with respect to the reference x axis. For the case $\sigma_x < \sigma_y$ the obtuse angle θ_{xp} denotes the position of σ_{P1} with respect to the reference x axis. The reason for this behavior will be illustrated in discussions of Mohr's circle.

In a similar manner we can determine the extreme values of the in-plane shear stresses and the orientations of the planes on which these stresses act. Differentiating Eq. (3.9b) with respect to θ_x and setting the resulting expression equal to zero yields

$$\frac{d\tau_{nt}}{d\theta_x} = -(\sigma_x - \sigma_y)\cos 2\theta_x - 2\tau_{xy}\sin 2\theta_x = 0$$

Solving for the angular orientations of the planes yields

$$\tan 2\theta_{xT} = -(\sigma_x - \sigma_y)/2\tau_{xy} \tag{3.12}$$

This equation defines two mutually perpendicular planes on which extreme values of the in-plane shear stresses occur. A comparison of Eqs. (3.10b) and (3.12) reveals they are negative reciprocals; therefore, the angles they define differ by 90°. This result indicates that planes on which extreme values of in-plane shear stress occur are located 45° from planes on which in-plane principal stresses occur. Substitution of θ_{xT} from Eq. (3.12) into Eq. (3.9b) yields the extreme values of in-plane shear stresses. Thus

$$\tau_P = \pm \sqrt{\left(\frac{\sigma_x - \sigma_y}{2}\right)^2 + \tau_{xy}^2} \tag{3.13}$$

Equation (3.13) indicates that shear stresses on the two planes defined by Eq. (3.12) are equal in magnitude. Since the two planes are perpendicular, this result is consistent with Eqs. (3.7). Selection of the sign to be associated with this extreme value of in-plane shear stress will depend on the angle selected to locate the n coordinate direction.

If the values of $\theta_{x\tau}$ obtained from Eq. (3.12) are substituted into Eq. (3.8b),

$$\sigma_n = (\sigma_x + \sigma_y)/2 = \sigma_{avg} \tag{3.14}$$

Eq. (3.14) indicates that planes on which the extreme values of shear stress occur are usually not free of normal stress. Extreme values of normal stress, on the other hand, occur on planes free of shear. Each of these conditions is illustrated in Fig. 3.8, where three different methods are used to illustrate the same state of stress at a point.

A comparison of Eqs. (3.11) with Eq. (3.13) indicates that

$$\tau_P = (\sigma_{P1} - \sigma_{P2})/2 \tag{3.15}$$

which means that the magnitude of the maximum in-plane shear

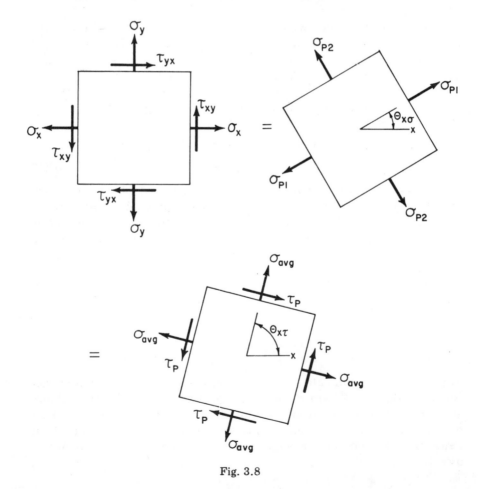

Fig. 3.8

stress equals one-half the difference between the two in-plane principal stresses.

In most design situations the quantities of interest are the maximum normal stress and the maximum shear stress on any plane through the point. For the two-dimensional or plane stress state, knowledge of the two in-plane principal stresses as given by Eqs. (3.11) as well as the out-of-plane principal stress $\sigma_z = 0$ provides the required maximum normal stress information since one of the principal stresses is always the maximum normal stress at the point. Unfortunately, the maximum in-plane shear stress as given by Eqs. (3.13) or (3.15) is not always the maximum shear stress at the point. Analysis of the general three-dimensional state of stress, however, provides us with a relationship involving principal stresses for determining the required maximum shear stress.

In a general three-dimensional stress state the principal stresses are usually denoted as $\sigma_1, \sigma_2, \sigma_3$ where $\sigma_1 \geqslant \sigma_2 \geqslant \sigma_3$. In this ordering of the principal stresses, positive stresses of any magnitude are considered greater than negative stresses (the stresses are ordered algebraically). In addition, σ_1 is considered the maximum principal stress and σ_3 the minimum principal stress regardless of their individual magnitudes. If we follow this convention for our special two-dimensional analysis, three cases arise which depend on the sign of the stresses as given by Eqs. (3.11). Consider first that σ_{P1} and σ_{P2} are both positive. Then

$$\sigma_1 = \sigma_{P1}, \qquad \sigma_2 = \sigma_{P2}, \qquad \sigma_3 = \sigma_z = 0 \qquad\qquad \text{(a)}$$

If σ_{P1} and σ_{P2} are of opposite sign,

$$\sigma_1 = \sigma_{P1}, \qquad \sigma_2 = \sigma_z = 0, \qquad \sigma_3 = \sigma_{P2} \qquad\qquad \text{(b)}$$

Finally, if σ_{P1} and σ_{P2} are both negative,

$$\sigma_1 = \sigma_z = 0, \qquad \sigma_2 = \sigma_{P1}, \qquad \sigma_3 = \sigma_{P2} \qquad\qquad \text{(c)}$$

From a general three-dimensional analysis it can be shown that the maximum shear stress at the point equals one-half the difference between the maximum and minimum principal stresses. Thus

$$\tau_{\max} = (\sigma_1 - \sigma_3)/2 \qquad\qquad (3.16)$$

An examination of Eqs. (a), (b), and (c) indicates that the in-plane maximum shear stress would be the maximum shear stress at the point only when the in-plane principal stresses are of opposite sign.

When σ_{P1} and σ_{P2} have the same sign, the maximum shear stress will occur on mutually perpendicular planes inclined $45°$ with respect to the reference xy plane. The planes of maximum shear for the various cases are illustrated in Fig. 3.9.

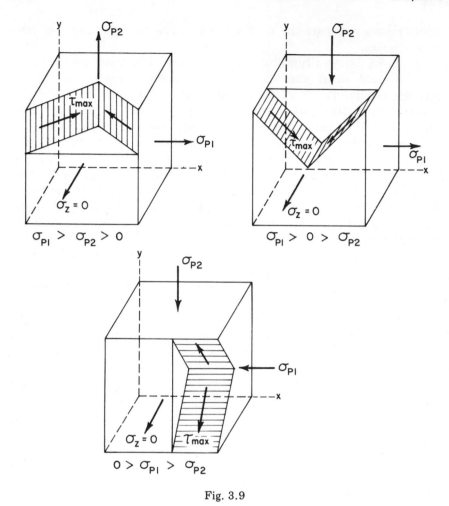

Fig. 3.9

EXAMPLE 3.1

At a point on the unloaded surface of a machine component a plane state of stress exists (Fig. E3.1a). Determine the normal and shear stresses which act on the inclined plane *a-a*.

SOLUTION

Normal and shear stresses on an inclined plane through a point can be determined by considering the equilibrium of a wedge-shaped element formed by the reference planes and the inclined plane. A two-dimensional sketch showing the stresses acting on the faces of the wedge is presented in Fig. E3.1b. If the area of the inclined face is ΔA_n, the horizontal and vertical faces have areas ΔA_n cos $30°$ and ΔA_n sin $30°$ respectively. A free-body diagram with forces involved

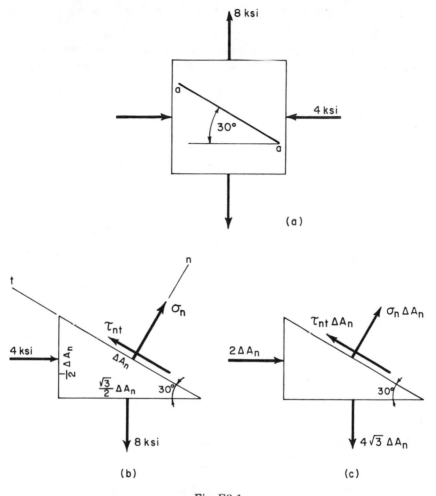

Fig. E3.1

in equilibrium considerations is shown in Fig. E3.1c. From a summation of forces in the n direction we obtain

$$\Sigma F_n = 0, \quad \sigma_n \Delta A_n + 2\Delta A_n \sin 30° - 4\sqrt{3}\,\Delta A_n \cos 30° = 0$$

$$\sigma_n = 5 \text{ ksi} \qquad\qquad Ans.$$

Similarly, we obtain from a summation of forces in the t direction

$$\Sigma F_t = 0, \quad \tau_{nt}\Delta A_n - 2\Delta A_n \cos 30° - 4\sqrt{3}\,\Delta A_n \sin 30° = 0$$

$$\tau_{nt} = 3\sqrt{3} \text{ ksi} \qquad\qquad Ans.$$

EXAMPLE 3.2

The Cartesian components of stress at a point in a plane stress field are shown in Fig. E3.2. Determine the normal and shear stresses on the plane having outer normal 30° from the x axis.

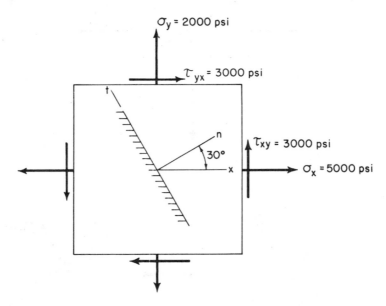

Fig. E3.2

SOLUTION

The normal stress σ_n is determined, using Eq. (3.8a), as

$$\sigma_n = \sigma_x \cos^2 \theta_x + \sigma_y \sin^2 \theta_x + 2 \, \tau_{xy} \sin \theta_x \cos \theta_x$$
$$= 5000 \, (\sqrt{3}/2)^2 + 2000 \, (1/2)^2 + 2(3000)(1/2)(\sqrt{3}/2)$$
$$= 6850 \text{ psi} \hspace{3cm} Ans.$$

In a similar manner we can determine the shear stress τ_{nt}, using Eq. (3.9a), as

$$\tau_{nt} = -(\sigma_x - \sigma_y) \sin \theta_x \cos \theta_x + \tau_{xy} (\cos^2 \theta_x - \sin^2 \theta_x)$$
$$= -(5000 - 2000)(1/2)(\sqrt{3}/2) + 3000 \, (\sqrt{3}/2)^2 - 3000 \, (1/2)^2$$
$$= 200 \text{ psi} \hspace{3cm} Ans.$$

EXAMPLE 3.3

The Cartesian components of stress at a point in a plane stress field are shown in Fig. E3.3. Determine the principal stresses and the maximum shear stress at the point.

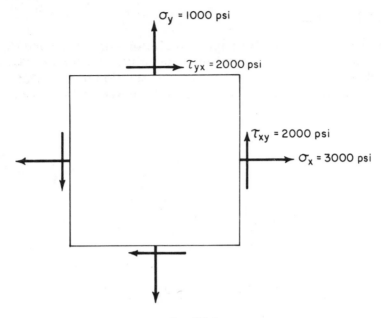

σ_y = 1000 psi

τ_{yx} = 2000 psi

τ_{xy} = 2000 psi

σ_x = 3000 psi

Fig. E3.3

SOLUTION

The in-plane principal stresses at the point are determined, using Eqs. (3.11), as

$$\sigma_{p1} = \frac{\sigma_x + \sigma_y}{2} + \sqrt{\left(\frac{\sigma_x - \sigma_y}{2}\right)^2 + \tau_{xy}^2}$$

$$= \frac{3000 + 1000}{2} + \sqrt{\left(\frac{3000 - 1000}{2}\right)^2 + (2000)^2} = 4235 \text{ psi}$$

Similarly,

$$\sigma_{p2} = \frac{\sigma_x + \sigma_y}{2} - \sqrt{\left(\frac{\sigma_x - \sigma_y}{2}\right)^2 + \tau_{xy}^2} = -235 \text{ psi}$$

Ordering the principal stresses then yields

$$\sigma_1 = \sigma_{p1} = 4235 \text{ psi}, \quad \sigma_2 = \sigma_z = 0, \quad \sigma_3 = \sigma_{p2} = -235 \text{ psi} \qquad Ans.$$

The maximum shearing stress τ_{max} is then computed, using Eq. (3.16), as

$$\tau_{max} = (\sigma_1 - \sigma_3)/2 = (4235 + 235)/2 = 2235 \text{ psi} \qquad Ans.$$

3.5 Mohr's Circle

A pictorial or graphical representation of Eqs. (3.8b) and (3.9b) is often used to provide insight into the nature of a two-dimensional state of stress at a point and to aid in the solution of problems. The form of representation may be developed by rewriting Eqs. (3.8b) and

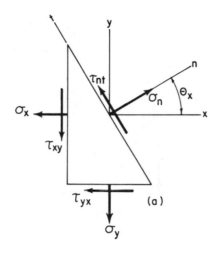

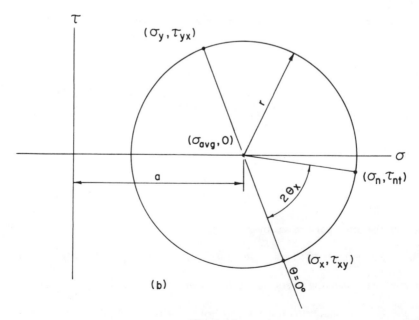

Fig. 3.10

(3.9b) in the following form:

$$\sigma_n - (1/2)(\sigma_x + \sigma_y) - (1/2)(\sigma_x \quad \sigma_y) \cos 2\theta_x + \tau_{xy} \sin 2\theta_x$$

$$\tau_{nt} = (1/2)(\sigma_x - \sigma_y) \sin 2\theta_x + \tau_{xy} \cos 2\theta_x$$

Squaring both equations, adding, and simplifying yields

$$\left(\sigma_n - \frac{\sigma_x + \sigma_y}{2}\right)^2 + \tau_{nt}^2 = \left(\frac{\sigma_x - \sigma_y}{2}\right)^2 + \tau_{xy}^2 \qquad (3.17)$$

Since σ_x, σ_y, and τ_{xy} are the known stress components associated with

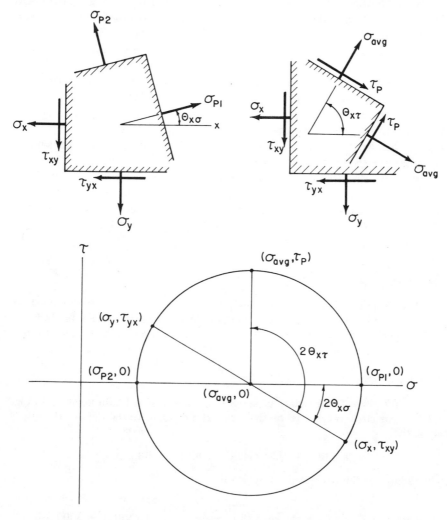

Fig. 3.11

the reference planes, Eq. (3.17) can be written in simplified form as $(\sigma_n - a)^2 + \tau_{nt}^2 = r^2$. This is the equation of a circle in terms of the variables σ_n and τ_{nt}. The circle is centered on the σ axis at a distance $a = (\sigma_x + \sigma_y)/2$ from the τ axis, and the radius of the circle is given by

$$r = \sqrt{\left(\frac{\sigma_x - \sigma_y}{2}\right)^2 + \tau_{xy}^2}$$

This circle in terms of stress components is known as *Mohr's circle.*

All the quantities present in Eq. (3.17) are shown plotted on the physical element in Fig. 3.10a and on Mohr's circle in Fig. 3.10b. On Mohr's circle normal stress components are plotted horizontally, while shear stress components are plotted vertically. Tensile stresses are plotted to the right of the τ axis. Compressive stresses are plotted to the left. Shear stress components that tend to produce a clockwise rotation are plotted above the σ axis, and those tending to produce a counterclockwise rotation are plotted below. When plotted in this manner, the stress components σ_n and τ_{nt} associated with each plane through the point are represented by a point on the circumference of the circle. An angle of rotation θ_x of the physical plane is represented by an angle $2\theta_x$ in the same direction on Mohr's circle.

We can also locate each of the quantities obtained by using Eqs. (3.10b), (3.11), (3.12), and (3.13) on Mohr's circle (Fig. 3.11). The student should verify that he can reproduce each of the equations mentioned above as well as Eqs. (3.8b) and (3.9b) from the data points for σ_x, σ_y, and $\tau_{xy} = \tau_{yx}$ in Figs. 3.10 and 3.11.

EXAMPLE 3.4

Use Mohr's circle to determine the principal stresses and the maximum shear stress associated with the Cartesian components of stress in Fig. E3.4a. Indicate on an appropriate sketch the orientation of the planes on which these stresses act.

SOLUTION

Mohr's circle for the state of stress indicated in Fig. E3.4a is shown in Fig. E3.4b. The circle is centered on the σ axis at a distance to the right of the origin equal to

$$a = (\sigma_x + \sigma_y)/2 = (7000 - 1000)/2 = 3000 \text{ psi}$$

The radius r of the circle is given by

$$r = \sqrt{\left(\frac{\sigma_x - \sigma_y}{2}\right)^2 + \tau_{xy}^2} = \sqrt{\left(\frac{7000 + 1000}{2}\right)^2 + (3000)^2} = 5000 \text{ psi}$$

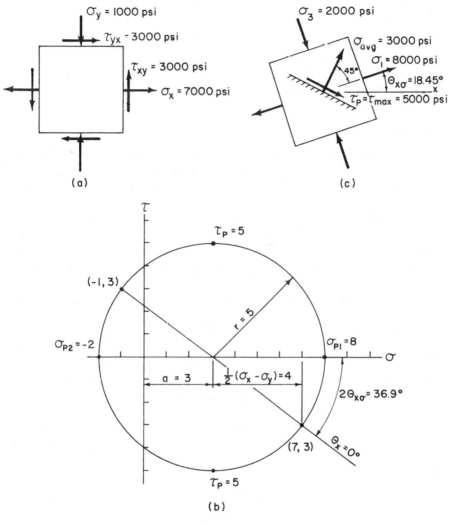

Fig. E3.4

From the circle it then follows that

$$\sigma_{p1} = a + r = 3000 + 5000 = 8000 \text{ psi}, \qquad \sigma_{p2} = a - r = 3000 - 5000 = -2000 \text{ psi}$$

$$\tau_p = r = 5000 \text{ psi}$$

Since the in-plane principal stresses are of opposite sign, the principal stresses are

$$\sigma_1 = \sigma_{p1} = 8000 \text{ psi}, \qquad \sigma_2 = \sigma_z = 0, \qquad \sigma_3 = \sigma_{p2} = -2000 \text{ psi} \qquad Ans.$$

Also, the extreme value of the in-plane shear stress τ_p is the maximum shear stress. Thus

$$\tau_{max} = \tau_p = 5000 \text{ psi} \qquad\qquad Ans.$$

From Fig. E3.4 we see that $\sin 2\theta_{x\sigma}$ equals 3/5; thus $\theta_{x\sigma}$ equals $18° 26'$ in the counterclockwise direction from the x axis. Figure E3.4c shows the planes with principal stresses and maximum shear.

___

3.6 Special States of Stress

The two-dimensional or plane state of stress is frequently encountered in engineering practice. Two other states often encountered are the *uniaxial* state of stress and the *hydrostatic* state of stress.

Uniaxial Stress—A *uniaxial* state of stress is one in which one normal stress component of the stress tensor exists and all other components are zero. This state of stress will be encountered in Chapter 9 when the tension test for determining material behavior is discussed.

For purposes of analysis consider σ_x to be the nonzero normal stress component; therefore, $\sigma_y = \sigma_z = \tau_{xy} = \tau_{yz} = \tau_{zx} = 0$. The uniaxial state of stress is obviously a special case of plane stress ($\sigma_z = \tau_{zx} = \tau_{zy} = 0$); therefore, all the relationships developed in Sections 3.4 and 3.5 will be valid. For example, the normal and shear stresses σ_n and τ_n can be determined on any plane oriented at an angle θ_x with respect to the x axis from Eqs. (3.8) and (3.9). Thus

$$\sigma_n = \sigma_x \cos^2 \theta_x \qquad\qquad (3.18a)$$

$$\tau_{nt} = -\sigma_x \sin \theta_x \cos \theta_x \qquad\qquad (3.19a)$$

or in double angle form

$$\sigma_n = (\sigma_x/2)(1 + \cos 2\theta_x) \qquad\qquad (3.18b)$$

$$\tau_{nt} = -(\sigma_x/2)\sin 2\theta_x \qquad\qquad (3.19b)$$

A plot of the variation of σ_n and τ_{nt} with change in angle θ_x is shown in Fig. 3.12. An examination of this figure shows that the normal stress σ_n always has the same sign and reaches a maximum ($\sigma_n = \sigma_x$) at $\theta_x = 0$ or π. The normal stress goes through a minimum ($\sigma_n = 0$) at $\theta_x = \pi/2$. The shear stress τ_{nt} also vanishes on planes where the normal stress is a maximum or a minimum. These planes, which are free of shear stress, have previously been defined as principal planes. For the case of uniaxial stress we have an infinite number of principal planes since planes having outer normals along or perpendicular to the x axis are free of shear (Fig. 3.12). From Fig. 3.12 it can also be

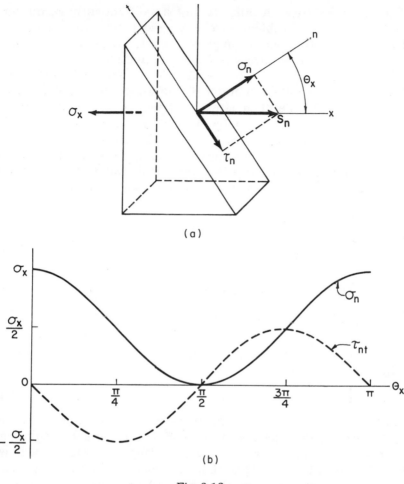

(a)

(b)

Fig. 3.12

seen that the shear stress τ_{nt} assumes extreme values ($\tau_{nt} = \pm\sigma_x/2$) at $\theta_x = \pi/4$ and $3\pi/4$. The magnitude of the normal stress on these planes of maximum shear is $\sigma_x/2$.

Both the normal stresses and the shear stresses that develop in a material as a result of uniaxial loading are important. Laboratory experiments indicate that brittle materials often fail in tension as a result of the maximum normal stress present on a transverse plane of a simple tension test specimen. However, ductile materials will fail in simple tension tests by developing slip planes (a shear type of failure) in the $45°$ directions where the shear stresses attain their maximum values.

   Hydrostatic Stress—A third state of stress frequently encountered
in fluid mechanics problems is characterized by a total absence of
shear stress on any plane through the point. Thus $\tau_{nt} = 0$, where n
and t are completely arbitrary orthogonal directions.

   Some insight into the nature of this stress distribution can be
gained by considering equilibrium of the wedge-shaped element (see
Fig. 3.7) used for the plane stress analysis. In the present case the
faces of the wedge having outer normals in the z direction will be
acted upon by a normal stress σ_z. Since this stress component does
not contribute to force equilibrium in the n and t directions shown
on the element, Eqs. (3.8) and (3.9) remain valid. Application of
these equations under the assumption of no shear stress on any plane
yields

$$\sigma_n = (1/2)(\sigma_x + \sigma_y) + (1/2)(\sigma_x - \sigma_y) \cos 2\theta_x \qquad \text{(a)}$$

$$0 = -(1/2)(\sigma_x - \sigma_y) \sin 2\theta_x \qquad \text{(b)}$$

Equation (b) can be satisfied for an arbitrary angle θ_x if and only if
$\sigma_x = \sigma_y$. Equation (a) then yields under this requirement

$$\sigma_n = \sigma_x = \sigma_y \qquad \text{(c)}$$

From a similar three-dimensional analysis, in which an element having the shape of a tetrahedron is used in place of the wedge, it is
possible to show that

$$\sigma_x = \sigma_y = \sigma_z = \sigma_n \qquad (3.20)$$

In this three-dimensional case the plane having outer normal in the n
direction can be oriented at arbitrary angles θ_x, θ_y, and θ_z with respect
to the reference directions. Since the plane on which σ_n acts was
selected arbitrarily, Eq. (3.20) indicates that for the case of no shear
stress on any plane the normal stress must have the same magnitude
on any plane that can be passed through the point. This stress situation is known as a *hydrostatic state of stress*. A hydrostatic state of
stress occurs in a fluid at rest and in a solid when loads are applied in
such a manner that $\sigma_x = \sigma_y = \sigma_z$ and $\tau_{xy} = \tau_{yz} = \tau_{zx} = 0$.

3.7   Triaxial Stress

   We have investigated the uniaxial, hydrostatic, and plane or two-
dimensional states of stress. Most practical design work deals with
these stress situations; however, occasions arise where stresses on an
arbitrary plane are required. We will attempt to illustrate how equilibrium considerations can be used to determine stresses on an arbi-

trary plane whose outer normal makes angles θ_x, θ_y, and θ_z with respect to the reference x, y, and z directions. For purposes of analysis consider a portion of the reference parallelepiped having the shape of a tetrahedron (Fig. 3.13). Three sides of the tetrahedron have the stress components associated with the sides of the reference parallelepiped. The fourth side represents the plane of interest and has the stress vector appropriate for its orientation. Normal and shear components of the stress vector represent the desired stress information for the arbitrary plane.

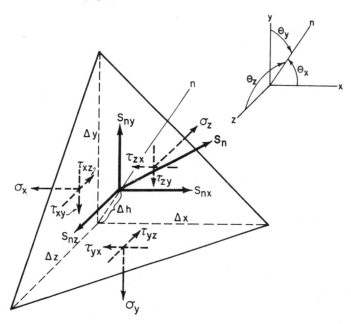

Fig. 3.13

For purposes of analysis consider the area of the inclined surface to be ΔA. The areas perpendicular to the reference x, y, and z axes are then $\Delta A \cos \theta_x$, $\Delta A \cos \theta_y$, and $\Delta A \cos \theta_z$ respectively.[1]

Relationships between the rectangular components of the stress vector on the inclined face (S_{nx}, S_{ny}, and S_{nz}) and components of the stress tensor on the other three faces (σ_x, τ_{xy}, τ_{xz}, σ_y, τ_{yx}, τ_{yz}, σ_z, τ_{zx},

---

1. These relationships can easily be proved by considering the volume of the tetrahedron in Fig. 3.13. Thus

$$V = (1/3) \, \Delta h \, \Delta A = (1/3) \, \Delta x \, \Delta A_x = (1/3) \, \Delta y \, \Delta A_y = (1/3) \, \Delta z \, \Delta A_z$$

but $\Delta h = \Delta x \cos \theta_x = \Delta y \cos \theta_y = \Delta z \cos \theta_z$; therefore

$$\Delta A_x = \Delta A \cos \theta_x, \qquad \Delta A_y = \Delta A \cos \theta_y, \qquad \Delta A_z = \Delta A \cos \theta_z$$

and τ_{zy}) can be obtained from a summation of forces in the coordinate directions. Again the tetrahedron is assumed to be extremely small so that the stresses can be assumed to be uniformly distributed over the faces. The higher order body force terms would again be negligible. From a summation of forces in the x direction we obtain

$$\Sigma F_x = S_{nx}(\Delta A) - \sigma_x(\Delta A \cos \theta_x) - \tau_{yx}(\Delta A \cos \theta_y) - \tau_{zx} \Delta A \cos \theta_z = 0$$

which yields

$$S_{nx} = \sigma_x \cos \theta_x + \tau_{yx} \cos \theta_y + \tau_{zx} \cos \theta_z \qquad (3.21a)$$

Summation of forces in the other two directions yields

$$S_{ny} = \sigma_y \cos \theta_y + \tau_{zy} \cos \theta_z + \tau_{xy} \cos \theta_x \qquad (3.21b)$$

$$S_{nz} = \sigma_z \cos \theta_z + \tau_{xz} \cos \theta_x + \tau_{yz} \cos \theta_y \qquad (3.21c)$$

The magnitude of the stress vector $\mathbf{S}_n$ is obtained from the rectangular components by using the equation

$$S_n = \sqrt{S_{nx}^2 + S_{ny}^2 + S_{nz}^2} \qquad (3.22)$$

We obtain the orientation of the stress vector $\mathbf{S}_n$ with respect to the reference axes from its magnitude and rectangular components since

$$S_{nx} = S_n \cos \phi_x, \qquad S_{ny} = S_n \cos \phi_y, \qquad S_{nz} = S_n \cos \phi_z$$

where ϕ_x, ϕ_y, and ϕ_z are the angles between the stress vector $\mathbf{S}_n$ and the reference axes. Since we know the magnitudes of the stress vector and its rectangular components from Eqs. (3.22) and (3.21), we can solve for the angles. Thus

$$\cos \phi_x = S_{nx}/S_n, \qquad \cos \phi_y = S_{ny}/S_n, \qquad \cos \phi_z = S_{nz}/S_n \quad (3.23)$$

Recall that the stress vector does not generally coincide with the outer normal to the plane.

The normal stress σ_n acting on the inclined plane can be determined as the algebraic sum of the projections of the components of the stress vector $\mathbf{S}_n$ onto the normal n. Thus

$$\sigma_n = S_{nx} \cos \theta_x + S_{ny} \cos \theta_y + S_{nz} \cos \theta_z \qquad (3.24a)$$

Substitution of Eqs. (3.7) and (3.21) into Eq. (3.24a) yields

$$\sigma_n = \sigma_x \cos^2 \theta_x + \sigma_y \cos^2 \theta_y + \sigma_z \cos^2 \theta_z + 2\tau_{xy} \cos \theta_x \cos \theta_y$$
$$+ 2\tau_{yz} \cos \theta_y \cos \theta_z + 2\tau_{zx} \cos \theta_z \cos \theta_x \qquad (3.24b)$$

The magnitude of the shear stress τ_n is then obtained as the vector difference between S_n and σ_n. Thus

$$\tau_n = \sqrt{S_n^2 - \sigma_n^2} \qquad (3.25)$$

Since normal and shear stresses on any plane can be specified in terms of the nine components of the stress tensor, these components have been shown to represent a complete specification of the state of stress at the point. Further analysis of the triaxial state of stress (beyond the scope of this book) would involve principal stress and maximum shear stress determinations. The significant results of this type of analysis have been mentioned at appropriate places. Students interested in a detailed analysis should consult a theory of elasticity text.

EXAMPLE 3.5

At a point in a stressed machine component the Cartesian components of stress are

$$\sigma_x = 9000 \text{ psi} \qquad \tau_{xy} = \tau_{yx} = 3000 \text{ psi}$$
$$\sigma_y = 6000 \text{ psi} \qquad \tau_{yz} = \tau_{zy} = 3000 \text{ psi}$$
$$\sigma_z = 3000 \text{ psi} \qquad \tau_{zx} = \tau_{xz} = 6000 \text{ psi}$$

Determine the normal and shear stresses on a plane whose normal is directed such that $\cos \theta_x = 1/3$, $\cos \theta_y = \cos \theta_z = 2/3$.

SOLUTION

From Eqs. (3.21) the components of the stress vector on the plane of interest are computed as

$$S_{nx} = \sigma_x \cos \theta_x + \tau_{yx} \cos \theta_y + \tau_{zx} \cos \theta_z$$
$$= 9000 \, (1/3) + 3000 \, (2/3) + 6000 \, (2/3) = 9000 \text{ psi}$$

Similarly,

$$S_{ny} = 6000 \, (2/3) + 3000 \, (2/3) + 3000 \, (1/3) = 7000 \text{ psi}$$
$$S_{nz} = 3000 \, (2/3) + 6000 \, (1/3) + 3000 \, (2/3) = 6000 \text{ psi}$$

Once the components of the stress vector are evaluated, the normal stress on the plane of interest is determined by using Eq. (3.24a). Thus

$$\sigma_n = S_{nx} \cos \theta_x + S_{ny} \cos \theta_y + S_{nz} \cos \theta_z$$
$$= 9000 \, (1/3) + 7000 \, (2/3) + 6000 \, (2/3) = 11{,}650 \text{ psi} \qquad \textit{Ans.}$$

The shear stress is evaluated by using Eq. (3.25) after the magnitude of the stress vector is determined from Eq. (3.22). Thus

$$S_n = \sqrt{S_{nx}^2 + S_{ny}^2 + S_{nz}^2} = \sqrt{(9000)^2 + (7000)^2 + (6000)^2} = 12{,}900 \text{ psi}$$
$$\tau_n = \sqrt{S_n^2 - \sigma_n^2} = \sqrt{(12{,}900)^2 - (11{,}650)^2} = 5500 \text{ psi} \qquad \textit{Ans.}$$

3.8 Stress-Force System Relationships

We have found it convenient to isolate a portion of the body by exposing an internal plane of interest with an imaginary cut. In the most general case the external forces acting on the body will produce an internal force system that can be represented on the plane of the cut by three force components F_x, F_y, and F_z and three moment components M_x, M_y, and M_z (Fig. 3.14a). On any small element of area on the surface of the cut the stress components σ_x, τ_{xy}, and τ_{xz} may act as shown in Fig. 3.14b. Since the stress components will

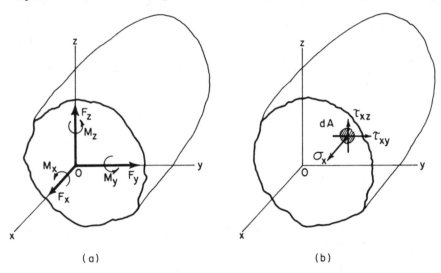

(a)   (b)

Fig. 3.14

vary from point to point, the following expressions must be used to relate the stress components to the internal force and moment components obtained from equilibrium considerations. Thus

$$F_x = \int_A \sigma_x \, dA, \qquad F_y = \int_A \tau_{xy} \, dA, \qquad F_z = \int_A \tau_{xz} \, dA \quad (3.26a)$$

$$M_x = \int_A \tau_{xz} \, y \, dA - \int_A \tau_{xy} \, z \, dA, \qquad M_y = \int_A \sigma_x z \, dA$$

$$(3.26b)$$

$$M_z = \int_A -\sigma_x y \, dA$$

Similar sets of equations may be developed for internal planes having other orientations.

PROBLEMS

3.1. The steel bar in Fig. P3.1 is subjected to a tensile load that produces a normal stress of 2000 psi on the cross section AB. Use the "wedge" method of analysis to determine the normal and shear stresses on plane CD.

Ans. $\sigma_n = 1700$ psi, $\quad \tau_{nt} = 710$ psi

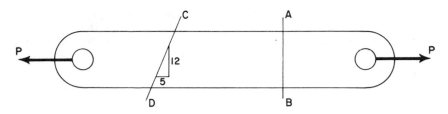

Fig. P3.1

3.2. At a point in a structural member the plane state of stress in Fig. P3.2 is known to exist. Use the "wedge" method of analysis to determine the normal and shear stresses on plane AB.

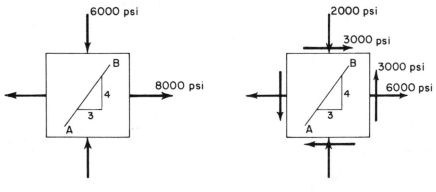

Fig. P3.2   Fig. P3.3

3.3. At a point in a machine component the plane state of stress shown in Fig. P3.3 is known to exist. Use the "wedge" method of analysis to determine the normal and shear stresses on plane AB.

Ans. $\sigma_n = 240$ psi, $\quad \tau_{nt} = 4680$ psi

3.4. At a point in a machine component the following plane state of stress is known to exist:

$$\sigma_x = 70 \text{ MN/m}^2, \quad \sigma_y = 10 \text{ MN/m}^2, \quad \tau_{xy} = 40 \text{ MN/m}^2$$

Use the "wedge" method of analysis to determine the normal and shear stresses on a plane whose normal is directed $30°$ from the x axis.

3.5. At a point in a machine component the following plane state of stress is known to exist:

$$\sigma_x = 8000 \text{ psi}, \qquad \sigma_y = -4000 \text{ psi}, \qquad \tau_{xy} = -3000 \text{ psi}$$

Use the "wedge" method of analysis to determine the normal and shear stresses on a plane whose normal is directed $-60°$ from the x axis.

Ans. $\sigma_n = 1598 \text{ psi}, \qquad \tau_{nt} = 6696 \text{ psi}$

3.6. The two wooden blocks in Fig. P3.6 have been glued together along joint AB. Use the "wedge" method of analysis to determine the axial force P that may be applied safely if the glue has a maximum allowable normal stress of 250 psi and a maximum allowable shear stress of 500 psi.

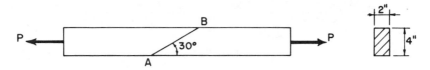

Fig. P3.6

3.7. At a point in a structural member the plane state of stress in Fig. P3.7 is known to exist. Use the stress transformation equations to determine the normal and shear stresses on plane AB.

Ans. $\sigma_n = 7000 \text{ psi}, \qquad \tau_{nt} = -1732 \text{ psi}$

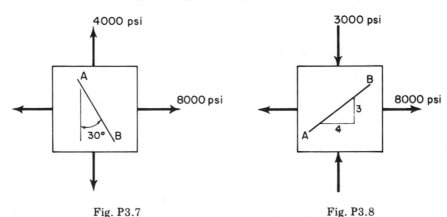

Fig. P3.7   Fig. P3.8

3.8. At a point in a structural member the plane state of stress in Fig. P3.8 is known to exist. Use the stress transformation equations to determine the normal and shear stresses on plane AB.

3.9. Use the stress transformation equations to determine the normal and shear stresses on plane AB in Fig. P3.9.

Ans. $\sigma_n = 48.6 \text{ MN/m}^2, \qquad \tau_{nt} = -24.7 \text{ MN/m}^2$

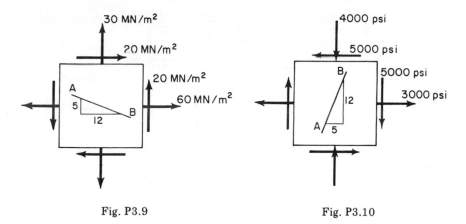

Fig. P3.9   Fig. P3.10

3.10. Use the stress transformation equations to determine the normal and shear stresses on plane AB in Fig. P3.10.

3.11. Use the stress transformation equations to determine the normal and shear stresses on plane AB in Fig. P3.11.

Ans. $\sigma_n = -6068$ psi, $\tau_{nt} = 6294$ psi

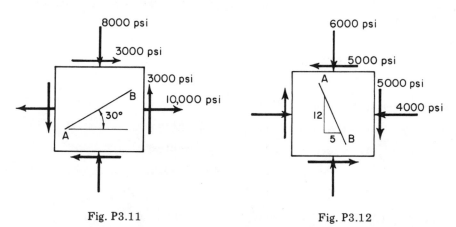

Fig. P3.11   Fig. P3.12

3.12. Use the stress transformation equations to determine the normal and shear stresses on plane AB in Fig. P3.12.

3.13. The Cartesian components of stress at a point in a plane stress field are shown in Fig. P3.13. Determine the principal stresses and the maximum shear stress at the point. Indicate on an appropriate sketch the orientation of the planes on which these stresses act.

Ans. $\sigma_1 = 22,000$ psi, $\sigma_2 = 0$, $\sigma_3 = -4000$ psi
$\theta_{x\sigma} = 33°$, $\tau_{max} = 13,000$ psi (in-plane)

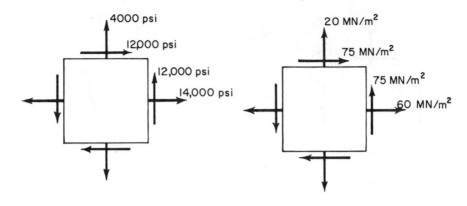

Fig. P3.13   Fig. P3.14

3.14. The Cartesian components of stress at a point in a plane stress field are shown in Fig. P3.14. Determine the principal stresses and the maximum shear stress at the point. Indicate on an appropriate sketch the orientation of the planes on which these stresses act.

In Problems 3.15–3.22 the Cartesian components of stress at a point in a plane stress field are given. Determine the principal stresses and the maximum shear stress at the point. Show on an appropriate sketch the planes on which these stresses act.

3.15. σ_x = 2000 psi, σ_y = −8000 psi, τ_{xy} = 12,000 psi

    Ans. σ_1 = 10,000 psi, σ_2 = 0, σ_3 = −16,000 psi

    $\theta_{x\sigma}$ = 33°, τ_{max} = 13,000 psi (in-plane)

3.16. σ_x = 7000 psi, σ_y = 1000 psi, τ_{xy} = −4000 psi

3.17. σ_x = 2400 psi, σ_y = 1000 psi, τ_{xy} = −2400 psi

    Ans. σ_1 = 4200 psi, σ_2 = 0, σ_3 = −800 psi

    $\theta_{x\sigma}$ = −36.9°, τ_{max} = 2500 psi (in-plane)

3.18. σ_x = 28 MN/m², σ_y = −20 MN/m², τ_{xy} = 7 MN/m²

3.19. σ_x = 12,000 psi, σ_y = 6000 psi, τ_{xy} = 4000 psi

    Ans. σ_1 = 14,000 psi, σ_2 = 4,000 psi, σ_3 = 0,

    $\theta_{x\sigma}$ = 26.6°, τ_{max} = 7000 psi (out-of-plane)

3.20. σ_x = −2000 psi, σ_y = −3000 psi, τ_{xy} = 1200 psi

3.21. σ_x = 31 MN/m², σ_y = 15MN/m², τ_{xy} = −15 MN/m²

    Ans. σ_1 = 40 MN/m², σ_2 = 6 MN/m², σ_3 = 0

    $\theta_{x\sigma}$ = −31.0°, τ_{max} = 20 MN/m² (out-of-plane)

3.22. σ_x = −2600 psi, σ_y = −4000 psi, τ_{xy} = 2400 psi

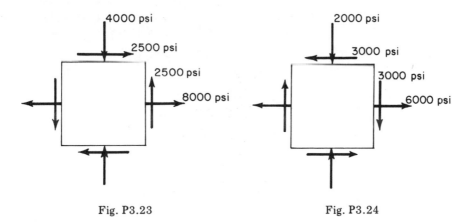

Fig. P3.23   Fig. P3.24

3.23. At a point in a structural member the plane state of stress in Fig. P3.23 is known to exist. Use Mohr's circle to determine the principal stresses and the maximum shear stress at the point. Show on an appropriate sketch the planes on which these stresses act.

Ans. $\sigma_1 = 8500$ psi, $\sigma_2 = 0$, $\sigma_3 = -4500$ psi
$\theta_{x\sigma} = 11.3°$, $\tau_{max} = 6500$ psi (in-plane)

3.24. At a point in a structural member the plane state of stress in Fig. P3.24 is known to exist. Use Mohr's circle to determine the principal stresses and the maximum shear stress at the point. Show on an appropriate sketch the planes on which these stresses act.

3.25. Solve Problem 3.16 using Mohr's circle.

3.26. Solve Problem 3.18 using Mohr's circle.

3.27. Solve Problem 3.20 using Mohr's circle.

3.28. Solve Problem 3.21 using Mohr's circle.

3.29. Solve Problem 3.12 using Mohr's circle.

3.30. Solve Problem 3.11 using Mohr's circle.

3.31. Solve Problem 3.9 using Mohr's circle.

3.32. On the cross section of the rectangular bar in Fig. P3.32 the following distribution of stress is known to exist:

$$\tau_{xy} = 0, \qquad \tau_{xz} = 0, \qquad \sigma_x = -By$$

where B is a constant. Determine (a) the net internal force and couple components acting at 0 that are statically equivalent to the given stress distribution and (b) express σ_x in terms of the bending couple M_z found in part (a).

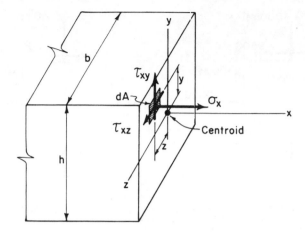

Fig. P3.32

3.33. On the cross section of the circular bar of radius C in Fig. P3.33 the following distribution of stress is known to exist:

$$\sigma_x = 0, \qquad \tau_{xr} = 0, \qquad \tau_{x\theta} = Br$$

where B is a constant. Determine (a) the shear stresses τ_{xy} and τ_{xz} in terms of y and z, (b) the net internal force and couple components acting at 0 which are statically equivalent to the given stress distribution, and (c) express $\tau_{x\theta}$ in terms of the twisting couple M_x found in part (b).

Ans. $\tau_{xy} = -Bz, \qquad \tau_{xz} = By, \qquad F_x = F_y = F_z = M_y = M_z = 0$

$$M_x = T = B \int_A r^2 \, dA = BJ, \qquad \tau_{x\theta} = Tr/J$$

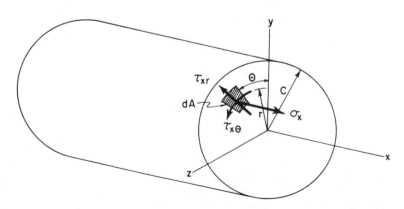

Fig. P3.33

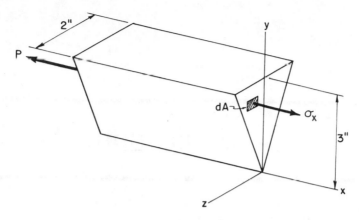

Fig. P3.34

3.34. The bar of triangular cross section in Fig. P3.34 has a load P applied parallel to the x axis. The load produces the following distribution of stress on the cross section:

$$\sigma_x = 10{,}000y, \qquad \tau_{xy} = 0, \qquad \tau_{xz} = 0$$

Determine the magnitude of the load P and the location of the intersection of its line of action with the cross section.

chapter 4

Equilibrium of Fluids

4.1 Pressure Variation in a Fluid

A fluid is a material set in motion under the action of any applied shearing stress. Thus when a fluid is at rest no shearing stresses are present. It was shown in Chapter 3 that for stress fields in which there are no shearing stresses the normal stresses at a point are the same in all directions. This result is known as *Pascal's law*. Since the normal stress is invariant at a point in a fluid at rest, it is common practice to call this stress the *pressure*. Usually the pressure is considered positive if it is a *compressive* stress, so that at any point

$$p = -\sigma_n \tag{4.1}$$

Recall that normal stresses were defined as positive if tensile.

Consider now a differential element of fluid as shown in Fig. 4.1. The pressure at the center of the element is p and the pressures on the faces of the element can be expressed in terms of this pressure, as illustrated. The resultant force in the x direction is

$$dF_x = -\frac{\partial p}{\partial x}\,dx\,dy\,dz$$

Similar expressions can be obtained for the differential forces in the y and z directions, and the resultant force is

$$d\mathbf{F} = -\left(\frac{\partial p}{\partial x}\mathbf{i} + \frac{\partial p}{\partial y}\mathbf{j} + \frac{\partial p}{\partial z}\mathbf{k}\right)dx\,dy\,dz$$

The force per unit volume $\mathbf{f}$ is

$$\mathbf{f} = \frac{d\mathbf{F}}{dx\,dy\,dz} = -\left(\mathbf{i}\frac{\partial}{\partial x} + \mathbf{j}\frac{\partial}{\partial y} + \mathbf{k}\frac{\partial}{\partial z}\right)p = -\nabla p \tag{4.2}$$

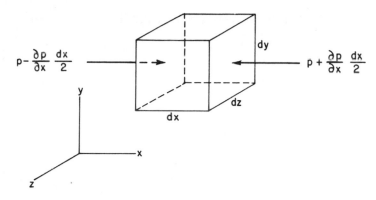

Fig. 4.1

where the symbol ∇ is the *"del"* or *gradient* vector operator; i.e.,

$$\nabla = \mathbf{i}\,\frac{\partial}{\partial x} + \mathbf{j}\,\frac{\partial}{\partial y} + \mathbf{k}\,\frac{\partial}{\partial z}$$

For a fluid at rest (Fig. 4.2) the resultant force due to pressure must be in equilibrium with the weight of the element; i.e.,

$$-\mathbf{j}\,\gamma\,dx\,dy\,dz = \nabla p\,dx\,dy\,dz \qquad (4.3)$$

where γ is the weight of the fluid per unit volume (specific weight). The *density* ρ of a fluid is its mass per unit volume and its *specific gravity* (sp. gr.) is the ratio of the density of the fluid to the density

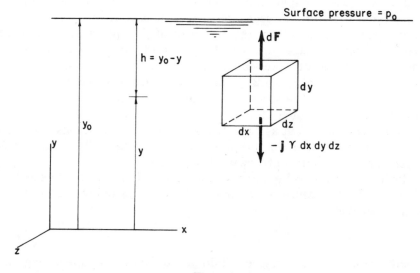

Fig. 4.2

of water at 4 C. It follows from Eq. (4.3) that

$$\frac{\partial p}{\partial x} = 0 \qquad (4.4)$$

$$\frac{\partial p}{\partial z} = 0 \qquad (4.5)$$

and

$$\frac{\partial p}{\partial y} = -\gamma \qquad (4.6)$$

Equations (4.4) and (4.5) indicate there is no pressure variation between points of equal elevation. For an *incompressible* fluid ($\gamma =$ constant) Eq. (4.6) can be integrated to give

$$p = -\gamma y + \text{constant} \qquad (4.7)$$

The constant of integration is determined by specifying a reference pressure p_0 at some elevation. In Fig. 4.2 the pressure at any depth can be determined from the relation

$$p = p_0 + \gamma(y_0 - y) = p_0 + \gamma h \qquad (4.8)$$

where h is the depth below the surface. This type of linear pressure variation is called a *hydrostatic* pressure variation.

Pressures can be specified in terms of either *gage* pressure or *absolute* (abs) pressure. A gage pressure is measured relative to atmospheric pressure; therefore, the pressure of p_0 (Fig. 4.2) is equal to zero (in terms of gage pressure) if the fluid surface is open to the atmosphere. Absolute pressure is measured relative to an absolute zero datum, thus in Fig. 4.2, p_0 is equal to atmospheric pressure for an open surface. Pressures are commonly expressed in terms of pounds per square inch (psi, lb/in.^{2}) or newtons per square metre (N/m^2). Notation such as gage or abs should in general also be used to indicate the datum for the pressure. In this text all pressures will be gage pressures unless otherwise designated.

Pressure may also be expressed in terms of an equivalent column of fluid; i.e., a pressure p can be developed by a column of fluid of height h so that

$$h = p/\gamma \qquad (4.9)$$

Specification of h for a given fluid is equivalent to specifying the pressure. For example a pressure of 5 psi can be expressed in terms of a column of water ($\gamma = 62.4 \ \text{lb/ft}^3$) through the relationship

$$h = 5(144)/62.4 = 5/0.433 = 11.55 \ \text{ft of water}$$

This equivalent height of fluid is the *pressure head.*

4.2 Manometry

The pressure distribution in a fluid at rest as given by Eq. (4.8) is utilized in the design of simple pressure measuring devices called *manometers.* Although manometers can take on a great variety of configurations, two common types are illustrated in Fig. 4.3.

The *piezometer tube* consists of a vertical open tube (Fig. 4.3a) and can be used to measure the pressure in the pressurized tank by simply measuring the height of fluid h above the point of interest. Since the pressure at the point of attachment of the piezometer tube is $p_0 + \gamma h_0$ and this pressure must be balanced by the height of fluid in the open tube, it follows that $p_0 + \gamma h_0 = \gamma h$ or $h = (p_0/\gamma) + h_0$. Thus the pressure head in the tank can be determined from a measurement of the height of fluid in the piezometer tube.

Although the piezometer tube represents the simplest type of manometer, it has the disadvantage that a long column of fluid is required to measure relatively small pressures. One method of over-

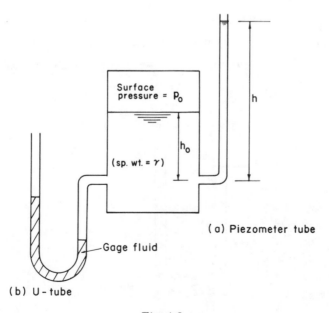

Fig. 4.3

coming this difficulty is to use the *U-tube manometer* (Fig. 4.3b). With this arrangement it is possible to utilize a fluid (*gage fluid*) in the manometer that is different from the fluid in the system in which the pressure is desired. Typical applications for manometers are illustrated in the following examples.

EXAMPLE 4.1

The pressure in pipe A, which contains water ($\gamma = 62.4$ lb/ft^3), is to be determined from the differential reading Δh of the U-tube manometer of Fig. E4.1. The gage fluid is mercury having a specific gravity of 13.6. Solve for (a) the gage pressure in A in terms of psi and (b) the pressure head in A in millimetres of mercury.

SOLUTION

Our basic procedure will be to start at one side of the manometer and work around to the other side, evaluating the pressure at certain key locations along the way. Let the unknown pressure in pipe A be p_A. Since the fluid in the manometer is at rest, the pressure at point 1 is $p_1 = p_A + \gamma_{H_2O}\Delta h$ from Eq. (4.8). This must also be equal to the pressure at point 2 since points 1 and 2 are at equal elevations in the same fluid. The pressure at point 3 is

$$p_3 = p_2 - \gamma_{Hg}\Delta h = p_A + \gamma_{H_2O}\Delta h - \gamma_{Hg}\Delta h$$

Thus

$$p_A - p_3 = \gamma_{H_2O}\left[(\gamma_{Hg}/\gamma_{H_2O})\Delta h - \Delta h\right] = \gamma_{H_2O}\Delta h\left[(\gamma_{Hg}/\gamma_{H_2O}) - 1\right]$$

(a) Since we wish to express p_A in terms of gage pressure, we set $p_3 = 0$ and

$$p_A = 62.4\ (\text{lb/ft}^3)(10/12)(\text{ft})(13.6 - 1) = 655\ \text{lb/ft}^2 = 4.55\ \text{psi (gage)} \quad Ans.$$

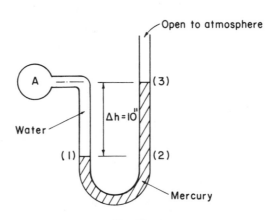

Fig. E4.1

(b) The equivalent pressure head of mercury is

$$h_A = p_A/\gamma_{Hg} = 655/[(13.6)(62.4)] = 0.772 \text{ ft Hg}$$
$$= 0.772 \ (304.8 \text{ mm/ft}) = 235 \text{ mm Hg} \qquad Ans.$$

EXAMPLE 4.2

The inclined-tube manometer of Fig. E4.2 is to be used to measure the pressure differential between the two points A and B. Determine the relationship between the differential pressure head and the reading Δh taken along the inclined tube.

SOLUTION

Since the readings taken from a manometer are frequently lengths of fluid columns, it is sometimes convenient to write the manometer equation in terms of pressure head. Thus we can start at point A where the pressure head is p_A/γ_1 (in terms of the fluid γ_1) and write the manometer equation in the form

$$\frac{p_A}{\gamma_1} + a - \frac{\gamma_g \, \Delta h \sin \theta}{\gamma_1} - (a - \Delta h \sin \theta) = \frac{p_B}{\gamma_1}$$

or

$$(p_A - p_B)/\gamma_1 = \Delta h \sin \theta \ [(\gamma_g/\gamma_1) - 1] \qquad Ans.$$

Note for a given differential pressure $\Delta p = p_A - p_B$

$$\Delta h = \frac{\Delta p/\gamma_1}{(\gamma_g/\gamma_1 - 1) \sin \theta}$$

and the differential reading Δh along the inclined tube can be increased by decreasing the angle θ. Thus inclined-tube manometers are commonly used to measure small pressures for which the inclined distance Δh may be easily read, whereas the corresponding reading on a U-tube manometer would be very small and subject to significant measurement errors. The differential reading is also strongly influenced by the ratio γ_g/γ_1, and the sensitivity of the manometer can

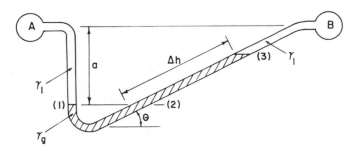

Fig. E4.2

be increased by using a gage fluid having a specific weight close to that of the working fluid.

4.3   Force on a Plane Submerged Area

Consider a plane surface immersed in a liquid at rest. The plane in which the surface lies intersects the free liquid surface at 0 and makes an angle θ with the free surface as in Fig. 4.4. The area can have an arbitrary shape as illustrated. We wish to determine the direction, magnitude, and line of action of the resultant force acting on one side of this area. The force arises from the fluid pressure acting on the surface; and since there are no shearing stresses present, the resultant force must be normal to the surface. At any given depth h the pressure is constant along a differential area dA so that the force acting on dA is $dF = \gamma h \, dA$ and the magnitude of the resultant force must be given by the equation

$$F_p = \int_A \gamma h \, dA = \int_A \gamma y \sin \theta \, dA$$

$$= \gamma \sin \theta \int_A y \, dA \qquad (4.10)$$

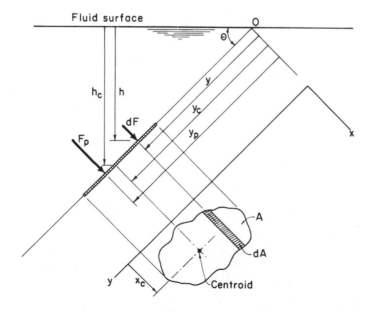

Fig. 4.4

The integral appearing in Eq. (4.10) is simply the first moment of the area with respect to the x axis, so we can write

$$\int_A y\, dA = y_c A$$

where y_c is the y coordinate of the centroid measured from 0. Equation (4.10) can thus be written $F_p = \gamma A y_c \sin \theta$ or

$$F_p = \gamma h_c A \qquad\qquad (4.11)$$

where h_c is the vertical distance from the fluid surface to the centroid of the area. Note that the magnitude of the force is independent of the angle θ and depends only on the specific weight, the total area, and the depth of the centroid below the surface.

The y coordinate y_p of the resultant force can be determined by summation of moments around the x axis; i.e.,

$$F_p y_p = \int_A y\, dF = \int_A y^2 \gamma \sin \theta\, dA$$

and

$$y_p = \frac{\displaystyle\int_A y^2\, dA}{y_c A} = \frac{I_x}{y_c A}$$

where I_x is the second moment of the area with respect to an axis formed by the intersection of the plane containing the surface and the free surface (x axis). Use can now be made of the parallel axis theorem to write

$$I_x = (I_x)_c + A\, y_c^2$$

where $(I_x)_c$ is the second moment of the area with respect to an axis passing through its centroid and parallel to the x axis. Thus

$$y_p = [(I_x)_c / y_c A] + y_c \qquad\qquad (4.12)$$

Equation (4.12) shows that the resultant force does not pass through the centroid but is always below it.

The x coordinate x_p for the resultant force can be determined in a similar manner by summing moments about the y axis. Thus

$$F_p x_p = \int_A xy\gamma \sin \theta\, dA$$

and

$$x_p = \frac{\displaystyle\int_A xy \, dA}{y_c A} = \frac{I_{xy}}{y_c A}$$

where I_{xy} is the product of inertia with respect to the x and y axes. Again using the parallel axis theorem,[1] we can write

$$x_p = [(I_{xy})_c / y_c A] + x_c \qquad (4.13)$$

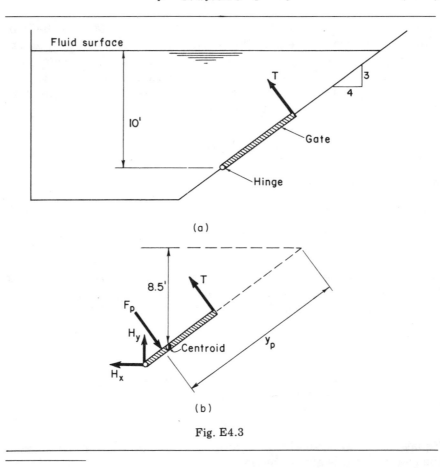

(a)

(b)

Fig. E4.3

1. Recall that the parallel axis theorem for the product of inertia of an area states that the product of inertia with respect to an orthogonal set of axes (xy coordinate system) is equal to the product of inertia with respect to an orthogonal set of axes parallel to the original set and passing through the centroid of the area plus the product of the area and the x and y coordinates of the centroid of the area. Thus $I_{xy} = (I_{xy})_c + A x_c y_c$.

where $(I_{xy})_c$ is the product of inertia with respect to an orthogonal coordinate system passing through the centroid and formed by a translation of the xy coordinate system. If the submerged area is symmetrical with respect to an axis passing through the centroid and parallel to the x or y axis, the resultant force must lie along the line $x = x_c$ since $(I_{xy})_c$ is identically zero in this case. The point through which the resultant force acts is called the *center of pressure*.

EXAMPLE 4.3

An area having the shape of an isosceles triangle with a base width of 4 ft and an altitude of 7.5 ft is located in the side of a tank (Fig. E4.3a). The base of the triangular area is horizontal and 10 ft below the surface of the fluid (sp.wt. = 50 lb/ft^3) contained in the tank. If the gate is hinged along its base, determine the force T required to hold the gate in position. Neglect friction at the hinge and the weight of the gate.

SOLUTION

A free-body diagram and the various pertinent dimensions are shown in Fig. E4.3b. The resultant fluid force is F_p = (50)(8.5)(15) = 6375 lb and the location of this force is obtained from the equation

$$y_p = [(I_x)_c/y_cA] + y_c = [46.8/(14.16)15] + 14.16 = 14.38 \text{ ft}$$

Summation of moments about the hinge gives

$$T(7.5) = (6375)(2.28), \qquad T = 1937 \text{ lb as shown} \qquad\qquad Ans.$$

4.4 Force on a Submerged Nonplanar Surface

The results obtained in Section 4.3 can be used to determine the resultant force acting on a curved surface subjected to fluid pressure. Consider, for example, the nonplanar section BC of a wall in a tank containing a liquid (Fig. 4.5a). We desire to find the resultant force acting on the curved section BC (having a unit length perpendicular to the plane of the paper). We first draw a free-body diagram of a section of fluid (Fig. 4.5b). The curved part of the tank forms one boundary, and the other two surfaces AB and AC are horizontal and vertical plane surfaces respectively. The forces F_1 and F_2 can be completely determined from the equations given in Section 4.3.

To maintain equilibrium, it follows that

$$(F_R)_H = F_2, \qquad (F_R)_V = F_1 + W$$

where W is the weight of the fluid contained in the volume forming the free-body diagram of Fig. 4.5b. Since the rectangular components

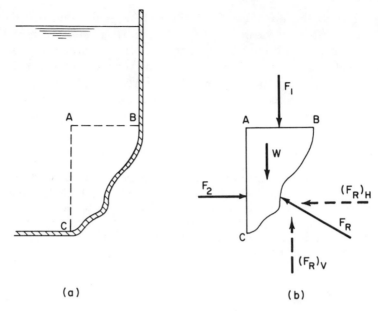

Fig. 4.5

of F_R are known, its magnitude can be determined; i.e.,

$$F_R = \sqrt{(F_R)_H^2 + (F_R)_V^2} \tag{4.14}$$

The location of F_R can be found by making use of the principle of moments. Note that the force the fluid exerts on the tank is equal and opposite to the force shown in the free-body diagram of Fig. 4.5b.

EXAMPLE 4.4

An 8-ft diameter cylinder restrains water at a depth of 4 ft (Fig. E4.4a). Determine the resultant force per unit length of cylinder due to the water.

SOLUTION

Draw a free-body diagram of a volume of fluid (Fig. E4.4b). The horizontal component F_H is obtained from the equation

$$F_H = F_2 = (62.4)(2)(4) = 499 \text{ lb}$$

The vertical component is given by

$$F_V = F_1 - W = (62.4)(4)(4) - (62.4)\,[16 - (\pi/4)(16)] = (62.4)(4\pi) = 784 \text{ lb}$$

and the magnitude of the resultant is $F_R = \sqrt{(499)^2 + (784)^2} = 931$ lb.

To determine the position of the line of action of the resultant, we can sum moments about some convenient axis. For this example, however, note that each elemental force acting on the surface of the cylinder passes through the center of the cylinder and thus the resultant must also pass through the center. The resultant force per unit length acting on the cylinder is therefore

931 lb

784

499

Ans.

through the center of the cylinder.

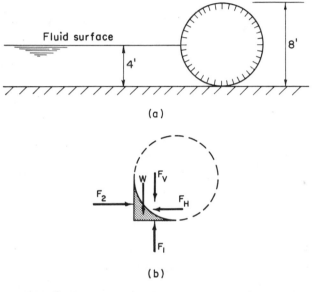

Fluid surface

4'

8'

(a)

F_2

W

F_v

F_H

F_1

(b)

Fig. E4.4

4.5 Buoyancy

The *buoyant force* acting on a body submerged or partially submerged in a fluid is defined as the resultant force due to the hydrostatic pressure distribution. Consider a body of arbitrary shape completely submerged in an incompressible fluid (Fig. 4.6a). Enclose the body within a rectangular parallelepiped as illustrated and draw a free-body diagram of the parallelepiped with the body removed (Fig. 4.6b). It is clear that the submerged body is subjected to the resultant of the forces shown in this diagram. The forces, such as F_3, on the four vertical faces are all equal and cancel so that the resultant fluid force must act vertically. Thus the buoyant force F_B is given by

$$F_B = F_2 - F_1 - W \qquad (4.15)$$

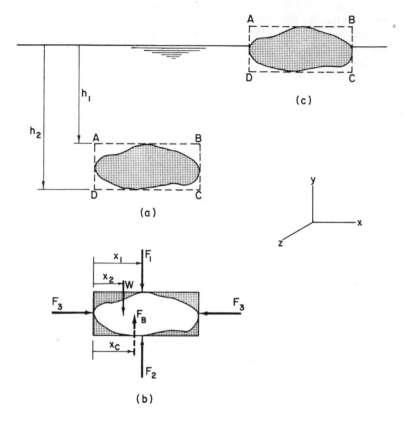

Fig. 4.6

where W is the weight of the fluid within the parallelepiped with the body removed. Equation (4.15) can be written as

$$F_B = \gamma h_2 A - \gamma h_1 A - \gamma [(h_2 - h_1)(A) - V] \qquad (4.16)$$

where A is the horizontal area of the upper or lower surface of the parallelepiped and V is the volume of the body. Equation (4.16) can be simplified to

$$F_B = \gamma V \qquad (4.17)$$

which reveals that the buoyant force is an upward force equal to the weight of the displaced fluid. This result is commonly referred to as *Archimedes' principle*.

The location of the line of action of the buoyant force can be determined by the principle of moments. Summing moments about an axis through A perpendicular to the paper we have

$$x_c F_B = x_1 (F_2 - F_1) - x_2 W$$

or

$$x_c \gamma V = x_1 \gamma (h_2 - h_1)A - x_2 \gamma [(h_2 - h_1)(A) - V]$$

$$x_c V = x_1 V_T - x_2 (V_T - V) \tag{4.18}$$

where V_T is the total volume $(h_2 - h_1)A$. The right-hand side of Eq. (4.18) is the first moment of the volume V with respect to a plane perpendicular to the plane of the paper and passing through AD so that x_c must be the x coordinate of the centroid of V. In a similar manner the z coordinate of the buoyant force is found to coincide with the z coordinate of the centroid. Thus we conclude that the buoyant force passes through the centroid of the volume of fluid displaced by the submerged body.

For bodies that are floating, i.e., partially submerged, as in Fig. 4.6c the same results can be obtained if the weight of the fluid (usually air) above the liquid surface is neglected.

EXAMPLE 4.5

A tank 3 m long with a cross-sectional area of $1/9$ m^2 and closed at one end is lowered into water ($\gamma = 98.03 \times 10^2$ N/m^3) with its open end down and held in an equilibrium position by a $1/10$ m^3 block of concrete ($\gamma = 236 \times 10^2$ N/m^3) (Fig. E4.5). Determine the final volume of air trapped in the tank when in the equilibrium position. Neglect the weight of the tank and assume that the air is compressed isothermally.

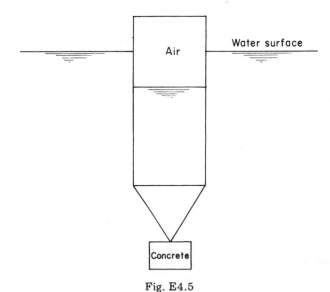

Fig. E4.5

SOLUTION

To maintain equilibrium, the force due to the pressure in the tank must balance the weight minus the buoyant force of the concrete. Thus

$$p(1/9) = (1/10)(23,600 - 9803), \qquad p = 12,420 \text{ N/m}^2 \text{ (gage)}$$

It is a well-known law of physics that a gas compressed isothermally obeys the relationship pV = constant, where p is the absolute pressure and V the volume. We assume that initially the tank is filled with air under atmospheric conditions $(p_{atm} = 1.013 \times 10^5 \text{ N/m}^2)$ so that

$$(0.124 \times 10^5 + 1.013 \times 10^5)V = (1.013 \times 10^5)(1/9)(3)$$

and

$$V = 0.298 \text{ m}^3 \qquad\qquad\qquad Ans.$$

PROBLEMS

4.1. A tank of fluid is resting on the floor of an elevator that is accelerating upward with a constant acceleration A_0. The fluid in the tank is incompressible, and there is no relative motion between the fluid and the tank. Use the procedure described in Section 4.1 to develop an expression for the pressure distribution in the tank.

   Ans. $p = -(\gamma + \rho A_0)y + \text{constant}$

4.2. Pipe B of Fig. P4.2 contains gasoline (sp. wt. = 42.4 pcf) and is connected by a mercury manometer to pipe A which contains water under a pressure of 20 psi. Determine the pressure in pipe B.

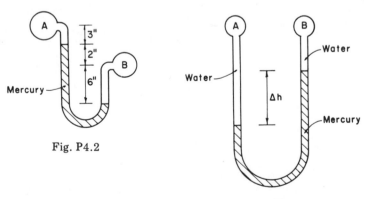

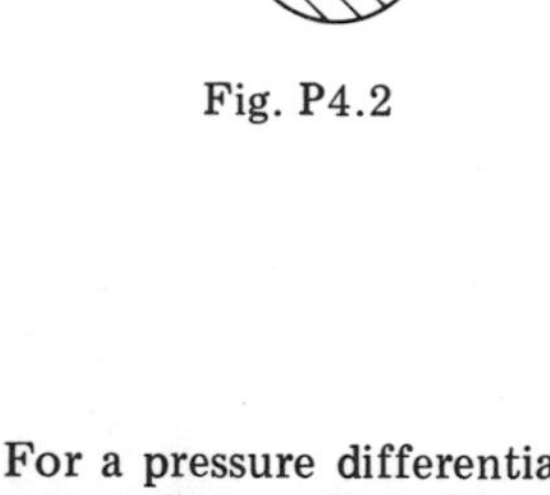

Fig. P4.2

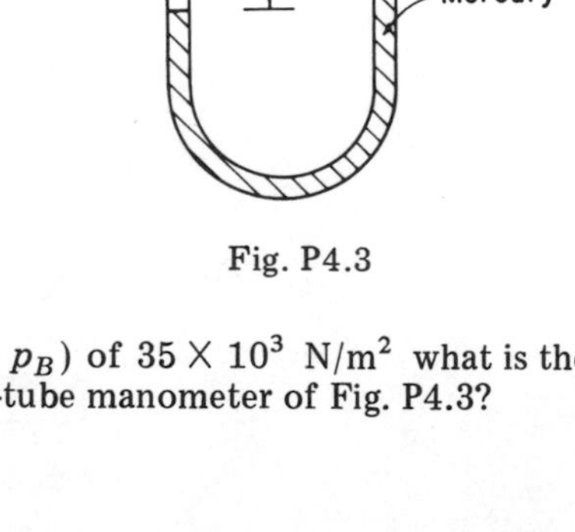

Fig. P4.3

4.3. For a pressure differential $(p_A - p_B)$ of 35×10^3 N/m^2 what is the corresponding reading Δh on the U-tube manometer of Fig. P4.3?

   Ans. 0.284 m

4.4. The pressure in pipe A, which contains water ($\gamma = 62.4$ pcf), is 10 psi (Fig. P4.4). Determine the pressure in pipe B, which also contains water, if the differential reading on the inclined-tube manometer is 12 in. as shown. The gage fluid is mercury (sp. gr. = 13.6).

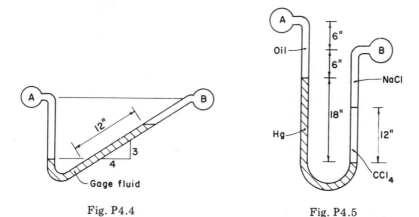

Fig. P4.4   Fig. P4.5

4.5. In Fig. P4.5 pipe A contains oil (sp. gr. = 0.8) and pipe B contains brine (sp. gr. = 1.1). The brine and mercury are separated by 12 in. of carbon tetrachloride (sp. gr. = 1.6). Determine the pressure differential between A and B.

   Ans.   8.01 psi, B greater than A

4.6. The differential mercury manometer of Fig. P4.6 is connected to pipe A containing oil (sp. gr. = 0.9) and pipe B containing water. Determine the differential reading h corresponding to a pressure in A of 3 psi and a vacuum of 6 in. of mercury in B.

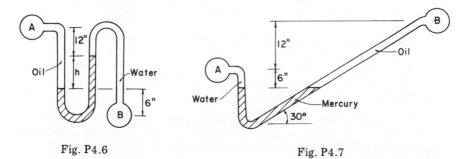

Fig. P4.6   Fig. P4.7

4.7. Determine the new differential reading on the inclined-tube manometer of Fig. P4.7 if the pressure in pipe A is increased 3 psi and the pressure in pipe B remains constant. Pipes A and B contain water and oil (sp. gr. = 0.9) respectively and the gage fluid is mercury.

   Ans.   1.095 ft along inclined tube

4.8. Determine the change in pressure in pipe B that will give a new differential reading of 6 in. on the manometer of Fig. P4.8. The pressure in pipe A remains constant.

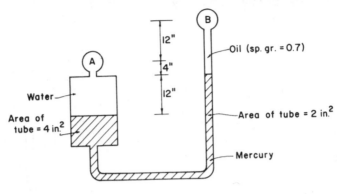

Fig. P4.8

4.9. For the inclined-tube manometer of Fig. P4.9 determine the angle θ so that a change in pressure of 0.1 psi between A and B will give a corresponding change in the reading h of 2 in.

Ans. 38.9°

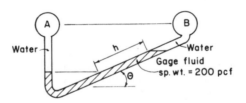

Fig. P4.9

4.10. The 3-ft wide rectangular gate of Fig. P4.10 is hinged at A and an 8-ft^3 block of concrete (sp. wt. = 150 pcf) is connected to the gate through its center of gravity. Determine the minimum force P required to hold the gate closed. The weight of the gate can be neglected and the hinge is smooth.

4.11. A gate 3 ft wide and having the rectangular cross section of Fig. P4.11 closes an opening in the side of a large tank. The gate is homogeneous and weighs 2000 lb. Determine the force P necessary to open the gate.

Ans. 3960 lb

4.12. A 2.0-m diameter circular gate is located in the side of a tank with an outward side slope of 4 vertical to 3 horizontal. The gate is hinged on a smooth horizontal axis along its diameter, and the water level in the tank is 4 m above the axis. Determine the vertical force applied at the bottom of the gate necessary to open it.

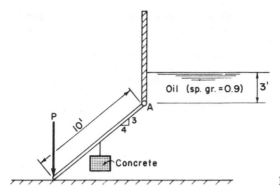

Oil (sp. gr. = 0.9)   3'

Fig. P4.10

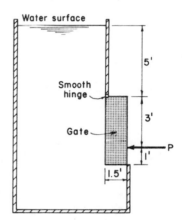

Fig. P4.11

4.13. An area having the shape of an isosceles triangle with a horizontal base of 4 ft and an altitude of 6 ft, the vertex being above the base, is submerged in a large open tank of water. The plane of the area is vertical. If the center of pressure is 4 in. below the centroid of the area, determine completely the resultant liquid force acting on one side of the area.

Ans. 4490-lb force perpendicular to the area at the center of pressure

4.14. A 6-ft square gate is located in the vertical side of a closed tank. The gate is hinged along a horizontal axis through the center of the gate and oil (sp. gr. = 0.8) stands to a depth of 10 ft above the hinge. The air pressure above the surface of the oil is 5 psi. Determine the magnitude of the torque that must be applied to the shaft to hold the gate in position if there are no stops at the edges and friction is neglected.

4.15. A rectangular tank is 6 ft wide and 12 ft long. The tank contains water to a depth of 6 ft and oil (sp. gr. = 0.8) on top of the water to a depth of 3 ft. Determine completely the resultant fluid force acting on one end of the tank.

Ans. 13,480 lb perpendicular to the tank end, acting along the vertical axis of symmetry 2.90 ft above the tank bottom.

4.16. An open tank with a rectangular vertical side 5 ft wide and 10 ft high is filled with a liquid of variable specific weight γ, where $\gamma = 30 + 4y$ pcf and y is measured vertically downward from the free surface. Determine the magnitude of the force exerted on the side by the liquid.

4.17. A timber 0.3 m^3 in volume (sp. gr. = 0.5) floats in a tank of oil (sp. wt. = 90×10^2 N/m^3). How many cubic metres of aluminum having a weight of 270×10^2 N/m^3 could be suspended beneath the timber without causing it to submerge?

Ans. 0.0683 m^3

4.18. The solid cylindrical body of Fig. P4.18 is rigidly attached to the wall of a tank and completely submerged in oil as shown. Determine the resultant force exerted by the oil on the cylinder.

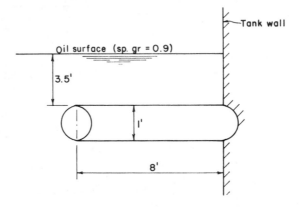

Fig. P4.18

4.19. A closed 10-ft diameter spherical tank initially contains air under atmospheric conditions. Water is forced into the tank until it is half filled. Determine the pressure of the fluid at the point where the depth of water is a maximum. Assume isothermal conditions.

Ans. 16.86 psi (gage)

4.20. A thin-walled hollow cylindrical tank is 6 ft long, has a cross-sectional area of 1.5 ft^2 and weighs 100 lb. The tank is forced into the water with its open end down until the top of the tank is level with the water surface. Determine the force required to hold the tank in this position. Assume isothermal conditions.

4.21. A thin-walled cylinder 5 ft long has a cross-sectional area of 3 ft^2 and weighs 200 lb. The cylinder is closed at both ends and placed in a large tank of water. Water is then pumped into the tank at the rate of 0.2 ft^3/s. How long a time will elapse before the tank is completely submerged?

Ans. 58.9 s

4.22. The thin-walled tank A of Fig. P4.22 weighs 100 lb, has a cross-sectional area of 1 ft^2 and is 8 ft long. One end is closed. The tank is pulled into the water with its open end down by a cable passing over a series of smooth pulleys and connected to a 1-ft^3 block of concrete B (sp. wt. = 150 pcf). Block B is submerged in gasoline (sp. gr. = 0.7). How much of the tank will project above the water surface when the system is in equilibrium? Assume isothermal conditions.

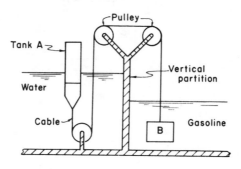

Fig. P4.22

part III

MOTION AND DEFORMATION

Kinematics

5.1 Introduction

Newton's second law of motion for a single particle relates the force required to produce a certain motion to the description of the motion itself. *Kinematics* is the study of the geometry of motion of particles, lines, and bodies without regard to the forces (and moments) required to maintain or produce this motion. *Kinetics* is the study of force systems required to produce accelerated motion of particles and rigid bodies. However, a working knowledge of the relationships between position, displacement, velocity, acceleration, and time is necessary before the effects of various force systems can be fully understood.

Two types of measurement are frequently used to describe positions with respect to a system of reference axes. Linear measurements are most useful in describing the motion of particles or points, whereas angular measurements are used to describe some motions associated with lines or bodies.

Motion of bodies or particles can be studied by using either fixed or moving reference axes. In the first case the motion is referred to as *absolute motion*, while in the second case it is called *relative motion*. This raises the concept as to what an observer sees, i.e., the motion observed by a fixed observer in his fixed reference axes as opposed to the motion observed by a moving observer in his moving reference axes. This is why the person standing on the corner sees what happens in an automobile accident entirely differently than either of the two drivers involved.

Newton's laws of motion require an absolute (inertial) reference coordinate system from which motion is observed rather than a relative one; this requires a nonrotating reference coordinate system fixed in space. In most engineering problems the motion of the earth can be neglected since the earth's accelerations are small; therefore,

159

motion measured with respect to a coordinate system attached to the earth is considered to be *absolute* or *inertial*. (An obvious exception is an inertial navigation system used for space navigation.)

A *particle* is defined as a body whose size is such that no significant errors are introduced in describing its motion by neglecting its dimensions. For example, the moon or the earth can be considered to be a particle when we are dealing with interplanetary distances, while the second hand on your watch may be treated as a body with significant dimensions in analysis of watch mechanisms.

If a particle moves along a straight-line path, the motion is referred to as rectilinear motion. If it moves along a curved path, the motion is known as *curvilinear motion* and may be either two or three dimensional. When a particle travels equal distances along its path in equal intervals of time, the motion is said to be *uniform*. If unequal distances are traveled in the same intervals of time, the motion is *nonuniform* or *variable*.

In this chapter the description of the motion of particles (points), lines, rigid bodies, and fluids is explored. Vector notation is employed where advantageous, but the scalar (component) terms must be used in obtaining numerical solutions, as with equilibrium in previous chapters.

The description of motion in a given problem can be expressed in any convenient reference coordinate system such as rectangular, curvilinear, cylindrical, polar, spherical, etc. The choice is for the engineer to make and is strongly influenced by the geometry of the problem so that mathematical difficulties are minimal. However, the choice of that to which we attach the coordinate system is solely governed by whether the description of motion is inertial as required by Newton's laws of motion.

5.2 Basic Concepts—Linear Motion

The term *linear motion* is frequently used to refer to the motion of points or particles. The basic concepts of linear motion are defined in terms of three-dimensional motion and then reduced to two special cases of motion in a plane (two-dimensional) and rectilinear motion (one-dimensional along a straight line).

Position Vector—The position of point P with respect to an observer attached to the xyz rectangular coordinate system at 0 is given by the *position vector* $\mathbf{OP} = \mathbf{r}$ in Fig. 5.1. As P moves along curve C, we see that $\mathbf{r} = \mathbf{r}(t)$; i.e., $\mathbf{r}$ is a function of time. The position vector is obtained by vector addition of its three components to give

$$\mathbf{r}(t) = x(t)\mathbf{i} + y(t)\mathbf{j} + z(t)\mathbf{k} \tag{5.1}$$

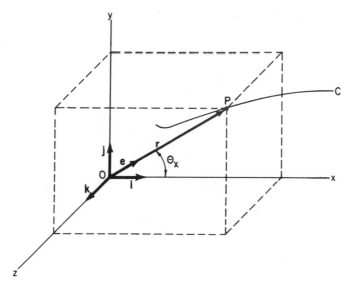

Fig. 5.1

where **i**, **j**, **k** are unit vectors along the x, y, z axes respectively. For convenience the position vector and each of its components are understood to be functions of time unless otherwise noted, so that Eq. (5.1) can be written as

$$\mathbf{r} = x\mathbf{i} + y\mathbf{j} + z\mathbf{k} \tag{5.2}$$

Observe that three scalar components are needed to describe the location of a point. Thus we say that a "freely" moving particle in space has three *degrees of freedom*.

The magnitude and direction of the position vector is given by

$$r = |\mathbf{r}| = \sqrt{x^2 + y^2 + z^2} \quad \text{(magnitude)} \tag{a}$$

and

$$\mathbf{e} = (x/r)\mathbf{i} + (y/r)\mathbf{j} + (z/r)\mathbf{k} \quad \text{(direction)} \tag{b}$$

since the scalar components are orthogonal. Recall from calculus that x/r, y/r, and z/r are the direction cosines between **r** and each scalar component that describes the orientation of **r**; i.e.,

$$x = \mathbf{r} \cdot \mathbf{i} = r \cos \theta_x, \text{ etc.}$$

Thus the position vector can also be written as

$$\mathbf{r} = r\mathbf{e} \tag{5.3}$$

where **e** is a unit vector as shown.

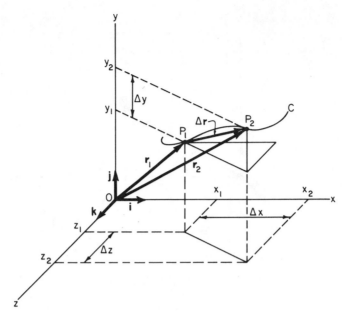

Fig. 5.2

Linear Displacement—To study the motion of point P as it moves along path C, consider the position vector at time t and $t + \Delta t$ (Fig. 5.2), where $r_1(t)$ and $r_2(t + \Delta t)$ are the two position vectors locating P_1 and P_2 respectively. From vector addition we obtain the *linear displacement* $\Delta r = u$ to be

$$\Delta r = u = r_2 - r_1 = (x_2 - x_1)i + (y_2 - y_1)j + (z_2 - z_1)k$$

or

$$\Delta r = \Delta x\, i + \Delta y\, j + \Delta z\, k \tag{5.4}$$

where Δx, Δy, Δz are the scalar components of the linear displacement. Note that the linear displacement (1) is independent of the location of the origin 0 and is solely dependent on the location of P_1, (2) is a valid definition for any time interval Δt, and (3) may be different in magnitude from the curvilinear distance Δs between points P_1 and P_2 along the curved path. The dimensions of linear displacement are those of length so that common units are feet, inches, and metres.

Velocity—The *average velocity* of P is defined as the linear displacement during time interval Δt divided by the time interval Δt, giving

$$V_{avg} = \frac{\Delta r}{\Delta t} = \frac{r(t + \Delta t) - r(t)}{\Delta t} \tag{5.5}$$

while the *instantaneous velocity* (or *velocity*) is obtained by taking the limit of Eq. (5.5) as Δt goes to zero, giving

$$v = \lim_{\Delta t \to 0} \frac{\Delta r}{\Delta t} = \frac{dr}{dt} = \dot{r} \tag{5.6}$$

where the dot over the position vector is a common notation indicating differentiation with respect to time. Thus velocity is the time rate of change of the position of the particle. Equation (5.6) is a valid definition regardless of the coordinate system employed to describe **r**. For rectangular coordinates we have

$$\mathbf{v} = \dot{\mathbf{r}} = \dot{x}\mathbf{i} + \dot{y}\mathbf{j} + \dot{z}\mathbf{k} \tag{5.7}$$

or

$$\mathbf{v} = \dot{\mathbf{r}} = v_x\mathbf{i} + v_y\mathbf{j} + v_z\mathbf{k}$$

since **i**, **j**, **k** are constant directions for an observer attached to the rectangular coordinates. The scalar components of the velocity are

$$v_x = \dot{x} = \frac{dx}{dt}, \qquad v_y = \dot{y} = \frac{dy}{dt}, \qquad v_z = \dot{z} = \frac{dz}{dt} \tag{5.8}$$

The direction of these velocity components is given by algebraic sign, depending on the component being in the positive or negative coordinate direction.

The velocity is a vector quantity with both magnitude and direction (Fig. 5.3a) from which we have

$$v^2 = v_x^2 + v_y^2 + v_z^2 \qquad \text{(magnitude)}$$

$$\mathbf{e}_t = (v_x/v)\mathbf{j} + (v_y/v)\mathbf{j} + (v_z/v)\mathbf{k} \qquad \text{(direction)}$$

where the magnitude v is called the *speed* and $\mathbf{e}_t$ is a unit vector

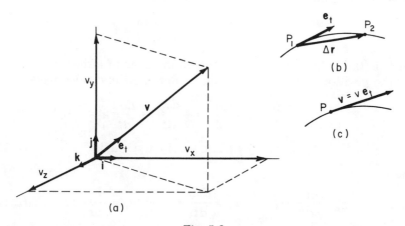

Fig. 5.3

(a)   (b)

Fig. 5.4

tangent to the curve. The direction is always tangent to the curve, as in Fig. 5.3b where in the limit as Δt goes to zero we see that Δr becomes tangent to the curve as P_2 approaches P_1. Thus the velocity can also be written as

$$\mathbf{v} = v\,\mathbf{e}_t \tag{5.9}$$

as shown in Fig. 5.3c.

The dimensions of velocity are LT^{-1}, while common engineering units are ft/s (fps), in./s (ips), and m/s (mps).

Acceleration—Consider the velocity of P as it moves along curve C. At point P_1 the velocity is $\mathbf{v}_1$, while Δt s later the velocity at P_2 is $\mathbf{v}_2$ (Fig. 5.4a). If the two velocity vectors are joined at $0'$ (Fig. 5.4b), we see that the change in velocity is $\Delta \mathbf{v}$, which represents both a change in magnitude $\Delta \mathbf{v}_{mag}$ and a change in direction $\Delta \mathbf{v}_{dir}$.

The average acceleration is defined as the change in velocity during time Δt divided by Δt; i.e.

$$\mathbf{a}_{avg} = \frac{\Delta \mathbf{v}}{\Delta t} \tag{5.10}$$

while the *instantaneous acceleration* (or simply *acceleration*) is obtained in the limit as Δt goes to zero, giving

$$\mathbf{a} = \lim_{\Delta t \to 0} \frac{\Delta \mathbf{v}}{\Delta t} = \frac{d\mathbf{v}}{dt} = \frac{d}{dt}\left(\frac{d\bar{\mathbf{r}}}{dt}\right) = \ddot{\mathbf{r}} \tag{5.11}$$

which is the definition of acceleration in terms of either the velocity $\mathbf{v}$ or the position vector $\mathbf{r}$. For rectangular coordinates we have

$$\mathbf{a} = \frac{d\mathbf{v}}{dt} = \ddot{\mathbf{r}} = \frac{dv_x}{dt}\,\mathbf{i} + \frac{dv_y}{dt}\,\mathbf{j} + \frac{dv_z}{dt}\,\mathbf{k}$$

$$= \ddot{x}\,\mathbf{i} + \ddot{y}\,\mathbf{j} + \ddot{z}\,\mathbf{k} = a_x\,\mathbf{i} + a_y\,\mathbf{j} + a_z\,\mathbf{k} \tag{5.12}$$

where

$$a_x = \frac{dv_x}{dt} = \ddot{x}, \qquad a_y = \frac{dv_y}{dt} = \ddot{y}, \qquad a_z = \frac{dv_z}{dt} = \ddot{z} \tag{5.13}$$

are the scalar components of acceleration relative to an observer attached to the xyz coordinate system. Similar to velocity, the direction of the acceleration components is given by algebraic sign.

Acceleration is a vector quantity with magnitude and direction that can be obtained from

$$a^2 = a_x^2 + a_y^2 + a_z^2 \quad \text{(magnitude)}$$

$$e_a = (a_x/a)i + (a_y/a)j + (a_z/a)k \quad \text{(direction)}$$

Generally, the direction of the acceleration e_a is not tangent to the curve since the velocity changes in both magnitude and direction so that there is one acceleration component associated with Δv_{dir} and a second component associated with Δv_{mag} (see Fig. 5.4b).

The dimensions of acceleration are LT^{-2} so that common engineering units are ft/s^2 (fps^2), in./s^2 (ips^2), and m/s^2 (mps^2).

EXAMPLE 5.1

A particle is moving along a curved path in space and its velocity is given by $v = 6ti + 1j - 8tk$ where v is in cm/s when t is in seconds. When $t = 2.0$ s, the particle is located at $r(2) = 16i - 6j$ cm. Determine (a) the acceleration, velocity, and position of the particle at $t = 1.0$ s and (b) the linear displacement of the particle during the time interval $t = 1.0$ s to $t = 2.0$ s.

SOLUTION

• (a) The acceleration can be obtained from direct differentiation of the velocity to obtain

$$a = \frac{dv}{dt} = 6i - 8k = 10 (0.6i - 0.8k) \text{ cm/s}^2 \qquad \textit{Ans.}$$

which is constant in magnitude and direction. The velocity at $t = 1.0$ s becomes

$$v = 6(1)i + j - 8(1)k = 10.5(0.572i + 0.0955j - 0.764k) \text{ cm/s} \qquad \textit{Ans.}$$

which is increasing linearly with time in the x and z directions.

To find the position vector, we integrate the velocity to obtain

$$r = \int v \, dt = 3t^2 i + tj - 4t^2 k + B$$

where B is a vector constant of integration that can be evaluated from knowing where the particle is at $t = 2.0$ s, giving

$$B = 16i - 6j - (12i + 2j - 16k)$$

or $B = 4i - 8j + 16k$ so that

$$r = (3t^2 + 4)i + (t - 8)j + (16 - 4t^2)k$$

When $t = 1.0$ s we have

$$r(1) = 7i - 7j + 12k \text{ cm} \qquad \textit{Ans.}$$

(b) The linear displacement becomes

$$u = \Delta r = r(2) - r(1) = 16i - 6j - (7i - 7j + 12k)$$

or

$$u = 9i + j - 12k \text{ cm}$$
$$= 15.03(0.598i + 0.0665j - 0.798k) \text{ cm} \qquad \textit{Ans.}$$

Motion in a Plane (two-dimensional)—In many problems encountered in engineering practice, the particle moves in a given plane and two coordinates are sufficient to describe the motion. In such cases the problem is often referred to as having two degrees of freedom. For rectangular coordinates it is convenient to have the xy plane correspond to the plane of motion (Fig. 5.5a) and to consider the motion of P along the curved path $y = f(x)$. The position of P is given by

$$r = xi + yj \qquad (5.14)$$

which is the same as Eq. (5.2) with z set equal to zero. It can be seen from Eq. (5.14) that a "freely" moving point has two degrees of freedom x and y.

However, when the point moves along a *fixed curve* such as $y = f(x)$, knowing $x(t)$ is sufficient to determine $y(t)$ for each value of the parameter t since $y = f(x) = f[x(t)] = y(t)$. In such cases two coordinates are necessary to locate P, but only one is independent, namely $x(t)$, since $y(t)$ is automatically specified by the function $y = f(x)$. Hence motion along a fixed curve is always a single degree of freedom situation regardless of whether the path is a straight line, a curve in a plane, or a curve in space.

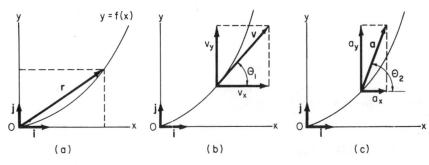

Fig. 5.5

The velocity can be obtained by differentiating Eq. (5.14) to give

$$\mathbf{v} = \dot{\mathbf{r}} = \dot{x}\mathbf{i} + \dot{y}\mathbf{j} = v_x\mathbf{i} + v_y\mathbf{j} \qquad (5.15)$$

where $\mathbf{v}$ and its components are shown in Fig. 5.5b. The magnitude and direction of $\mathbf{v}$ are given by

$$v^2 = v_x^2 + y_y^2 \qquad \text{(magnitude)} \qquad (c)$$

$$\mathbf{e}_t = (v_x/v)\mathbf{i} + (v_j/v)\mathbf{j} \qquad \text{(direction)} \qquad (d)$$

as before. The direction can also be expressed in terms of a single angle θ_1 for motion in a plane since

$$v_x/v = \cos\theta_1, \qquad v_y/v = \sin\theta_1$$

so that

$$\tan\theta_1 = \frac{v_y}{v_x} = \frac{\dot{y}}{\dot{x}} = \frac{dy}{dx}$$

which is also the slope (tangent) of the curved path in Fig. 5.5b.

The acceleration can be obtained by differentiating Eq. (5.15) to give

$$\mathbf{a} = \frac{d\mathbf{v}}{dt} = \ddot{\mathbf{r}} = \ddot{x}\mathbf{i} + \ddot{y}\mathbf{j} = a_x\mathbf{i} + a_y\mathbf{j} \qquad (5.16)$$

where $\mathbf{a}$ and its components are shown in Fig. 5.5c. The magnitude and direction of $\mathbf{a}$ are given by expressions similar to Eqs. (c) and (d). The direction of $\mathbf{a}$ can also be expressed in terms of the angle θ_2 by $\tan\theta_2 = a_y/a_x$ (Fig. 5.5c).

EXAMPLE 5.2

A particle is moving along a fixed path in the xy plane given by

$$y = x^2/3 \qquad (a)$$

and the horizontal position is given by $x(t) = (t^3/3) - t^2 + 2$, where x is in feet when t is in seconds. Determine the position, velocity, and acceleration of the particle when $t = 3.0$ s.

SOLUTION

The position can be obtained directly from

$$x = (t^3/3) - t^2 + 2 = [(3)^3/3] - (3)^2 + 2 = 2 \text{ ft} \qquad (b)$$

and Eq. (a), $y = x^2/3 = 4/3 = 1.33$ ft. Thus

$$\mathbf{r} = 2\mathbf{i} + 1.33\mathbf{j} \text{ ft} \qquad \qquad Ans.$$

The velocity can be obtained either by substituting $x(t)$ into Eq. (a) and differentiating, or by differentiating both Eq. (a) and $x(t)$ separately. Following the latter procedure gives

$$\frac{dy}{dt} = \frac{2}{3} x \frac{dx}{dt}$$

or

$$v_y = (2/3)x \, v_x \tag{c}$$

and

$$v_x = \frac{dx}{dt} = t^2 - 2t = (3)^2 - 2(3) = 3 \text{ fps} \tag{d}$$

so that from Eqs. (b) and (c) we have $v_y = (2/3)(2)(3) = 4$ fps. The velocity becomes

$$\mathbf{v} = v_x \mathbf{i} + v_y \mathbf{j} = 3\mathbf{i} + 4\mathbf{j} = 5(0.6\mathbf{i} + 0.8\mathbf{j}) \text{ fps} \qquad Ans.$$

from which we see that the speed is 5 fps.

The acceleration can be evaluated from differentiation of Eqs. (c) and (d). From Eq. (d) we have

$$a_x = \frac{dv_x}{dt} = 2t - 2 = 2(3) - 2 = 4 \text{ fps}^2$$

and from Eq. (c)

$$a_y = \frac{dv_y}{dt} = \frac{2}{3} x \frac{dv_x}{dt} + \frac{2}{3} \frac{dx}{dt} v_x$$

or

$$a_y = (2/3)x \, a_x + (2/3)v_x^2 = (2/3)(2)(4) + (2/3)(3)^2 = (34/3) = 11.33 \text{ fps}^2$$

Thus

$$\mathbf{a} = a_x \mathbf{i} + a_y \mathbf{j} = 4\mathbf{i} + 11.33\mathbf{j} \text{ fps}^2 \qquad Ans.$$

or

$$\mathbf{a} = 12.02(0.332\mathbf{i} + 0.944\mathbf{j}) \text{ fps}^2 \qquad Ans.$$

Rectilinear Motion (straight-line motion)—A very special but important type of motion occurs when a particle moves along a straight-line path. For this case it is convenient to select one of the rectangular coordinates (say the x axis) to be along the line of motion as in Fig. 5.6a, so that the position vector becomes

$$\mathbf{r} = x\mathbf{i} \tag{5.17}$$

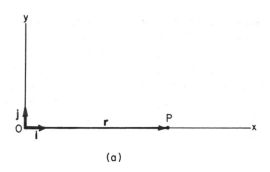

(a)

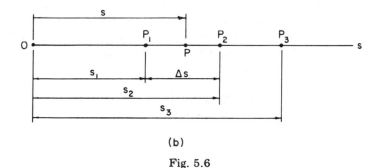

(b)

Fig. 5.6

The velocity and acceleration simply become

$$\mathbf{v} = \dot{\mathbf{r}} = \dot{x}\mathbf{i} = v_x\mathbf{i} \qquad (5.18)$$

$$\mathbf{a} = \frac{d\mathbf{v}}{dt} = \ddot{x}\mathbf{i} = a_x\mathbf{i} \qquad (5.19)$$

so that all the kinematic quantities are dependent on only a single coordinate x and have the same direction $\mathbf{i}$. Note that these vector quantities can change only in magnitude. Thus it is often convenient to drop the vector notation for this case and use a single coordinate s to locate P from the origin 0 (Fig. 5.6b). In this case we can write the velocity and acceleration as

$$v = \frac{ds}{dt} = \dot{s} \qquad (5.20)$$

$$a = \frac{dv}{dt} = \ddot{s} \qquad (5.21)$$

and let the algebraic sign determine the direction. Thus positive velocity and acceleration are in the same direction as positive displacement s. When the magnitude of the velocity is decreasing, the sense of the acceleration is opposite to that of the velocity, and this condition is sometimes referred to as *deceleration*.

If during a time interval Δt the particle moves from P_1 to P_2, the change in position Δs is the linear displacement u of the particle as before. If during the interval, however, the particle moves first from P_1 to P_3 and then back to P_2, the linear displacement is still u, but the *total distance traveled* S is the accumulated length of path traversed; namely,

$$S = (s_3 - s_1) + (s_3 - s_2) \tag{5.22}$$

Note that total distance traveled is a scalar quantity, while linear displacement generally is a vector quantity.

An alternate expression for acceleration that is useful when the acceleration is a function of the position s can be obtained from Eqs. (5.20) and (5.21) through application of the chain rule of differentiation to Eq. (5.21). This gives

$$a = \frac{dv}{dt} = \frac{dv}{ds}\frac{ds}{dt} = v\frac{dv}{ds} \tag{5.23}$$

EXAMPLE 5.3

The position of a particle as it moves along a straight-line path is given by the expression $s = t^4 - 8t^3 + 16t^2$, where s and t are expressed in feet and seconds respectively. When $t = 1.0$ s, the velocity is to the left. Determine (a) the displacement of the particle during the first 3 s, (b) the total distance traveled by the particle during this 3-s interval, and (c) the acceleration when $t = 1.0$ s.

SOLUTION

(a) The displacement of the particle during the time interval from $t = 0$ s to $t = 3$ s is computed as

$$u = s_3 - s_0 = 9 - 0 = 9 \text{ ft} \leftarrow \qquad\qquad Ans.$$

since positive velocity of 12 fps is to the left, which establishes to the left as the positive direction.

(b) The total distance traveled may be greater than the displacement if the particle motion has reversed directions in the time interval. Since a change in direction will occur at points of zero velocity, we can solve for these times of change. Thus

$$v = \frac{ds}{dt} = 4t^3 - 24t^2 + 32t = 4t(t - 2)(t - 4) = 0 \tag{a}$$

From this expression it is obvious that a change in direction has occurred in the time interval of interest at $t = 2$. From the positions of the particle at $t = 0$ s, $t = 2$ s, and $t = 3$ s we can compute the total distance traveled as

$$s_0 = 0, \qquad s_2 = 16 \text{ ft}, \qquad s_3 = 9 \text{ ft}$$

$$S = (s_2 - s_0) + (s_2 - s_3) = (16 - 0) + (16 - 9) = 23 \text{ ft} \qquad \qquad Ans.$$

(c) The acceleration can be obtained from Eq. (a) to be

$$a = 12t^2 - 48t + 32 = 12 - 48 + 32 = -4$$

or

$$\mathbf{a = 4 \text{ fps}^2} \rightarrow \qquad \qquad Ans.$$

EXAMPLE 5.4

A particle moving along a straight-line path has an initial velocity of 24 fps to the right. The particle is subjected to a constant acceleration until its velocity is 6 fps in the opposite direction. At this point the velocity remains constant for 3 s. During the entire motion, the particle moves 120 ft. Determine (a) the acceleration of the particle during the first part of the motion and (b) the displacement of the particle at the end of the second part of the motion.

SOLUTION

Problems of this type can frequently be solved by using a semigraphical motion curve approach rather than formal mathematical integration procedures. The most useful motion curve, especially for problems involving constant acceleration, is a plot of linear velocity versus time (*vt diagram*). From the definition of velocity, Eq. (5.20), we may write $ds = v\, dt$ from which

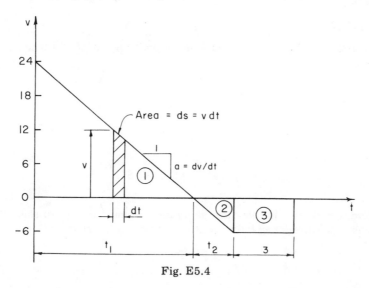

Fig. E5.4

$$\int_{s_1}^{s_2} ds = \int_{t_1}^{t_2} v \, dt$$

or $s_2 - s_1$ = area under a vt diagram. The vt diagram for the example under consideration is shown in Fig. E5.4. The slope of the vt curve represents the magnitude and direction of the acceleration (negative) from Eq. (5.21). The area above the t axis (area 1) represents distance traveled in one direction (to the right), while the areas below the axis (areas 2 and 3) represent distance traveled in the opposite direction (to the left). The total distance traveled during any time interval is the sum of the areas above and below the t axis. The displacement during a given time interval, however, is the difference between areas on the two sides of the t axis since the area above the t axis is positive for positive velocity while the area below is negative for negative velocity.

From the similar triangles of the diagram it can be seen that

$$t_1 = 24/a, \qquad t_2 = 6/a$$

$$S = A_1 + A_2 + A_3 = 120, \qquad (1/2)(24)(24/a) + (1/2)(6)(6/a) + 6(3) = 120$$

$$\mathbf{a} = 3 \text{ ft/s}^2 \leftarrow \qquad\qquad\qquad\qquad\qquad Ans.$$

$$\mathbf{u} = A_1 - A_2 - A_3 = (1/2)(24)(24/3) - (1/2)(6)(6/3) - 6(3) = 72 \text{ ft} \rightarrow \quad Ans.$$

EXAMPLE 5.5

The acceleration of a particle in rectilinear motion is given by $a = -9s$ where a is in f/s² when s is in feet. The particle is 3 ft to the right of the origin and is released from rest. Determine the velocity of the particle when it passes through the origin.

SOLUTION

Since the acceleration is a function of position, Eq. (5.23) is most useful and gives

$$\int_0^v v \, dv = - \int_3^0 9s \, ds$$

or

$$v^2 = -9s^2 \, \bigg|_3^0 = 9(3)^2 = 81$$

Therefore, $v = \pm 9$ fps. In this case the minus sign must be chosen since the acceleration is negative for an initial displacement of +3 ft. Thus

$$\mathbf{v} = 9 \text{ fps} \leftarrow \qquad\qquad\qquad\qquad\qquad Ans.$$

What would be the velocity if the particle is initially 3 ft to the left of the origin?

$$v = 9 \text{ fps} \rightarrow \qquad\qquad Ans.$$

5.3   Basic Concepts—Angular Motion of a Line in a Plane

A second type of measurement, widely used in kinematics, denotes the angular position of a line with respect to some fixed reference axis. Since the measurement is associated with a line, it can have no significance for points or particles. It is used either as a measure of the rotation of a rigid body or as the orientation of a position vector, as in polar coordinates. Consider the case of a body of arbitrary shape rotating about a fixed axis thru 0 (Fig. 5.7) so that lines $0X$ and $0P$ always lie in the same plane. The angular position of a line $0P$ at any instant of time t can be specified by a single angle θ measured with respect to some convenient fixed reference axis such as $0X$. Since the body is rigid, specification of the angle θ as a function of time completely describes the motion of the body as a whole. Therefore, angular motion of a line for a rigid body rotating around a fixed axis is analogous to the rectilinear motion of a particle that can be completely described in terms of a single coordinate as a function of time.

If the angular position of the line changes by an amount $\Delta\theta$ during a time interval Δt, the change in position $\Delta\theta$ is the *angular displacement* of the line. The angular displacement is positive if the angle increases and negative if it decreases. The choice of reference axis and positive sense of measurement is arbitrary; however, a counterclockwise direction of rotation is customarily taken as posi-

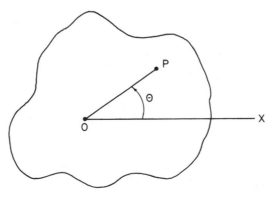

Fig. 5.7

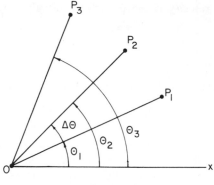

Fig. 5.8

tive. As in the case for rectilinear motion, if the line moves first from θ_1 to θ_3 and then back to θ_2 as in Fig. 5.8, the angular displacement remains $\Delta\theta$, but the *total angle turned* Θ is the accumulated angular movement; i.e.,

$$\Theta = (\theta_3 - \theta_1) + (\theta_3 - \theta_2) \qquad (5.24)$$

The total angle turned is a scalar quantity, while infinitesimal angular displacements are vector quantities.

The concepts of angular velocity and angular acceleration for a line can be developed in the same manner as for the linear velocity and linear acceleration. Thus an *average angular velocity* ω_{avg} can be computed as the ratio of angular displacement to the time interval; i.e.,

$$\omega_{avg} = \frac{\theta_2 - \theta_1}{t_2 - t_1} = \frac{\Delta\theta}{\Delta t}$$

An *instantaneous angular velocity* ω is obtained by taking shorter and shorter time intervals. In the limit

$$\omega = \lim_{\Delta t \to 0} \frac{\Delta\theta}{\Delta t} = \frac{d\theta}{dt} \qquad (5.25)$$

By a similar process an *average angular acceleration* α_{avg} can be determined as the ratio of change in angular velocity to the elapsed time. Thus

$$\alpha_{avg} = \frac{\omega_2 - \omega_1}{t_2 - t_1} = \frac{\Delta\omega}{\Delta t}$$

The same limit process can then be applied to obtain an *instantaneous angular acceleration* α. Thus

$$\alpha = \lim_{\Delta t \to 0} \frac{\Delta \omega}{\Delta t} = \frac{d\omega}{dt} \qquad (5.26a)$$

Therefore, angular velocity can be defined as the time rate of change of angular position. Similarly, angular acceleration can be defined as the time rate of change of angular velocity. Angles are expressed in units of radians, which are dimensionless. Thus the angular velocity has units of rad/s, and angular acceleration has units of rad/s².

An alternate expression for the angular acceleration can be obtained by substituting Eq. (5.25) into Eq. (5.26a) to yield

$$\alpha = \frac{d}{dt}\left(\frac{d\theta}{dt}\right) = \frac{d^2\theta}{dt^2} \qquad (5.26b)$$

Also, by the chain rule,

$$\alpha = \frac{d\omega}{d\theta}\frac{d\theta}{dt} = \omega\frac{d\omega}{d\theta} \qquad (5.27)$$

Angular displacement, total angle turned, angular velocity, and angular acceleration have an equivalent linear quantity associated with rectilinear motion. As in the case of linear velocity and acceleration, angular velocity and acceleration are vector quantities; the magnitudes are given by Eqs. (5.25) and (5.26), while the direction is along a line perpendicular to the plane of the motion and in a direction given by the right-hand screw rule (Fig. 5.9). In the case of displacement, infinitesimal angular displacements are vector quanti-

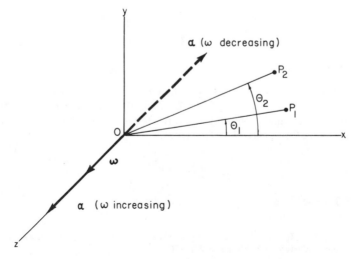

Fig. 5.9

ties but finite angular displacements are not.[1]   Even though finite angular displacements have magnitude and direction (quantities associated with vectors), they do not add according to the parallelogram law and thus do not qualify as vectors. For the frequently encountered special case of plane motion where the line element remains in a given plane, finite angular displacements can be treated as vector quantities.

EXAMPLE 5.6

A racing car is traveling along a straight-line track at a constant velocity of 408 mph. Determine the angular velocity and angular acceleration of the line of sight when the car is in the position shown in Fig. E5.6. The observation tower is located 4000 ft from the track.

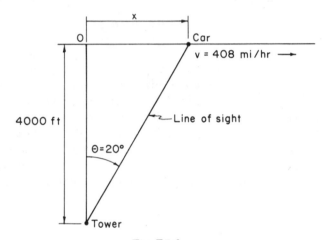

Fig. E5.6

SOLUTION

From the geometry of the situation a relationship between x and θ can be expressed as $x = 4000 \tan \theta$. From Eqs. (5.20) and (5.25) it is observed that linear velocity and angular velocity are time derivatives of the linear and angular coordinates x and θ respectively. Therefore,

$$\frac{dx}{dt} = 4000 \sec^2 \theta \, \frac{d\theta}{dt}$$

or $v = 4000 \, \omega \sec^2 \theta$ and $\omega = (v/4000) \cos^2 \theta$ since

---

1. See T. C. Huang, *Engineering Mechanics*, Vol. 2, *Dynamics* (Reading, Mass.: Addison-Wesley, 1967), pp. 530–33.

$$v = 408 \text{ mph} = 408 \text{ mph} (88 \text{ fps}/60 \text{ mph}) = 600 \text{ fps}$$

$$\omega = (600/4000) \cos^2 20° = 0.133 \text{ rad/s} \quad\quad\quad\quad Ans.$$

From Eq. (5.27) it is observed that

$$\alpha = \omega \frac{d\omega}{d\theta} = \frac{v}{4000} \cos^2 \theta \left(- \frac{v}{4000} 2 \cos \theta \sin \theta \right)$$

$$= - \frac{v^2}{(4000)^2} \cos^2 \theta \sin 2\theta$$

$$= - \frac{600^2}{4000^2} \cos^2 20° \sin 40° = -0.0128 \text{ rad/s}^2$$

or

$$\alpha = 0.0128 \text{ rad/s}^2 \quad\quad\quad\quad Ans.$$

where the positive direction for θ, ω, and α is clockwise for this problem (Fig. E5.6).

Example 5.6 illustrates how a position function ($x = 4000 \tan \theta$) can be set up from the geometry of the problem and then differentiated to obtain a relationship between linear and angular velocities and accelerations. Example 5.7 illustrates the same type of procedure for a body moving in a fixed plane.

EXAMPLE 5.7

A ladder of length ℓ has its ends in contact with a wall and floor (Fig. E5.7). If point A moves to the right with a constant velocity v_A, determine (a) the velocity and acceleration of point B and (b) the angular velocity and angular acceleration of the ladder.

SOLUTION

(a) The positions of points A and B are related by the expression

$$x^2 + y^2 = \ell^2$$

which states that the length of the ladder is constant. Since $v_A = dx/dt$, $v_B = dy/dt$, and $a_B = d^2 y/dt^2$, the above expression can be differentiated to yield the desired quantities as follows:

$$2x \frac{dx}{dt} + 2y \frac{dy}{dt} = 0, \quad x v_A + y v_B = 0, \quad v_B = -(x/y) v_A = -\cot \theta \, v_A \quad Ans.$$

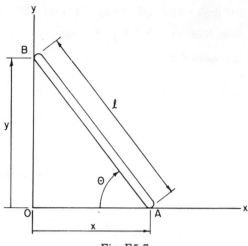

Fig. E5.7

Differentiating a second time gives

$$2x \frac{d^2 x}{dt^2} + 2 \left(\frac{dx}{dt}\right)^2 + 2y \frac{d^2 y}{dt^2} + 2 \left(\frac{dy}{dt}\right)^2 = 0$$

and

$$x a_A + v_A^2 + y a_B + v_B^2 = 0$$

Since v_A = constant, $a_A = 0$ so that

$$a_B = -(1/y)(v_A^2 + v_B^2) = -v_A^2/(\ell \sin^3 \theta) \qquad\qquad Ans.$$

where $y = \ell \sin \theta$.

   (b) The angular velocity and acceleration of the ladder can be found by expressing the position of A in terms of the angle θ. Thus

$$x = \ell \cos \theta, \qquad \frac{dx}{dt} = -\ell \sin \theta \frac{d\theta}{dt}$$

and

$$v_A = -\ell \omega \sin \theta, \qquad \omega = -v_A/(\ell \sin \theta) \qquad\qquad Ans.$$

where clockwise is positive. Differentiating a second time gives

$$\frac{d^2 x}{dt^2} = -\ell \sin \theta \frac{d^2 \theta}{dt^2} - \ell \cos \theta \left(\frac{d\theta}{dt}\right)^2$$

and

$$a_A = -\ell \alpha \sin \theta - \ell \omega^2 \cos \theta$$

Since $a_A = 0$,

$$\alpha = -(\omega^2 \cos \theta)/\sin \theta = -(v_A^2 \cos \theta)/(\ell^2 \sin^3 \theta) \qquad \qquad Ans.$$

where clockwise is positive.

5.4 Curvilinear Coordinates

In this section, curvilinear coordinates are considered for describing the motion of a point along a curved path that lies in a plane. Consider the motion of particle P along the plane curve in Fig. 5.10. In this case the path may be described by the equation $y = f(x)$, while the position of the particle along the path may be specified by $s = s(t)$. Hence s is called the curvilinear distance.

The curvilinear coordinates are oriented so that the tangential direction is always taken to be locally tangent to the curve at P with the unit vector $\mathbf{e}_t$ in the direction of increasing s. The normal direction is always perpendicular to the tangential direction with a unit vector $\mathbf{e}_n$ directed from P toward the instantaneous center of curvature C as shown. Frequently, curvilinear coordinates are also called *normal-tangential coordinates*. The position of P can also be expressed in terms of $\mathbf{r}$ and its components x and y (Fig. 5.10).

The velocity of P can be determined from Fig. 5.11a as P moves to P' in the time interval Δt, during which it moves through a curvilinear distance Δs while the instantaneous radius of curvature ρ sweeps out an angle $\Delta \theta$. From the geometry the linear displacement

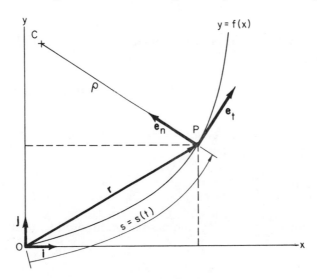

Fig. 5.10

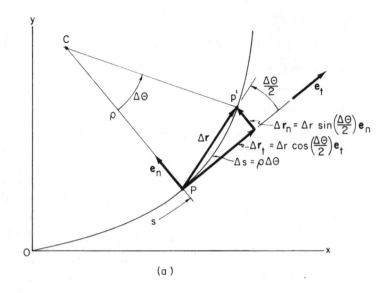

(a)

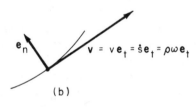

(b)

Fig. 5.11

$\Delta \mathbf{r}$ can be resolved into components along the tangential and normal directions giving

$$\Delta \mathbf{r} = \Delta \mathbf{r}_t + \Delta \mathbf{r}_n = \Delta r \left[\cos \left(\frac{\Delta \theta}{2} \right) \mathbf{e}_t + \sin \left(\frac{\Delta \theta}{2} \right) \mathbf{e}_n \right] \qquad \text{(a)}$$

where $\Delta r = |\Delta \mathbf{r}|$. As Δt becomes smaller and smaller, we see that $\Delta \theta$ becomes very small so that

$$\cos \left(\frac{\Delta \theta}{2} \right) \approx 1.0$$

$$\sin \left(\frac{\Delta \theta}{2} \right) \approx \frac{\Delta \theta}{2} \text{ which is small compared to unity}$$

and

$$\Delta r \approx \Delta s = \rho \Delta \theta \qquad \text{(b)}$$

Hence the velocity becomes

$$\mathbf{v} = \lim_{\Delta t \to 0} \frac{\Delta r}{\Delta t} \mathbf{e}_t - \lim_{\Delta t \to 0} \frac{\Delta s}{\Delta t} \mathbf{e}_t = \frac{ds}{dt} \mathbf{e}_t = v\mathbf{e}_t$$

or

$$\mathbf{v} = v\mathbf{e}_t = \dot{s}\, \mathbf{e}_t \tag{5.28}$$

which shows the velocity to be tangent to the curve as before. In view of Eq. (b) we see that $ds = \rho\, d\theta$ so that the speed can also be written as

$$v = \frac{ds}{dt} = \rho\, \frac{d\theta}{dt} = \rho\omega \tag{5.29}$$

where ω is the angular velocity of the instantaneous radius of curvature ρ as well as the unit vectors $\mathbf{e}_t$ and $\mathbf{e}_n$. The results of Eqs. (5.28) and (5.29) are shown in Fig. 5.11b.

The acceleration can be obtained by differentiating Eq. (5.28) to obtain

$$\mathbf{a} = \frac{d\mathbf{v}}{dt} = \frac{dv}{dt} \mathbf{e}_t + v\, \frac{d\mathbf{e}_t}{dt} \tag{c}$$

In Fig. 5.12a the orientation of the unit vectors $\mathbf{e}_t$ and $\mathbf{e}_n$ and the instantaneous radius of curvature is shown in terms of angle θ. The angular orientation of the unit vectors $\mathbf{e}_t$ and $\mathbf{e}_n$ can be seen more clearly from the unit circle of Fig. 5.12b from which we obtain

$$\mathbf{e}_t = \cos\theta\, \mathbf{i} + \sin\theta\, \mathbf{j}, \qquad \mathbf{e}_n = -\sin\theta\, \mathbf{i} + \cos\theta\, \mathbf{j} \tag{d}$$

Differentiating Eqs. (d) by the chain rule gives

$$\frac{d\mathbf{e}_t}{dt} = \frac{d\mathbf{e}_t}{d\theta} \frac{d\theta}{dt} = \dot{\theta}\, (-\sin\theta\, \mathbf{i} + \cos\theta\, \mathbf{j})$$

or

$$\frac{d\mathbf{e}_t}{dt} = \omega\mathbf{e}_n \tag{5.30}$$

where $\omega = \dot{\theta}$ is the angular velocity of the instantaneous radius of curvature and the unit vectors $\mathbf{e}_t$ and $\mathbf{e}_n$. Similarly, the time derivative of $\mathbf{e}_n$ becomes

$$\frac{d\mathbf{e}_n}{dt} = -\dot{\theta}\, \mathbf{e}_t = -\omega\mathbf{e}_t \tag{e}$$

Equations (5.30) and (e) represent the velocity of the tip of the unit

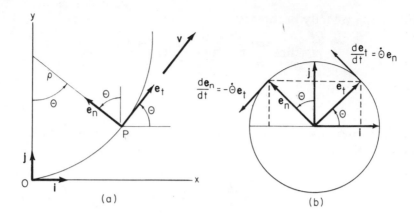

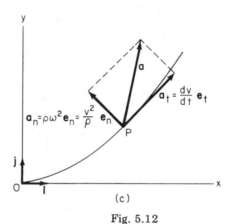

Fig. 5.12

vectors (Fig. 5.12b). Substitution of Eq. (5.30) into Eq. (c) yields

$$a = \frac{dv}{dt} \mathbf{e}_t + v\omega \, \mathbf{e}_n = \ddot{s} \, \mathbf{e}_t + \rho\omega^2 \, \mathbf{e}_n = \mathbf{a}_t + \mathbf{a}_n \tag{5.31}$$

since $v = \dot{s}$ and $v = \rho\omega$. Equation (5.31) shows that the acceleration along a curved path can be thought of in terms of two components. The tangential acceleration is given by

$$a_t = \frac{dv}{dt} = \ddot{s} \tag{5.32}$$

which is due to the *change in magnitude* of the velocity (speed v) and is always directed tangent to the curve. The normal acceleration is given by

$$a_n = v\omega = v^2/\rho = \rho\omega^2 \tag{5.33}$$

which is due to the *change in direction* of the velocity and is directed toward the center of curvature. These acceleration components are illustrated in Fig. 5.12c from which it is seen that the normal acceleration is zero only when either (1) the velocity is zero or (2) the radius of curvature is infinitely large (i.e., a straight line). A comparison of Eqs. (5.28) and (5.32) with Eqs. (5.20) and (5.21) for rectilinear motion shows that rectilinear motion is simply a special case of curvilinear motion where the radius of curvature is infinitely large so that a_n is always zero.

EXAMPLE 5.8

A particle is moving along a curved path in such a manner that its speed is changing at a constant rate. At a certain point on the path the rectangular components of velocity and acceleration are given as

$$v_x = 9.6 \text{ ft/s} \qquad a_x = 3 \text{ ft/s}^2$$
$$v_y = 7.2 \text{ ft/s} \qquad a_y = -9 \text{ ft/s}^2$$

Determine (a) the normal and tangential components of acceleration, (b) the radius of curvature of the path at the point, and (c) the distance traveled during the next 6 s.

SOLUTION

(a) The velocity and acceleration can be written in vector form as

$$\mathbf{v} = 9.6\mathbf{i} + 7.2\mathbf{j} = 12(0.8\mathbf{i} + 0.6\mathbf{j}), \qquad \mathbf{a} = 3\mathbf{i} - 9\mathbf{j}$$

Since the velocity is always tangent to the path and the normal component of acceleration is toward the center of curvature, unit vectors in the tangential and normal directions respectively can be expressed as

$$\mathbf{e}_t = 0.8\mathbf{i} + 0.6\mathbf{j}, \qquad \mathbf{e}_n = 0.6\mathbf{i} - 0.8\mathbf{j}$$

Therefore

$$\mathbf{a}_t = (\mathbf{a} \cdot \mathbf{e}_t)\mathbf{e}_t = (2.4 - 5.4)\mathbf{e}_t = -3\mathbf{e}_t = -2.4\mathbf{i} - 1.8\mathbf{j} \text{ ft/s}^2$$
$$\mathbf{a}_n = (\mathbf{a} \cdot \mathbf{e}_n)\mathbf{e}_n = (1.8 + 7.2)\mathbf{e}_n = 9\mathbf{e}_n = 5.4\mathbf{i} - 7.2\mathbf{j} \text{ ft/s}^2 \qquad Ans.$$

(b) The radius of curvature is related to the velocity and normal component of acceleration (Eq. 5.33) so that

$$\rho = v^2/a_n = 12^2/9 = 16 \text{ ft} \qquad Ans.$$

(c) The speed of the particle is changing at a constant rate. Thus

$$a_t = \frac{dv}{dt} = -3, \qquad v = \frac{ds}{dt} = -3t + c_1, \qquad s = -(3t^2/2) + c_1 t + c_2$$

At $t = 0$, $v = 12$; therefore $c_1 = 12$. At $t = 0$, $s = 0$; therefore $c_2 = 0$. The velocity and position of the particle with respect to the original point are thus given by

$$v = -3t + 12, \qquad s = -1.5t^2 + 12t$$

The distance traveled can now be evaluated after it is determined whether a change in direction of motion has occurred in the time interval of interest. The velocity will be zero at the point where the direction changes. For this example $v = -3t + 12 = 0$. Thus at $t = 4$ the direction changes. Therefore

$$S = (s_4 - s_0) + (s_4 - s_6)$$

Since

$$s_4 = -1.5(4)^2 + 12(4) = -24 + 48 = 24 \text{ ft}$$

$$s_6 = -1.5(6)^2 + 12(6) = -54 + 72 = 18 \text{ ft}$$

$$S = (24 - 0) + (24 - 18) = 30 \text{ ft} \qquad\qquad Ans.$$

Motion of a Point on a Circular Path—A special but extremely important case of curvilinear motion occurs when point P moves on a circular path of constant radius r (Fig. 5.13a). This circular motion is important because the motion of P is typical of the motion of any point in the rigid body B that is rotating about a fixed axis perpendicular to the page through 0. (Recall that the distance between any two points in a rigid body is constant.)

For this case the radius of curvature is a constant, and the curvilinear distance s is related to the angle θ by $s = r\theta$ so that the velocity becomes

$$\mathbf{v} = \frac{ds}{dt}\mathbf{e}_t = r\frac{d\theta}{dt}\mathbf{e}_t = r\omega\mathbf{e}_t \qquad (5.34)$$

as in Fig. 5.13b where ω is the angular velocity of the radius of curvature. Note that $\mathbf{e}_t$ is always in the direction of increasing s so that if ω is clockwise, it is negative and the velocity is opposite to that shown. The tangential acceleration becomes

$$a_t = \frac{dv}{dt} = r\frac{d\omega}{dt} = r\alpha \qquad (5.35)$$

which is in the direction of $\mathbf{e}_t$ for positive α (same direction as θ which is counterclockwise for this case) and in the negative direction of $\mathbf{e}_t$ for negative values (clockwise for this case) of α. The tangential acceleration is shown for α being counterclockwise (positive) in Fig. 5.13c. The normal acceleration becomes

$$a_n = r\omega^2 = v^2/r \qquad (5.36)$$

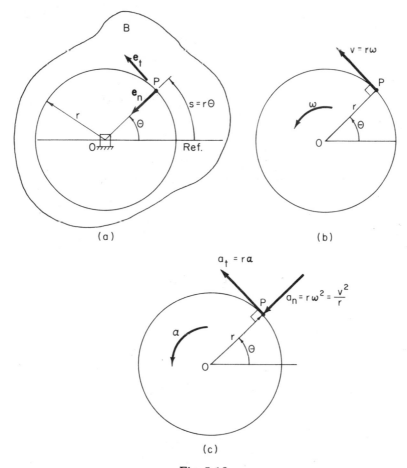

$$s = r\Theta$$

$$v = r\omega$$

$$a_t = r\alpha$$

$$a_n = r\omega^2 = \frac{v^2}{r}$$

Fig. 5.13

in the direction of $\mathbf{e}_n$, which is always directed from P toward 0 regardless of the direction of v or ω (Fig. 5.13c). Equations (5.34), (5.35), and (5.36) are valid for circular motion of points in rigid bodies rotating about a fixed axis.

EXAMPLE 5.9

The 20-cm rigid bar $0A$ in Fig. E5.9a is rotating about a fixed axis through 0 with an angular velocity given by $\omega = 1 + 8t - 3t^2$ where ω is in rad/s for t in seconds. The positive direction of ω is the same as θ. Determine (a) the velocity distribution of point P as a function of its distance from 0 and plot this function and (b) the velocity and acceleration of point A when $t = 2.0$ s if $\theta = 60°$. Express the answer of part b in both normal-tangential and rectangular coordinates.

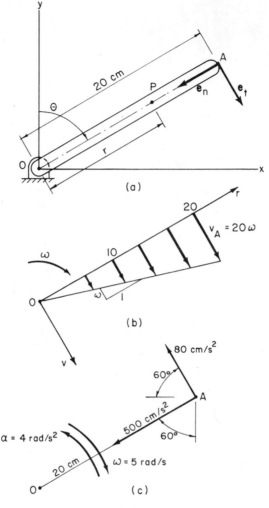

Fig. E5.9

SOLUTION

(a) The velocity for any point P located r cm from 0 is given by

$$\mathbf{v} = r\omega \, \mathbf{e}_t \tag{a}$$

which is a linear function of r for a given ω (Fig. E5.9b). Note that the slope of the line is ω.

(b) The velocity of A can be determined from Eq. (a) with ω evaluated at $t = 2.0$ as follows

$$\omega = 1 + 8(2) - 3(2)^2 = 5 \text{ rad/s}$$

$$r = 20 \text{ cm}$$

$$v_A = 20(5)e_t = 100 \text{ cm/s} \quad \bigvee_{60°} \qquad \qquad Ans.$$

and

$$v_A = 100(\cos 60°i - \sin 60°j) = 50i - 86.6j \text{ cm/s} \qquad Ans.$$

The acceleration can be determined from Eqs. (5.35) and (5.36), once α is known, as follows:

$$\alpha = \frac{d\omega}{dt} = 8 - 6t = 8 - 12 = -4 \text{ rad/s}^2 = 4 \text{ rad/s}^2 \Big)$$

since θ, ω, and α are positive clockwise for this problem. Thus

$$a_t = r\alpha \, e_t = 20(-4) = 80 \text{ cm/s}^2 \quad {}_{60°}\!\!\nearrow \qquad \qquad Ans.$$

and

$$a_n = r\omega^2 \, e_n = 20(5)^2 \, e_n = 500 \, e_n = 500 \text{ cm/s}^2 \quad \swarrow_{30°} \qquad Ans.$$

as shown in Fig. E5.9c. In rectangular coordinates the answer is

$$a_A = 80(-\cos 60° \, i + \sin 60° \, j) + 500(-\cos 30° \, i - \sin 30° \, j)$$
$$= -473i - 180.7j \text{ cm/s}^2 \qquad \qquad Ans.$$

The student should check that the magnitude and direction of a_A is the same regardless of whether the answer is expressed in normal-tangential or rectangular coordinates.

5.5   Polar and Cylindrical Coordinates

It is often convenient to describe motion in terms of polar coordinates for motion of points in a plane and of cylindrical coordinates for three-dimensional motion.   Polar and cylindrical coordinate systems are simply alternative methods for expressing position vector, velocity, and acceleration.

Polar Coordinates—Consider the motion of point P along the plane curved path in Fig. 5.14a.   The coordinates (r, θ) are called *polar coordinates*, where r measures how far P is from 0 while θ measures the orientation (direction) of r from the reference line 0x. The position vector can be expressed in terms of

$$r = re_r \qquad \qquad (5.37)$$

where e_r is a unit vector in the direction of r. The unit vector e_θ is

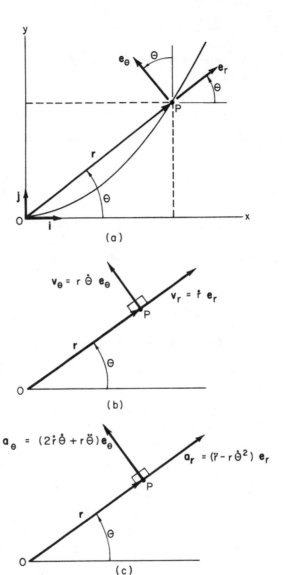

Fig. 5.14

perpendicular to $\mathbf{e}_r$ and in the direction of increasing θ as shown. The relationship between rectangular and polar coordinates is seen to be

$$x = r \cos \theta, \qquad y = r \sin \theta, \qquad \tan \theta = y/x \qquad (5.38)$$

and $r = \sqrt{x^2 + y^2}$, while the unit vectors are related by

$$\mathbf{e}_r = \cos \theta \; \mathbf{i} + \sin \theta \; \mathbf{j} \qquad (5.39)$$

and

$$\mathbf{e}_\theta = -\sin\,\theta\,\mathbf{i} + \cos\,\theta\,\mathbf{j} \qquad (5.40)$$

where we see that the unit vectors $\mathbf{e}_r$ and $\mathbf{e}_\theta$ are functions of θ and hence rotate as P moves.

The velocity can be obtained by differentiating Eq. (5.37) to obtain

$$\mathbf{v} = \dot{\mathbf{r}} = \dot{r}\,\mathbf{e}_r + r\,\dot{\mathbf{e}}_r \qquad (a)$$

where $\dot{\mathbf{e}}_r$ is obtained from Eqs. (5.39) and (5.40) as follows:

$$\dot{\mathbf{e}}_r = \frac{d\mathbf{e}_r}{d\theta}\frac{d\theta}{dt} = \dot{\theta}\,(-\sin\,\theta\,\mathbf{i} + \cos\,\theta\,\mathbf{j}) = \dot{\theta}\,\mathbf{e}_\theta \qquad (b)$$

so that Eq. (a) becomes

$$\mathbf{v} = \dot{\mathbf{r}} = \dot{r}\,\mathbf{e}_r + r\dot{\theta}\,\mathbf{e}_\theta \qquad (5.41)$$

or

$$= v_r\,\mathbf{e}_r + v_\theta\,\mathbf{e}_\theta \qquad (5.42)$$

where $\quad v_r = \dot{r}$ is the radial velocity component due to a change in magnitude of $\mathbf{r}$

$v_\theta = r\dot{\theta}$ is the tangential velocity component due to a change in direction of $\mathbf{r}$

These velocity components are shown in Fig. 5.14b. Compare Eqs. (b) and (5.41) and note what happens to Eq. (5.41) when $\mathbf{r}$ is replaced by $\mathbf{e}_r$.

The acceleration can be obtained by differentiating Eq. (5.41), where each term has a time derivative including $\dot{\mathbf{e}}_\theta$, which can be obtained from Eqs. (5.40) and (5.39) to be

$$\dot{\mathbf{e}}_\theta = \frac{d\mathbf{e}_\theta}{d\theta}\frac{d\theta}{dt} = \dot{\theta}\,(-\cos\,\theta\,\mathbf{i} - \sin\,\theta\,\mathbf{j}) = -\dot{\theta}\,\mathbf{e}_r \qquad (c)$$

so that the acceleration becomes

$$\begin{aligned}\mathbf{a} = \ddot{\mathbf{r}} = \dot{\mathbf{v}} &= \ddot{r}\,\mathbf{e}_r + \dot{r}\,\dot{\mathbf{e}}_r + \dot{r}\,\dot{\theta}\,\mathbf{e}_\theta + r\ddot{\theta}\,\mathbf{e}_\theta + r\dot{\theta}\,\dot{\mathbf{e}}_\theta \\ &= \ddot{r}\,\mathbf{e}_r + \dot{r}\dot{\theta}\,\mathbf{e}_\theta + \dot{r}\dot{\theta}\,\mathbf{e}_\theta + r\ddot{\theta}\,\mathbf{e}_\theta - r\dot{\theta}^2\,\mathbf{e}_r\end{aligned} \qquad (d)$$

on using Eqs. (b) and (c) for $\dot{\mathbf{e}}_r$ and $\dot{\mathbf{e}}_\theta$ respectively. The five acceleration terms can be interpreted physically as follows:

$$\ddot{r}\,\mathbf{e}_r = \text{change in magnitude of } v_r$$
$$\dot{r}\dot{\theta}\,\mathbf{e}_\theta = \text{change in direction of } v_r$$

$\dot{r}\dot{\theta}\ \mathbf{e}_\theta$ = change in magnitude of v_θ due to change in r $\left.\right\}$ ²

$r\ddot{\theta}\ \mathbf{e}_\theta$ = change in magnitude of v_θ due to change in $\dot{\theta}$

$-r\dot{\theta}^2\ \mathbf{e}_r$ = change in direction of v_θ

Collecting terms in Eq. (d) gives

$$\mathbf{a} = \ddot{\mathbf{r}} = \dot{\mathbf{v}} = (\ddot{r} - r\dot{\theta}^2)\ \mathbf{e}_r + (2\dot{r}\dot{\theta} + r\ddot{\theta})\ \mathbf{e}_\theta \qquad (5.43)$$

or

$$= a_r \mathbf{e}_r + a_\theta\ \mathbf{e}_\theta \qquad (5.44)$$

where $a_r = \ddot{r} - r\dot{\theta}^2$ is the radial acceleration component
$a_\theta = 2\dot{r}\dot{\theta} + r\ddot{\theta}$ is the tangential acceleration component

These acceleration components are illustrated in Fig. 5.14c, where we should note that θ, $\dot{\theta}$, and $\ddot{\theta}$ are all positive in the counterclockwise direction. The term $2\dot{r}\dot{\theta}$ is commonly called the Coriolis acceleration component.

Cylindrical Coordinates—The position vector $\mathbf{r}_P$ can be expressed in terms of the three cylindrical coordinate variables (r, θ, z) as in Fig. 5.15 where we see that the projection of the position of P onto the xy plane gives P', which can be easily described in terms of r and θ. The z coordinate gives the location of P above the xy plane. Note that the unit vectors $\mathbf{e}_r$ and $\mathbf{e}_\theta$ rotate with the OPP' plane while the unit vector $\mathbf{k}$ always points in the z direction. For this case, the position vector $\mathbf{r}_P$ can be expressed as

$$\mathbf{r}_P = r\mathbf{e}_r + z\mathbf{k} \qquad (5.45)$$

from which the velocity becomes

$$\mathbf{v} = \dot{\mathbf{r}}_P = \dot{r}\ \mathbf{e}_r + r\dot{\theta}\ \mathbf{e}_\theta + \dot{z}\mathbf{k}$$
$$= v_r \mathbf{e}_r + v_\theta\ \mathbf{e}_\theta + v_z \mathbf{k} \qquad (5.46)$$

where v_z is the velocity component in the z direction. The acceleration becomes

$$\mathbf{a} = \dot{\mathbf{v}} = \ddot{\mathbf{r}}_P = (\ddot{r} - r\dot{\theta}^2)\mathbf{e}_r + (2\dot{r}\dot{\theta} + r\ddot{\theta})\mathbf{e}_\theta + \ddot{z}\mathbf{k}$$
$$= a_r\mathbf{e}_r + a_\theta\ \mathbf{e}_\theta + a_z \mathbf{k} \qquad (5.47)$$

where a_z is the acceleration component in the z direction. Comparison of Eqs. (5.45), (5.46), and (5.47) with Eqs. (5.37), (5.41), and

---

2. Note $v_\theta = r\dot{\theta}$ so that the magnitude of v_θ can be varied by a change in either r or $\dot{\theta}$ or both. Hence v_θ generates two acceleration terms when it changes in magnitude.

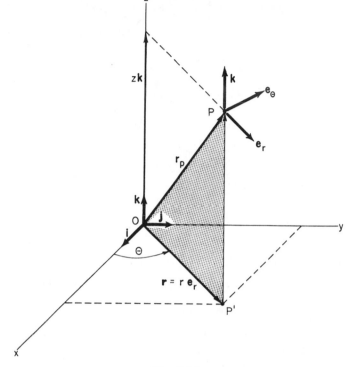

Fig. 5.15

(5.43) shows that polar coordinates are a special case of cylindrical coordinates with z equal to zero.

EXAMPLE 5.10

Particle P moves in the slot AB of arm $0AB$ (Fig. E5.10). Arm $0AB$ rotates with a uniform angular velocity $\dot{\theta}$ in a clockwise direction and completes one revolution each second. Particle P follows the stationary cam C with a profile given by $r = 6 + 2 \sin \theta$, where r is in centimetres when θ is in radians. Determine the velocity and acceleration of P when $\theta = \pi/3$ rad.

SOLUTION

From the data on angular motion we conclude that $\dot{\theta}$ is -2π rad/s since arm $0AB$ rotates with uniform angular velocity. Since the cam profile is given as a function of θ, the chain rule of differentiation gives

$$r = 6 + 2 \sin \theta \text{ cm} \qquad\qquad \theta = -2\pi t + \theta_0 = \pi/3 \text{ rad}$$

$$\dot{r} = 2(\cos \theta)\dot{\theta} \text{ cm/s} \qquad\qquad \dot{\theta} = -2\pi \text{ rad/s}$$

$$\ddot{r} = -2(\sin \theta)\dot{\theta}^2 + 2\cos \theta \ \ddot{\theta} \text{ cm/s}^2 \qquad \ddot{\theta} = 0 \text{ rad/s}^2$$

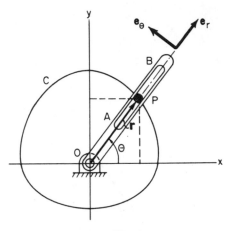

Fig. E5.10

Substituting these results into Eq. (5.41) for the velocity, we have

$$\mathbf{v} = \dot{\mathbf{r}} = 2[\cos{(\pi/3)}]\,(-2\pi)\mathbf{e}_r + [6 + 2\sin{(\pi/3)}]\,(-2\pi)\mathbf{e}_\theta$$

$$\mathbf{v} = -6.28\mathbf{e}_r - 48.6\mathbf{e}_\theta \text{ cm/s} \qquad\qquad Ans.$$

where we see that both velocity components are negative because r is becoming smaller as θ decreases and $\dot{\theta}$ is negative.

Similarly, for acceleration from Eq. (5.43) we find

$$\mathbf{a} = \ddot{\mathbf{r}} = \{-2\,[\sin{(\pi/3)}]\,(2\pi)^2 - [6 + 2\sin{(\pi/3)}]\,(2\pi)^2\}\mathbf{e}_r$$

$$+ \{[6 + 2\sin{(\pi/3)}]\,(0) + 2[2\cos{(\pi/3)}]\,(-2\pi)\}\mathbf{e}_\theta$$

$$\mathbf{a} = -373\,\mathbf{e}_r + 78.8\,\mathbf{e}_\theta \text{ cm/s}^2 \qquad\qquad Ans.$$

EXAMPLE 5.11

Particle P moves along a circular path of radius r, which lies in the xy plane (Fig. E5.11a). Determine the velocity and acceleration of P at any point on the circle. Compare this answer to that obtained using normal and tangential components of motion.

SOLUTION

Note that r is a constant radius of r units and that the curvilinear distance s, the radius r, and the angle θ are related through the arc length relationship $s = r\theta$. Thus we can write that

$$r = r, \qquad \dot{r} = 0, \qquad \ddot{r} = 0$$
$$s = r\theta, \qquad \dot{s} = r\dot{\theta}, \qquad \ddot{s} = r\ddot{\theta}$$

must hold for this type of motion. Substitution of the data on r into Eq. (5.41)

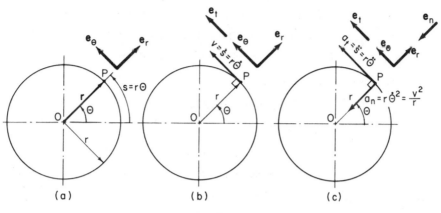

Fig. E5.11

for velocity gives

$$\mathbf{v} = \dot{\mathbf{r}} = \dot{r}\mathbf{e}_r + r\dot{\theta}\ \mathbf{e}_\theta = r\dot{\theta}\ \mathbf{e}_\theta \qquad (a)\ Ans.$$

where we see that the velocity is always tangent to the circular path and its speed is given by $r\dot{\theta}$. A comparison of Eq. (a) with Eq. (5.28), $\mathbf{v} = \dot{s}\ \mathbf{e}_t = v\ \mathbf{e}_t$, shows that the speed v (magnitude of the velocity) can be expressed as $v = \dot{s} = r\dot{\theta}$ and that the unit vectors $\mathbf{e}_\theta$ and $\mathbf{e}_t$ are coincident for the case in Fig. E5.11b.

The acceleration becomes

$$\mathbf{a} = -r\dot{\theta}^2\ \mathbf{e}_r + r\ddot{\theta}\ \mathbf{e}_\theta \qquad (b)\ Ans.$$

from Eq. (5.43). Here we see that one acceleration component is always directed at the center of the circle, while the other is always tangent to the circle. If we compare Eq. (b) with Eq. (5.31),

$$\mathbf{a} = \frac{v^2}{\rho}\ \mathbf{e}_n + \ddot{s}\ \mathbf{e}_t = r\dot{\theta}^2\ \mathbf{e}_n + r\ddot{\theta}\ \mathbf{e}_t$$

we see that the magnitudes and directions are the same. The difference in sign is due to $\mathbf{e}_r$ being positive in the direction of 0 of P while $\mathbf{e}_n$ is positive in the direction of P to 0. These acceleration components and the unit vectors are shown in Fig. E5.11c.

EXAMPLE 5.12

A particle is moving along a curved path given by $r = t^3$, $\theta = \pi t^2$, $z = 2t^2$ where r and z are in feet, θ is in radians, and t is in seconds. Determine the position, velocity, and acceleration when $t = 0.5$ s.

SOLUTION

The values needed to express the position, velocity, and acceleration can be obtained from

$$r = t^3 \text{ ft} \qquad \theta = \pi t^2 \text{ rad} \qquad z = 2t^2 \text{ ft}$$
$$\dot{r} = 3t^2 \text{ fps} \qquad \dot{\theta} = 2\pi t \text{ rad/s} \qquad \dot{z} = 4t \text{ fps}$$
$$\ddot{r} = 6t \text{ fps}^2 \qquad \ddot{\theta} = 2\pi \text{ rad/s}^2 \qquad \ddot{z} = 4 \text{ fps}^2$$

The position is obtained from Eq. (5.45) to be

$$\mathbf{r}_p = r\mathbf{e}_r + z\mathbf{k} = 0.125\mathbf{e}_r + 0.50\mathbf{k} \qquad\qquad Ans.$$

The velocity is obtained from Eq. (5.46) to be

$$\mathbf{v} = \dot{\mathbf{r}}_p = \dot{r}\,\mathbf{e}_r + r\dot{\theta}\,\mathbf{e}_\theta + \dot{z}\mathbf{k} = 3t^2\,\mathbf{e}_r + 2\pi t^4\,\mathbf{e}_\theta + 4t\mathbf{k}$$
$$= 0.75\mathbf{e}_r + 0.393\mathbf{e}_\theta + 2\mathbf{k} \text{ fps} \qquad\qquad Ans.$$

and the acceleration from Eq. (5.47) to be

$$\mathbf{a} = \ddot{\mathbf{r}}_p = (\ddot{r} - r\dot{\theta}^2)\mathbf{e}_r + (r\ddot{\theta} + 2\dot{r}\dot{\theta})\mathbf{e}_\theta + \ddot{z}\mathbf{k}$$
$$= (6t - 4\pi^2 t^5)\mathbf{e}_r + (2\pi t^3 + 12\pi t^3)\mathbf{e}_\theta + 4\mathbf{k}$$
$$= (3 - \pi^2/8)\mathbf{e}_r + [(\pi/4) + (3\pi/2)]\mathbf{e}_\theta + 4\mathbf{k}$$
$$= 1.768\mathbf{e}_r + 5.50\mathbf{e}_\theta + 4.0\mathbf{k} \text{ fps}^2 \qquad\qquad Ans.$$

5.6 Relative Motion of Particles

In this section, the concept of relative motion is considered using two observers, one attached to the earth and the other moving. Consider the situation in Fig. 5.16a where the motion of particle P along the curved path C is to be measured by two observers. The first is attached to the earth at $0'$ and can describe motion in terms of rectangular coordinates X, Y, Z with unit vectors $\mathbf{i}'$, $\mathbf{j}'$, $\mathbf{k}'$. The second is attached to point 0 which moves along curve C_0 and can describe motion relative to himself in terms of the nonrotating rectangular coordinates x, y, z with unit vectors $\mathbf{i}$, $\mathbf{j}$, $\mathbf{k}$. For convenience the axes of the two coordinate systems are parallel so that $\mathbf{i}' = \mathbf{i}$, $\mathbf{j}' = \mathbf{j}$, and $\mathbf{k}' = \mathbf{k}$. Hence the second observer is a nonrotating but translating observer.

The first observer can describe the absolute motion of both P and 0 in terms of position vectors $\mathbf{r}_P$ and $\mathbf{r}_0$. The second observer at 0 can describe the motion of P relative to himself (i.e., point 0) in terms of the *relative position vector* $\mathbf{r}_{P/0}$. The three position vectors are related by vector addition giving

$$\mathbf{r}_P = \mathbf{r}_0 + \mathbf{r}_{P/0} \qquad\qquad (5.48)$$

which states that the position of P is equal to the position of 0 plus the position of P relative to 0. Equation (5.48) can be easily expressed in terms of the two rectangular coordinate systems giving three scalar equations.

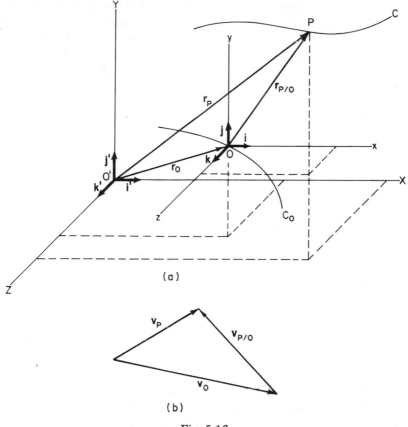

(a)

(b)

Fig. 5.16

The velocity relationship can be obtained by differentiating Eq. (5.48) with respect to time, giving

$$\dot{\mathbf{r}}_P = \dot{\mathbf{r}}_0 + \dot{\mathbf{r}}_{P/0} \tag{5.49}$$

or

$$\mathbf{v}_P = \mathbf{v}_0 + \mathbf{v}_{P/0} \tag{5.50}$$

which simply states that the absolute velocity of P is equal to the vector sum of the absolute velocity of 0 plus the velocity of P with respect to 0. The *relative velocity* $\mathbf{v}_{P/0}$ is the velocity of P as seen by a nonrotating observer moving with 0. The subscripts of Eq. (5.50) can be remembered by the scheme that the product of the subscripts on the right-hand side (0 times $P/0$) is equal to the subscript on the left-hand side (P). The vector addition required by Eqs. (5.49) or (5.50) is illustrated in Fig. 5.16b.

We have experienced this physical situation many times. For example, every time we walk along a sidewalk and see someone else either walking toward or away from us, we see the velocity of the other person relative to ourselves; i.e., we see $v_{P/0}$. To obtain the absolute velocity of the other person, we must add our own velocity to the relative velocity we observe, which Eq. (5.50) states for the most general case.

The acceleration relationship can be obtained from Eq. (5.49) by differentiation to obtain

$$\ddot{r}_P = \ddot{r}_0 + \ddot{r}_{P/0} \qquad\qquad (5.51)$$

or

$$a_P = a_0 + a_{P/0} \qquad\qquad (5.52)$$

The interpretation of Eqs. (5.51) and (5.52) for accelerations is precisely the same as that of Eq. (5.50) for velocities where $a_{P/0}$ is the acceleration of P relative to 0, which is called the *relative acceleration*. The accelerations add vectorially in a manner identical to the velocities (Fig. 5.16b).

From Eq. (5.52) it is seen that the absolute acceleration of P is equal to the acceleration of P relative to 0 when the acceleration of 0 is zero; i.e., when 0 moves with constant velocity along a straight line. Hence a point moving with constant velocity ($\dot{r}_0$ = constant) is an inertial reference point insofar as calculation of acceleration and force through Newton's second law is concerned.

The entire concept of relative motion is based on the use of both a stationary and a nonrotating moving observer and the use of vector addition for position, velocity, and acceleration.

EXAMPLE 5.13

A pilot is flying a light airplane with a cruising speed of 100 mph and is making a 260-mi cross-country flight with a magnetic compass heading of $120°$ (Fig. E5.13a). His weather briefing indicates that a 40-mph south wind is blowing. Determine (a) his ground speed and his compass heading to follow a straight-line flight path and (b) how long it should take him to make the flight.

SOLUTION

The airplane flies at a speed of 100 mph relative to the air mass in which it is operating while following the flight path from A to B. This means that the absolute velocity of the airplane (ground speed) must be along the direction AB. Let the x axis be in the direction of the flight path from A to B (Fig. E5.13b) where the relative velocity equation

$$v_p = v_{p/w} + v_w \qquad\qquad (a)$$

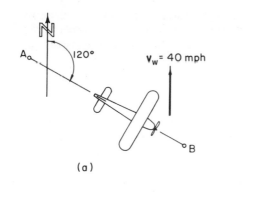

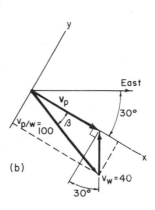

Fig. E5.13

is also shown. The angle β is called the crab angle, and the fixed and moving co-ordinates are parallel to the x and y axes.

(a) The ground speed and crab angle can be found from Eq. (a) and Fig. E5.13b to give

$$v_p i = (100 \cos \beta - 40 \sin 30°)i + (40 \cos 30° - 100 \sin \beta)j$$

or

$$i : v_p = 100 \cos \beta - 20 \qquad (b)$$

$$j : \sin \beta = (40/100)(0.867) = 0.347 \qquad (c)$$

From Eq. (c) we find that $\beta = 20.3°$ and from Eq. (b) that the ground speed is

$$v_p = 100 \cos (20.3) - 20 = 73.6 \text{ mph} \qquad Ans.$$

The magnetic compass heading is the flight path heading plus the crab angle β; thus

$$\text{Magnetic compass heading} = 120 + 20.3 = 140°{}^3 \qquad Ans.$$

(b) Providing that the wind, airplane, and pilot perform consistently, the time required to make the flight is

$$t = \text{distance/ground speed} = 260/73.6 = 3.53 \text{ hr} \qquad Ans.$$

or

$$t = 3 \text{ hr } 32 \text{ min} \qquad Ans.$$

---

3. The pilot would fly a compass heading of $139°$–$141°$ for a while and would keep checking his progress since he knows that the wind does not blow at 40.0 mph for long.

Pilots of light airplanes actually perform these calculations in planning a flight. The ground speed and crab angle are determined by use of a special flight computer, where the velocity diagram in Fig. E5.13b is actually constructed to scale and the values are read directly.

EXAMPLE 5.14

Two blocks A and B of the mechanism in Fig. E5.14a move in horizontal and vertical slots respectively and are connected by bar AB, $\ell = 4$ ft. The motion of block A is given by $X = a \sin \Omega t = 3 \sin \pi t$, where X is in feet, Ω is in rad/s, and t is in seconds. When the angle between bar AB and the vertical axis is $30°$, determine (a) the relative velocity of B with respect to A and (b) the relative acceleration of B with respect to A.

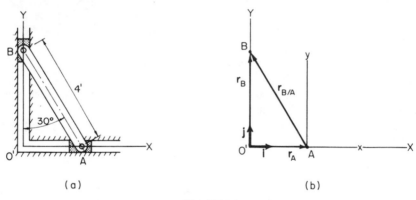

(a)   (b)

Fig. E5.14

SOLUTION

Attach the moving (nonrotating) rectangular coordinates x, y to block A (Fig. E5.14b) so that the position vectors can be expressed as

$$\mathbf{r}_A = X\mathbf{i} = 3 \sin \pi t\,\mathbf{i}, \qquad \mathbf{r}_B = Y\mathbf{j}, \qquad \mathbf{r}_{B/A} = -x\mathbf{i} + y\mathbf{j} \tag{a}$$

To proceed, we must know the position, velocity, and acceleration of block A. From the geometry we find

$$X = 4 \sin 30° = 3 \sin \pi t, \qquad \sin \pi t = 2/3 = 0.667$$

and

$$\cos \pi t = \sqrt{1 - (2/3)^2} = \sqrt{5/9} = 0.745$$

for the instant under consideration. Thus $\mathbf{r}_A$, $\dot{\mathbf{r}}_A$, $\ddot{\mathbf{r}}_A$, and $\mathbf{r}_B$ become

$$\mathbf{r}_A = 2\mathbf{i} \tag{b}$$
$$\dot{\mathbf{r}}_A = \dot{X}\mathbf{i} = 3\pi \cos \pi t\,\mathbf{i} = 7.03\mathbf{i} \text{ fps} \tag{c}$$
$$\ddot{\mathbf{r}}_A = \ddot{X}\mathbf{i} = -3\pi^2 \sin \pi t\,\mathbf{i} = -19.75\mathbf{i} \text{ fps}^2 \tag{d}$$

and

$$r_B = Yj = 4 \cos 30° \, j = 3.46j \text{ ft} \qquad (e)$$

(a) The relative velocity can be obtained from Eq. (5.50) if the velocity of B is known, since the velocity of A is already known. We can find the velocity of B by using the fact that the length of the bar is constant. Thus

$$X^2 + Y^2 = \ell^2 = 16 \qquad (f)$$

Differentiation gives

$$\dot{Y} = -(X/Y)\dot{X} = -(2/3.46)(7.03) = -4.06 \text{ fps} \qquad (g)$$

Substitution of Eqs. (c) and (g) into

$$v_{B/A} = \dot{r}_{B/A} = \dot{r}_B - \dot{r}_A = \dot{x}i + \dot{y}j$$

gives

$$v_{B/A} = -4.06j - 7.03i \text{ fps} \qquad\qquad Ans.$$

(b) The relative acceleration can be obtained from Eq. (5.52) if we know the acceleration of B, since the acceleration of A is known. Differentiation of Eq. (f) twice gives

$$\ddot{Y} = -\frac{(\dot{X}^2 + \dot{Y}^2)}{Y} - \frac{X}{Y}(\ddot{X})$$

$$= -\frac{(7.03)^2 + (4.06)^2}{3.46} - \left(\frac{2.0}{3.46}\right)(-19.75) = -19.08 + 11.42 = -7.66 \text{ fps}^2 \qquad (h)$$

Substitution of Eqs. (d) and (h) into

$$a_{B/A} = \ddot{r}_{B/A} = \ddot{r}_B - \ddot{r}_A = \ddot{x}i + \ddot{y}j$$

gives

$$a_{B/A} = -7.66j - (-19.75)i = 19.75i - 7.66j \text{ fps}^2 \qquad Ans.$$

These answers reflect what a nonrotating observer attached to block A would see if he were looking at block B.

5.7 Position and Velocity of a Rigid Body in Plane Motion

We next consider (a) the description of the various types of rigid-body motion that can occur, (b) the position and velocity relationships for a rigid body in plane motion, and (c) the special case of the rolling wheel.

Types of Rigid-Body Motion—In solving rigid-body dynamics problems, it is important to recognize the type of motion the body has since this recognition usually indicates the best kinematic de-

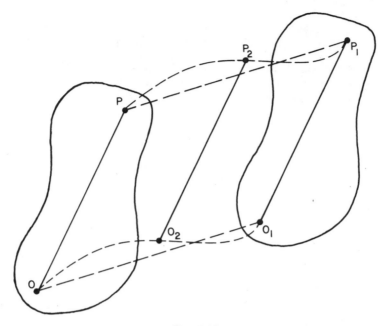

Fig. 5.17

scription of body motion for that problem. Five types of rigid-body motion can be described.

Translation is a rigid-body motion where any straight line $0P$ in the body remains parallel to its original direction throughout the motion (Fig. 5.17). Thus all the particles of the body must move along parallel paths. If these paths are straight lines (such as P to P_1) the motion is defined to be *rectilinear translation;* while if these paths are curved (such as P, P_2, P_1), the motion is defined to be *curvilinear translation.*

Pure rotation (rotation about a fixed axis) is a rigid-body motion that occurs when the body rotates about the fixed axis AB in Fig. 5.18. In this case points P and P_1 travel in circular paths around points 0 and 0_1 respectively, which are located on the *axis of rotation.* The circular paths of P and P_1 lie in parallel planes that are perpendicular to the axis of rotation AB, while points 0 and 0_1 have zero velocity and acceleration. The lines $0P$ and 0_1P_1 have a common angular velocity ω and acceleration α, which are parallel vectors as shown.

The difference between curvilinear translation and pure rotation is illustrated in Fig. 5.19 where all points in the body of Fig. 5.19a move in circular curvilinear paths but the line $0P$ remains parallel throughout the motion. For the case in Fig. 5.19b all points in the

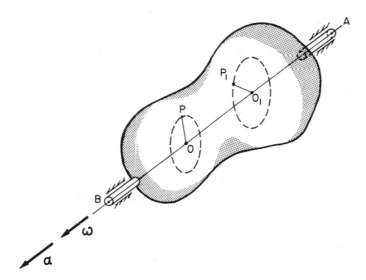

Fig. 5.18

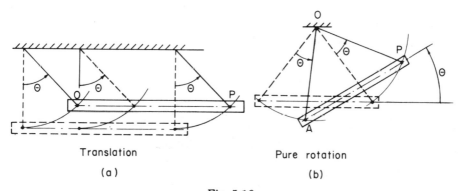

Translation

(a)

Pure rotation

(b)

Fig. 5.19

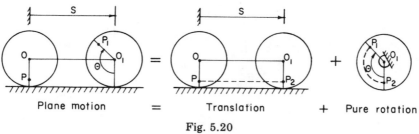

Plane motion   =   Translation   +   Pure rotation

Fig. 5.20

body move in circular paths about point 0, and lines $0P$, PA, and $0A$ rotate through the same angle θ.

Plane motion is the superposition of the two simpler motions of translation and pure rotation, where each point in the body moves in a plane parallel to the plane of motion for any other point in the body. This superposition concept is shown in Fig. 5.20 where a disk is moving along a horizontal surface. The center of the disk moves a distance s to 0_1 while P moves to P_1. The line $0P$ is seen to rotate through an angle θ.

The same final position can be obtained from the two simpler motions of translation, where 0 moves to 0_1 and P moves to P_2, plus pure rotation of the disk around 0_1 through an angle θ during which P_2 moves to P_1 along the circular path. Note that the same final position can be obtained by having the body rotate first, then translating so that the order of superposition is of little consequence.

The following two motions are more general and are beyond the scope of this text. We are describing them only to illustrate the limitation of the types of motion associated with translation, pure rotation, and plane motion. The first type is called *fixed point rotation* (Fig. 5.21), where one point in the body (point 0, for example) is fixed. There is no preferred axis of rotation since the instantaneous axis of rotation changes with time. For this case the angular velocity can change in magnitude and direction since ω and α are nonparallel

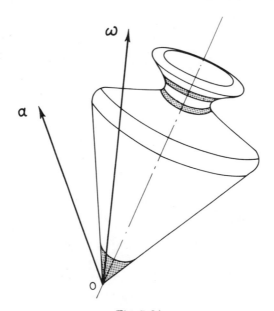

Fig. 5.21

as shown. Thus fixed axis rotation is a special case of fixed point rotation. The motion of a top on a rough floor is an example of this type. The second type is called *general space motion*, which is the superposition of translation and fixed point rotation types of motion. General motion includes all other types as a special case. A space capsule or satellite is an example of this type.

Position and Velocity Relationships Consider a rigid body to be in plane motion with instantaneous angular velocity ω and instantaneous angular acceleration α (recall that ω and α are parallel for plane motion) (Fig. 5.22a). Let the xy plane be parallel to the plane of motion and intersect the body forming the shaded area, let 0 be any point of the body in the shaded area as well as the origin for the non-

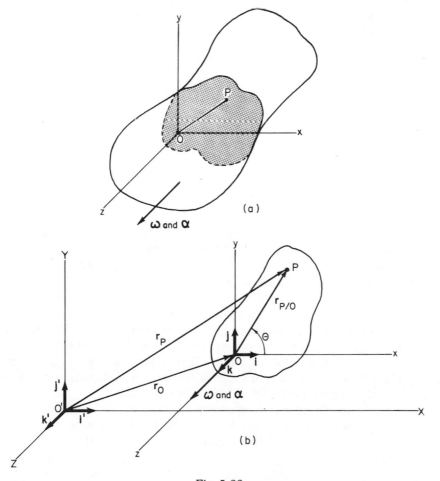

Fig. 5.22

rotating xyz coordinate system which is to move with point 0, and let P be any other point in the shaded area.

To describe the absolute motion of points P and 0, let us remove the shaded area (called a lamina) and establish a fixed reference at $0'$ (Fig. 5.22b). Here the absolute position of P and 0 is given by r_P and r_0 respectively. The position of P relative to 0 is given by the *relative position vector* $r_{P/0}$ which forms an angle θ with respect to the moving x axis. Note that an observer at $0'$ (attached to the XYZ coordinate system) observes the absolute position of both points P and 0, while a nonrotating observer moving with point 0 (attached to the xyz coordinate system) can see only the position of P relative to himself, i.e., $r_{P/0}$. (See Section 5.6 for greater detail on what these observers see in general.)

The relationship of the positions of points P and 0 is obtained by vector addition to give

$$r_P = r_0 + r_{P/0} \qquad (5.53)$$

which states that the position of P is equal to the position of 0 plus the position of P with respect to 0.

The velocity relationship can be obtained by differentiating Eq. (5.53) with respect to time to obtain

$$\dot{r}_P = \dot{r}_0 + \dot{r}_{P/0} \qquad (a)$$

or

$$v_P = v_0 + v_{P/0} \qquad (b)$$

which states that the absolute velocity of P is equal to the vector sum of the absolute velocity of 0 plus the velocity of P with respect to 0. The *relative velocity* $v_{P/0}$ is the velocity the nonrotating (but moving) observer at 0 sees relative to the xyz coordinate system.

To evaluate the meaning of the velocity relationship described by Eqs. (a) or (b) for a rigid body in plane motion at any instant, consider what the velocity relationship is for each of the simpler motions of translation and pure rotation separately as indicated in Figs. 5.23a, b, and c. The plane motion situation is shown in Fig. 5.23a where the velocities v_P and v_0, the angular velocity ω, and the orientation of $r_{P/0}$ (i.e., the angle θ) are shown for a given instant.

Suppose the body were translating only as in Fig. 5.23b. Then θ is constant, which implies $\omega_T = 0$ so that the magnitude and direction of $r_{P/0}$ are also constant. Hence $\dot{r}_{P/0} = v_{P/0} = 0$ and Eq. (b) becomes

$$v_{P_T} = v_0 \qquad (c)$$

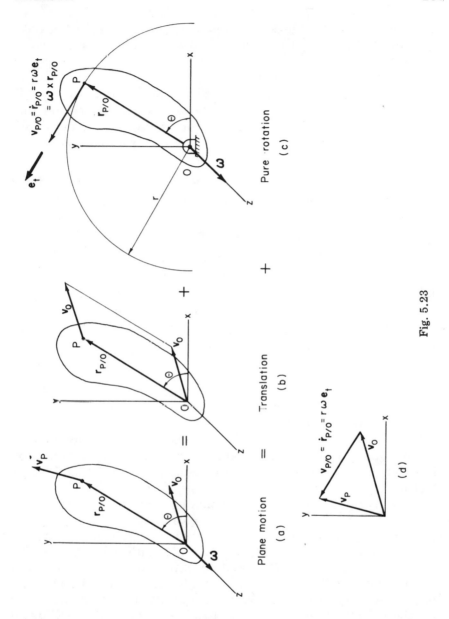

Fig. 5.23

as shown where $\mathbf{v}_{P_T}$ is used to denote $\mathbf{v}_P$ due to translation only. Equation (c) shows that the velocity of every point in a translating body is the same.

Suppose the body is rotating only about the fixed z axis through point 0 with an angular velocity of ω (Fig. 5.23c). In this case $\mathbf{v}_{0_R} = 0$, where subscript R denotes rotation terms. The only motion that point P can have is that of *circular motion with radius r*, and we see that $\mathbf{r}_{P/0}$ can change direction only. We have shown that the velocity of a point moving on a circular path is given by

$$\mathbf{v}_{P/0} = r\,\omega\,\mathbf{e}_t \tag{d}$$

where $\mathbf{e}_t$ is a unit vector tangent to the path. Also, we see that the cross product $(\omega \times \mathbf{r}_{P/0})$ is the relative velocity $\mathbf{v}_{P/0}$ (recall that the direction of ω is given by the right-hand screw rule and the order $(\omega \times \mathbf{r}_{P/0} \neq \mathbf{r}_{P/0} \times \omega)$ is important). Thus Eq. (d) can be written as

$$\mathbf{v}_{P/0} = \dot{\mathbf{r}}_{P/0} = r\,\omega\,\mathbf{e}_t = \omega \times \mathbf{r}_{P/0} \tag{e}$$

Superimposing the results of Eqs. (c) and (e) according to Eq. (b) gives the desired relationship between velocities of any two points in a rigid body to be

$$\mathbf{v}_P = \mathbf{v}_0 + \mathbf{v}_{P/0} = \mathbf{v}_0 + r\,\omega\,\mathbf{e}_t \tag{5.54}$$

$$= \mathbf{v}_0 + \omega \times \mathbf{r}_{P/0} \tag{5.55}$$

Either of these equations is called the *rigid-body relative velocity equation*, and the vector addition is shown in Fig. 5.23d where this vector addition can be done in terms of any convenient set of axes such as x and y. The concept of being able to compute the absolute velocity of a point in a rigid body by the process of superposition (Figs. 5.23b, c) is important for visualizing the use of Eqs. (5.54) or (5.55).

There is often concern about which point to select for 0 in using Eq. (5.54) to obtain the velocity of a given point P. The governing rule is to select any point in the body for which we know or can find the velocity and call this 0. The velocity of P with respect to 0 is due to the circular motion of P around point 0, as though 0 is on a fixed axis of rotation (Fig. 5.23c) and computed from Eq. (d). Then the absolute velocity of P is the vector sum of the absolute velocity of 0 and the velocity of P with respect to 0 as shown in Fig. 5.23d and computed from Eq. (5.54).

EXAMPLE 5.15

The 10-m ladder in Fig. E5.15 is sliding down and away from the wall. At the instant shown, point A is moving to the left with a velocity of 4 m/s. Determine the angular velocity of the ladder and the velocity of point B along the wall.

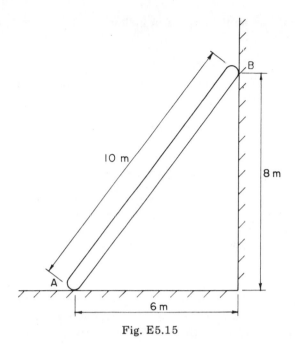

Fig. E5.15

SOLUTION

Write Eq. (5.54) in terms of points A and B and draw the corresponding figure, which represents the translation of A and pure rotation components of B about the fixed axis at A as follows:

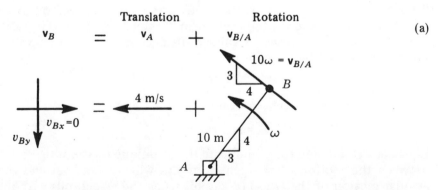

(a)

Here v_{By} is assumed positive downward and ω of AB is assumed counterclockwise, giving $\mathbf{v}_{B/A}$ as directed. Observe that $v_{Bx} = 0$ since point B moves along the vertical wall. Summing velocity components of Eq. (a) in the horizontal direction, where to the right is positive, gives

$$\overset{+}{(\rightarrow)}\ 0 = -4 - (4/5)(10\omega) = -4 - 8\omega$$

or $\omega = -0.5$ rad/s, where the minus sign indicates ω is opposite to the assumed direction. Thus

$$\omega = 0.5 \text{ rad/s} \; \rangle \qquad\qquad Ans.$$

Summing velocity components in the vertical direction with downward as positive gives

$$\left(\downarrow\right) v_{By} = 0 - (3/5)(10\omega) = -6(-0.5) = 3 \text{ m/s}$$

or

$$v_B = \vec{0} + 3 \downarrow = 3 \text{ m/s} \downarrow \qquad\qquad Ans.$$

The same answer would result if Eq. (a) were written as $v_A = v_B + v_{A/B}$, where point B becomes the fixed point for relative motion. Also, the cross product ($\omega \times r_{B/A}$) could have been used, but it is usually much simpler to sketch the velocity diagram of Eq. (a) as shown here.

The rolling wheel—A special but important engineering element is that of a wheel that rolls along a straight surface, which is typical of vehicle wheels as well as some gearing arrangements. There are two conditions of importance; the wheel either rolls without slipping or rolls and slips.

Consider the wheel of radius r that rolls without slipping (Fig. 5.24a), where 0 is the wheel center and moves along the straight-line path a distance s from 0 to 0_1. Let P be the initial point of contact of the wheel, which is in contact with point A on the horizontal surface. We see that P moves to P_1 as the point of contact moves from A to A_1. From the geometry, the rectilinear distance s from 0 to 0_1 (also A to A_1) is the same as the curvilinear distance from A_1 to P_1, so that s and θ are related by

$$s = r\theta \qquad\qquad (5.56)$$

The velocity of the wheel center is

$$v_0 = \dot{s}i = r\dot{\theta}i = r\omega i \qquad\qquad (5.57)$$

Equations (5.56) and (5.57) indicate that a unique relationship exists between the motion of the center of the wheel (s and v_0) and the angular motion of the wheel (θ and ω) when the wheel rolls without slipping. Hence the motion of a wheel that rolls without slipping is a single degree of freedom problem since knowledge of either $s(t)$ or $\theta(t)$ is sufficient to specify all the motion that takes place.

When the wheel rolls and slips, it is impossible to relate s and θ (or v_0 and ω) since the word slip implies that relative motion exists at the point of contact between points P' and A_1 (Fig. 5.24a). This may

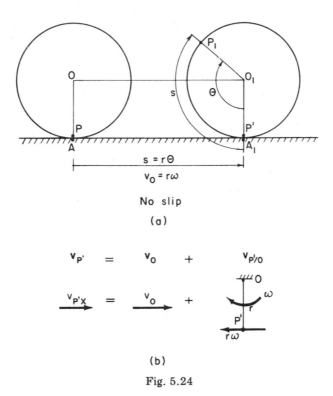

s = r\Theta

v_0 = r\omega

No slip

(a)

$\mathbf{v}_{P'} = \mathbf{v}_0 + \mathbf{v}_{P'/0}$

$\mathbf{v}_{P'x} = \mathbf{v}_0 + \quad$

(b)

Fig. 5.24

be demonstrated from the velocity diagram of Fig. 5.24b from which we find

$$\mathbf{v}_{P'} = v_{P'x}\mathbf{i} = (v_0 - r\omega)\mathbf{i} \qquad (5.58)$$

is the velocity of point P' of the wheel at the point of contact. For the no-slip condition, substitution of Eq. (5.57) into Eq. (5.58) shows that $\mathbf{v}_{P'} = 0$. For the rolling and slipping condition, Eq. (5.58) shows that P' is moving relative to A_1 (note $\mathbf{v}_{A_1} = 0$) and that the magnitude and direction of $\mathbf{v}_{P'}$ is dependent on *both* v_0 and ω. Hence a wheel that rolls and slips is a two degree of freedom problem since both $s(t)$ and $\theta(t)$, or $v_0(t)$ and $\omega(t)$, are needed to specify the motion of the wheel. A set of equations similar to Eqs. (5.56) and (5.57) can be developed for a wheel that rolls without slipping on a curved path.

EXAMPLE 5.16

The wheel in Fig. E5.16 is attached to body C by an inextensible cord that remains taut throughout the motion and is wrapped around the wheel at B. The position of 0 is given by $s = 2t^2$ (s in centimetres for t in seconds) and the angular

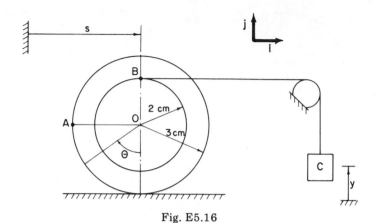

Fig. E5.16

position of the wheel is given by $\theta = 16t - 4t^2$ (θ in radians for t in seconds). Determine (a) the velocity of point A of the wheel and body C when $t = 4.0$ s and (b) a general relationship between the vertical position of body C and s and θ of the wheel.

SOLUTION

The wheel is generally rolling and slipping except at the moment $t = 1.714$ s. See Eq. (5.58).

(a) To find $\mathbf{v}_A$, we must know v_0 and ω, which are

$$\mathbf{v_0} = \dot{s} = 4t = 16 \text{ cm/s} \rightarrow$$

and

$$\boldsymbol{\omega} = \dot{\theta} = 16 - 8t = -16 \text{ rad/s} = 16 \text{ rad/s} \curvearrowright$$

where the minus sign means ω is in the negative θ direction. Thus we have

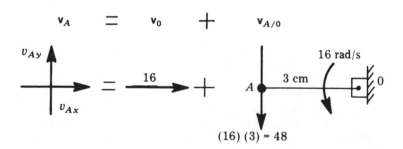

or

$(\leftrightarrow)\, v_{Ax} = 16 + 0 = 16, \quad (\updownarrow)\, v_{Ay} = 0 - 48 = -48, \quad \mathbf{v}_A = 16\mathbf{i} - 48\mathbf{j} \text{ cm/s}$

$Ans.$

The velocity of body C can be obtained by noting that the cord velocity at B is the same as the tangential velocity of a corresponding point in the wheel. Thus we have

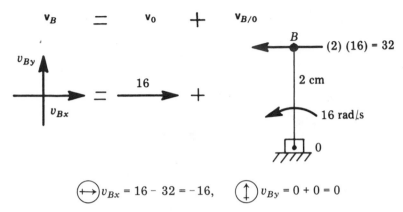

$$\left(\overset{+}{\longleftrightarrow}\right) v_{Bx} = 16 - 32 = -16, \qquad \left(\overset{+}{\uparrow}\right) v_{By} = 0 + 0 = 0$$

and

$$\mathbf{v}_C = |\mathbf{v}_B|\mathbf{j} = 16\mathbf{j} \text{ cm/s} \qquad\qquad Ans.$$

where we note body C moves upward when the cord at point B moves to the left.

(b) The relationship between s and θ and the vertical position of body $C(y)$ can be obtained by superposition of translation and pure rotation as follows. First, let the wheel *translate* a distance s to the right as shown, during which we see that body C moves s units downward; i.e., $y_T = -s$. Second, let the wheel *rotate* through an angle θ about an axis through 0, during which we see that body C moves downward a distance of 2θ units; i.e., $y_R = -2\theta$. Thus we have that

$$\mathbf{r}_C = (y_T + y_R)\mathbf{j} = -(s + 2\theta)\mathbf{j}$$

or

$$\mathbf{r}_C = (-2t^2 - 32t + 8t^2)\mathbf{j} = (6t^2 - 32t)\mathbf{j} \text{ cm} \qquad\qquad Ans. \text{ (a)}$$

where y is positive upward, s positive to the right, and θ is positive clockwise for this relationship. We can check the velocity of body C from Eq. (a) to obtain

$$\mathbf{v}_C = \dot{\mathbf{r}}_C = (12t - 32)\mathbf{j} = [12(4) - 32]\mathbf{j} = 16\mathbf{j} \text{ cm/s} \qquad\qquad Ans.$$

as before.

EXAMPLE 5.17

A gear of radius r is attached to arm $0A$ which rotates about a fixed axis at A (Fig. E5.17). This mechanical arrangement is a simplification of gearing used in some automatic transmissions. The gear is sometimes called a planetary gear. Determine (a) the relationship between angles ϕ (arm $A0$) and θ (gear); (b) the velocity of point D when in the position shown if $\dot{\phi} = 6.0$ rad/s counterclockwise,

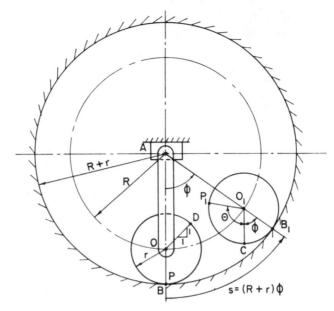

Fig. E5.17

$R = 6.0$ in., and $r = 2.0$ in.; and (c) the number of revolutions the gear makes for each revolution of arm $0A$ ($R = 6$ in., $r = 2$ in.).

SOLUTION

(a) To determine the angular relationship, let arm $0A$ rotate through an angle ϕ (Fig. E5.17), where 0 moves to 0_1, P moves to P_1, the point of contact moves from B to B_1, and the gear rotates through an angle θ (measured from the vertical reference line). Since we are dealing with a gear, the no-slip condition holds so that the arc lengths B to B_1 and $B_1 CP_1$ must be equal, giving

$$s = (R + r)\phi = r(\theta + \phi)$$

which reduces to

$$R\phi = r\theta \qquad\qquad \textit{Ans. (a)}$$

Equation (a) states that arc lengths 00_1 and CP_1 are equal, which is not obvious from the geometry.

(b) The velocity of point D can be found as soon as the angular velocity of the gear is known. This can be obtained from Eq. (a) to give

$$\boldsymbol{\omega} = \dot{\theta} = (R/r)\dot{\phi} = (R/r)\omega_{A0} = (6/2)6 = 18 \text{ rad/s} \ \big\downarrow$$

Thus

$$\mathbf{v}_D = \mathbf{v}_0 + \mathbf{v}_{D/0} = \mathbf{v}_A + \mathbf{v}_{0/A} + \mathbf{v}_{D/0}$$

where $v_A = 0$, so that

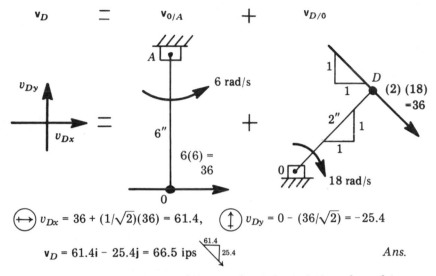

$$\overset{\longleftrightarrow}{\oplus} \ v_{Dx} = 36 + (1/\sqrt{2})(36) = 61.4, \quad \uparrow \ v_{Dy} = 0 - (36/\sqrt{2}) = -25.4$$

$$v_D = 61.4i - 25.4j = 66.5 \text{ ips} \quad \overset{61.4}{\underset{25.4}{\diagdown}} \qquad Ans.$$

(c) The number of revolutions of the gear for each revolution of arm $0A$ can be obtained directly from Eq. (a) to give

$$N_\theta/N_\phi = R/r = 6/2 = 3 \qquad Ans.$$

5.8 Instantaneous Centers of Zero Velocity

Every rigid body moving with plane motion has an instantaneous center of zero velocity that lies either in the body or the body extended, about which the body appears to rotate at a given instant in time. Consider the rigid body moving with plane motion (Fig. 5.25a). Let A and B be any two points in the body with nonparallel velocities v_A and v_B. Now, construct a line ad perpendicular to v_A and a line be perpendicular to v_B that intersect at point C either in the body (Fig. 5.25a) or the body extended (Fig. 5.25b). Velocities v_A and v_B can be written as

$$v_A = v_C + r_A \, \omega \, e_{tA} \tag{a}$$

and

$$v_B = v_C + r_B \, \omega \, e_{tB} \tag{b}$$

where e_{tA} and e_{tB} are unit vectors in the direction of v_A and v_B respectively. See Eq. (5.54).

From Eq. (a) and Fig. 5.25a we must conclude that v_C is parallel to v_A since e_{tA} is parallel to v_A. Similarly, from Eq. (b) it is con-

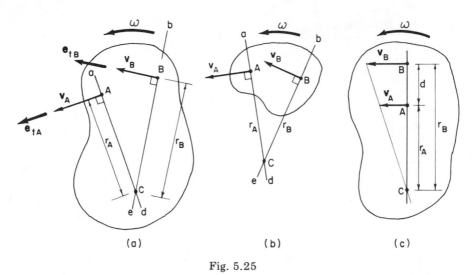

Fig. 5.25

cluded that $\mathbf{v}_C$ is parallel to $\mathbf{v}_B$. The only way $\mathbf{v}_C$ can be parallel to two nonparallel directions simultaneously is for $\mathbf{v}_C$ to be zero. Thus point C is called an *instantaneous center of zero velocity*, and the velocity of any point P in the body can be obtained from

$$\mathbf{v}_P = r\omega\mathbf{e}_t \tag{5.59}$$

where $\mathbf{e}_t$ is a unit vector perpendicular to the line from C to P and in the direction of increasing θ; i.e., the same direction ω is turning around point C.

Note that while the absolute velocity of the instantaneous center is zero, this does not imply either (1) that the acceleration of the instantaneous center is zero or (2) that the instantaneous center remains fixed relative to the rigid body; thus the name *instantaneous center of rotation*.

The nonparallel velocities of points A and B in Fig. 5.25a made it easy to locate point C by constructing lines perpendicular to the velocities. When the two velocities are parallel and perpendicular to the line joining the points (Fig. 5.25c), the instantaneous center can be located by the fact that C must lie on line ABC and Eqs. (5.54) and (5.59) must hold so that

$$\omega = v_A/r_A = v_B/r_B = (v_B - v_A)/d \tag{c}$$

From Eq. (c) it is seen that ω is zero when $v_A = v_B$ so that r_A and r_B become infinite. This means that the instantaneous center for a translating rigid body is at infinity.

EXAMPLE 5 18

The 10-in. diameter wheel in Fig. E5.18a has an angular velocity of 12 rad/s counterclockwise and is rolling without slipping on the horizontal surface. Bar AB is pinned to the periphery of the wheel at A and has end B sliding along the horizontal surface. Determine (a) the velocity of point A, (b) the angular velocity of bar AB, and (c) the velocity of point B by the method of instantaneous centers.

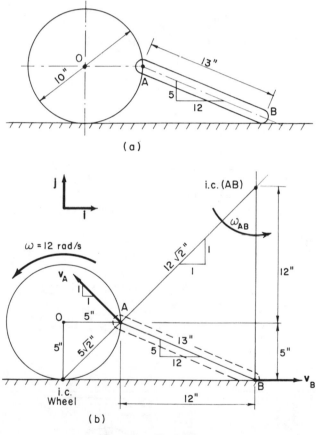

Fig. E5.18

SOLUTION

(a) The velocity of point A can be obtained immediately from recognizing that the point of contact is the instantaneous center for the nonslipping wheel (Fig. E5.18b). Thus

$$\mathbf{v}_A = 5\sqrt{2}\,(12)\mathbf{e}_{tA} = 84.9 \text{ ips} \qquad \textit{Ans.}$$

(b) To find ω_{AB}, we must know where the instantaneous center (i.c.) of body AB is located. This location is obtained by constructing lines perpendicular to $\mathbf{v}_A$ and $\mathbf{v}_B$ (which must be horizontal) (Fig. E5.18b). Thus the velocity of A can be given by two expressions of

$$\mathbf{v}_A = 5\sqrt{2}\omega\,\mathbf{e}_{tA} = -12\sqrt{2}\omega_{AB}\mathbf{e}_{tA}$$

so that

$$\omega_{AB} = (-5/12)\,\omega = 5 \text{ rad/s} \,\rangle \qquad \textit{Ans.}$$

(c) The velocity of B is then obtained from the instantaneous center of AB and ω_{AB} to be

$$\mathbf{v}_B = (5 + 12)\,\omega_{AB}\mathbf{i} = 17(-5)\mathbf{i} = -85\mathbf{i} = 85 \text{ ips} \leftarrow \qquad \textit{Ans.}$$

EXAMPLE 5.19

The mechanism in Fig. E5.19a consists of two bars AD and BDC pinned together at D. When in the position shown, the velocity of point C is observed to be $\mathbf{v}_C = 24\mathbf{i} + 18\mathbf{j}$ fps. Determine for this position (a) the angular velocity of member BDC, (b) the velocity of D, and (c) the angular velocity of member AD.

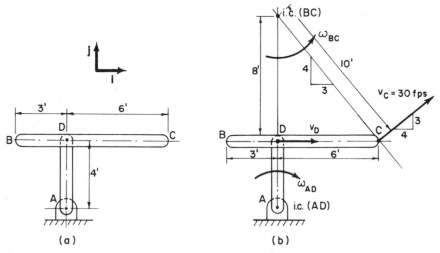

Fig. E5.19

SOLUTION

(a) From the given data

$$v_C = 30 \text{ fps} \quad \overset{3}{\underset{4}{\diagup}}$$

and v_D is horizontal so that the instantaneous center of member BDC can be located as shown in Fig. E5.19b. Thus

$$\omega_{BC} = v_C/r_C = 30/10 = 3.0 \text{ rad/s} \, \rotatebox{0}{)} \qquad\qquad Ans.$$

(b) The velocity of point D becomes

$$v_D = r_D \omega_{BC} \mathbf{i} = 8(3)\mathbf{i} = 24.0 \text{ fps} \rightarrow \qquad\qquad Ans.$$

(c) Recognizing that A is an instantaneous center for member AD,

$$\omega_{AD} = v_D/r_{DA} = 24.0/4.0 = 6 \text{ rad/s} \, \rotatebox{0}{)} \qquad\qquad Ans.$$

An alternative method of solving for velocities in rigid bodies is to employ the concept of instantaneous centers of rotation as demonstrated here. However, we must emphasize that instantaneous centers of rotation are accelerating points and hence can never be used in calculating accelerations. They may be used only for calculating velocities.

5.9   Acceleration of a Rigid Body in Plane Motion

Consider again the rigid body in plane motion in Fig. 5.22a, where points 0 and P are any two points of the body lying in the shaded plane, which is perpendicular to ω and α and parallel to the plane of motion. The position relationship for these points is illustrated in Fig. 5.22b from which vector addition gave

$$\mathbf{r}_P = \mathbf{r}_0 + \mathbf{r}_{P/0} \qquad\qquad (a)$$

from which the rigid-body relative velocity equation was obtained to be

$$\mathbf{v}_P = \mathbf{v}_0 + \mathbf{v}_{P/0} = \mathbf{v}_0 + r\omega\mathbf{e}_t \qquad\qquad (b)$$

where $\mathbf{v}_0$ = the velocity of point 0
  $\mathbf{v}_{P/0}$ = the velocity of P with respect to 0 and is due to the rotation of P around 0 in a circular path of radius r
  ω = the angular velocity of the body
  $\mathbf{e}_t$ = a unit vector tangent to the circular path of radius r and in the direction of ω

The acceleration of point P can be obtained by differentiating Eq. (b) with respect to time, giving

$$\mathbf{a}_P = \mathbf{a}_0 + \mathbf{a}_{P/0} = \mathbf{a}_0 + r\dot{\omega}\mathbf{e}_t + r\omega\dot{\mathbf{e}}_t \qquad (c)$$

where $\dot{\omega} = \alpha$, and from Eq. (5.30) we have $\dot{\mathbf{e}}_t = \omega\mathbf{e}_n$, so that Eq. (c) can be written as

$$\mathbf{a}_P = \mathbf{a}_0 + \mathbf{a}_{P/0} = \mathbf{a}_0 + r\alpha\mathbf{e}_t + r\omega^2\mathbf{e}_n \qquad (5.60)$$

Equation (5.60) simply states that the acceleration of P is equal to the acceleration of 0 plus the acceleration of P with respect to 0 and is called the *rigid body relative acceleration equation.*

A physical interpretation of Eq. (5.60) can be obtained by considering plane motion to be the superposition of translation and pure rotation as with the rigid-body relative velocity equation. Figure 5.26a illustrates the rigid body in plane motion at a given instant when point 0 has an acceleration of $\mathbf{a}_0$; the body has an angular velocity and acceleration of ω and α respectively, the relative position vector $\mathbf{r}_{P/0}$ has orientation θ with respect to the nonrotating x axis, and $\mathbf{a}_P$ is the acceleration of point P.

If the rigid body is translating only as shown in Fig. 5.26b, θ is constant so that $\omega_T = \alpha_T = 0$ and the orientation and magnitude of $\mathbf{r}_{P/0}$ is constant, which implies $\dot{\mathbf{r}}_{P/0} = \ddot{\mathbf{r}}_{P/0} = (\mathbf{a}_{P/0})_T = 0$. Thus

$$\mathbf{a}_{PT} = \mathbf{a}_0 \qquad (d)$$

which states that the acceleration of every point in a translating rigid body is the same.

If the rigid body has pure rotation only about the fixed z axis through point 0 (Fig. 5.26c), $(\mathbf{a}_0)_R$ is zero and point P travels in a circular path of radius r around point 0. For this case we have shown that the acceleration of P relative to 0 is composed of a tangential component ($\mathbf{a}_t = r\alpha\mathbf{e}_t$ due to a change in magnitude of $\mathbf{v}_{P/0}$) and a normal component ($\mathbf{a}_n = r\omega^2\mathbf{e}_n$ due to a change in direction of $\mathbf{v}_{P/0}$). Thus the acceleration of P with respect to 0 due to rotation only (as seen by the nonrotating observer attached at 0) is composed of the two terms

$$\mathbf{a}_{PR} = \mathbf{a}_{P/0} = r\alpha\mathbf{e}_t + r\omega^2\mathbf{e}_n \qquad (5.61)$$

as shown in Fig. 5.26c. The vector addition of terms as required by Eqs. (5.60) and (5.61) and shown in Fig. 5.26d can be done in terms of any convenient coordinate system such as the xy coordinates shown.

The relative acceleration given by Eq. (5.61) may also be written

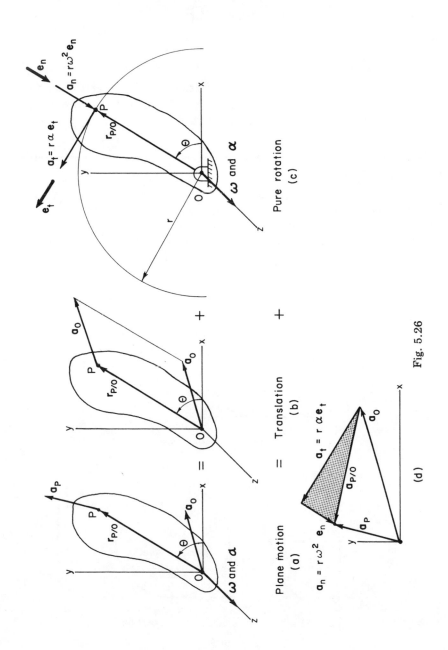

Plane motion
(a)

= Translation
(b)

+ Pure rotation
(c)

$a_n = r \omega^2 e_n$

$a_t = r \alpha e_t$

$a_n = r\omega^2 e_n$

$a_t = r \alpha e_t$

(d)

Fig. 5.26

in terms of vector cross products; i.e.,

$$\mathbf{a}_t = r\alpha\mathbf{e}_t = \alpha \times \mathbf{r}_{P/0} \qquad (e)$$

and

$$\mathbf{a}_n = r\omega^2\mathbf{e}_n = \omega \times (\omega \times \mathbf{r}_{P/0}) = \omega \times \mathbf{v}_{P/0} \qquad (f)$$

as seen from Fig. 5.26c. There is no particular advantage in using the vector cross product forms given by Eqs. (e) and (f) for plane motion problems. The following examples illustrate the use of Eq. (5.60) for calculating accelerations.

EXAMPLE 5.20

The position of the center of the gear in Fig. E5.20a is given by $s = 10t^3 - 40t^2$ (s in centimetres when t is in seconds). The gear has a velocity to the right when $t = 1.0$ s and is in the position shown. Determine the acceleration of points A and P (point of contact) at this instant. Assume the gear rolls without slipping.

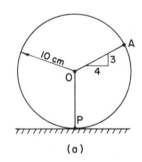

(a)

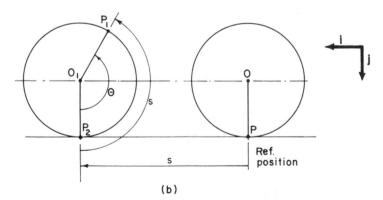

(b)

Fig. E5.20

SOLUTION

To calculate acceleration of points A and P, we must know the acceleration of point 0 as well as the angular velocity and acceleration of the gear. However, before these quantities can be established, we also must know whether s is positive to the right or the left. This can be done by knowing that the gear is moving to the right at $t = 1.0$ s. From the velocity,

$$v_0 = \frac{ds}{dt} = 30t^2 - 80t = -50 \text{ cm/s} \qquad (a)$$

which has a negative sign. Thus to the left is positive for s, v_0, and a_0. The acceleration of 0 is obtained from Eq. (a) to be

$$a_0 = \frac{dv_0}{dt} = 60t - 80 = 60 - 80 = -20 \text{ cm/s}^2 \qquad (b)$$

at $t = 1.0$ s.

From Fig. E5.20b the quantities s and θ are related by $s = r\theta$ so that $v_0 = r\omega$ and $a_0 = r\alpha$ from which

$$\omega = v_0/r = 3t^2 - 8t = 3 - 8 = -5 \text{ rad/s} = 5 \text{ rad/s} \,\big)$$

and

$$\alpha = a_0/r = 6t - 8 = 6 - 8 = -2 \text{ rad/s}^2 = 2 \text{ rad/s}^2 \,\big)$$

for $t = 1.0$ s and $r = 10$ cm since θ is positive counterclockwise.

The acceleration of point A becomes

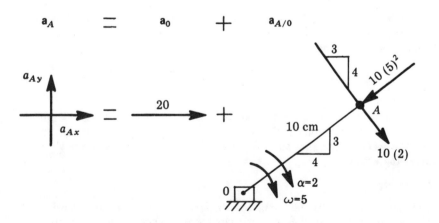

$$(\rightarrow) \quad a_{Ax} = 20 + (3/5)(10)(2) - (4/5)(10)(25) = 20 + 12 - 200 = -168$$

$$a_{Ax} = 168 \text{ cm/s}^2 \leftarrow$$

$\uparrow$ $a_{Ay} = 0 - (4/5)(10)(2) - (3/5)(10)(25) = -16 - 150 = -166$

$a_{Ay} = 166 \text{ cm/s}^2 \downarrow$

or

$\mathbf{a}_A = 168\mathbf{i} + 166\mathbf{j} \text{ cm/s}^2$ *Ans.*

The acceleration of point P becomes

$\mathbf{a}_p \quad = \quad \mathbf{a}_0 \quad + \quad \mathbf{a}_{p/0}$

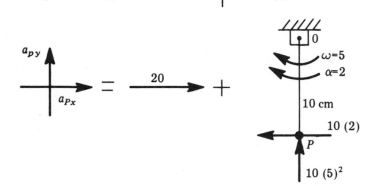

$\xrightarrow{}$ $a_{Px} = 20 - 20 = 0$, $\uparrow$ $a_{Py} = 0 + 250 = 250\uparrow$

or

$\mathbf{a}_P = 0\mathbf{i} - 250\mathbf{j} \text{ cm/s}^2$ *Ans.*

Note that point P is an instantaneous center of zero velocity and is accelerating; however, the acceleration is zero in the horizontal direction since the wheel is rolling without slipping.

EXAMPLE 5.21

Body B in Fig. E5.21 is attached to a cord wrapped around the wheel. Determine (a) the relationship between (y, s_0, θ), (v_B, v_0, ω) and (a_B, a_0, α) when the wheel rolls and slips and (b) the acceleration of point A where the cord is leaving the wheel.

SOLUTION

(a) Let us adopt the following convention: y is positive downward, s_0 is positive to the right, and θ is positive clockwise. Then y can be obtained as follows. Let the wheel translate s_0 units to the right, during which B moves downward s_0 units so that $y_T = s_0$. Now, let the wheel rotate through an angle θ about an axis through 0 so that $r\theta$ units of cord unwind and B moves downward $r\theta$ units so $y_R = r\theta$. Thus for a rolling and slipping wheel we have the position, velocity, and acceleration relationships of B given by

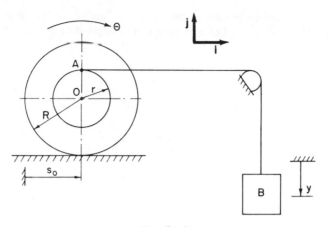

Fig. E5.21

$$y = y_T + y_R = s_0 + r\theta, \qquad v_B = \dot{y} = v_0 + r\omega \qquad\qquad Ans.\ (a)$$

and

$$a_B = \ddot{y} = a_0 + r\alpha \qquad\qquad Ans.\ (b)$$

When the wheel rolls without slipping, as shown before,

$$s_0 = R\theta, \qquad v_0 = R\omega, \qquad a_0 = R\alpha \quad \text{(no-slip relationships)} \qquad (c)$$

must hold.

(b) The acceleration of point A can be obtained from

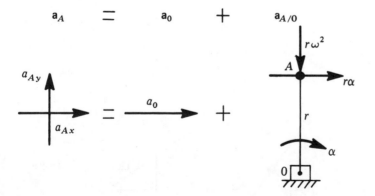

$$\mathbf{a}_A = (a_0 + r\alpha)\mathbf{i} - r\omega^2\mathbf{j} \qquad\qquad Ans.$$

from which a comparison with Eq. (b) shows that the acceleration of B is the same as the horizontal (tangential) acceleration component of point A on the wheel. When the wheel rolls without slipping, the acceleration of 0 is given by $R\alpha$.

The relationships developed in this example apply to the geometry described. However, the method of approach can be applied to other situations.

EXAMPLE 5.22

The mechanism in Fig. E5.22 consists of two bars pinned together at B. When in the position shown, the velocity of point C is to the left, bar AB has an angular velocity of 12 rad/s counterclockwise and an angular acceleration of 2 rad/s² clockwise, and bar BC has an angular acceleration of 10 rad/s² counterclockwise. Determine the acceleration of point C for this instant.

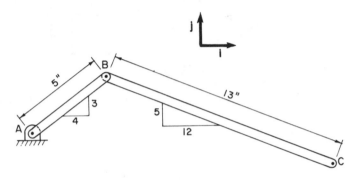

Fig. E5.22

SOLUTION

To determine the acceleration of point C, we must know the angular velocity of bar BC, which can be determined from

$$\mathbf{v}_C = \mathbf{v}_B + \mathbf{v}_{C/B} = \mathbf{v}_A + \mathbf{v}_{B/A} + \mathbf{v}_{C/B} \tag{a}$$

where $\mathbf{v}_A = 0$. Thus

$$\mathbf{v}_C \quad = \quad \mathbf{v}_{B/A} \quad + \quad \mathbf{v}_{C/B}$$

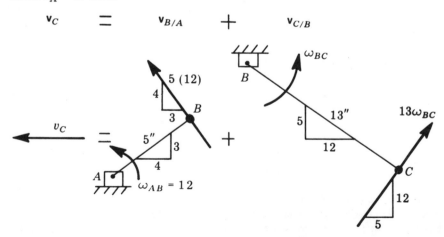

$$\updownarrow \; 0 = (4/5)(5)(12) + (12/13)(13)\omega_{BC}$$

or

$$\omega_{BC} = -48/12 = -4 \text{ rad/s} = 4 \text{ rad/s} \downharpoonright$$

The acceleration of point C can be obtained from

$$\mathbf{a}_C = \mathbf{a}_B + \mathbf{a}_{C/B} = \mathbf{a}_A + \mathbf{a}_{B/A} + \mathbf{a}_{C/B} \tag{b}$$

where $\mathbf{a}_A = 0$ so that

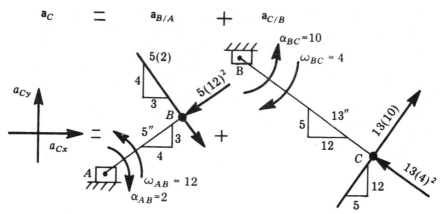

$$\longleftrightarrow a_{Cx} = -(4/5)(5)(12)^2 + (3/5)(5)(2) + (5/13)(13)(10) - (12/13)(13)(4)^2$$
$$= -576 + 6 + 50 - 192 = -712 \text{ ips}^2$$

and

$$\updownarrow a_{Cy} = -(3/5)(5)(12)^2 - (4/5)(5)(2) + (12/13)(13)(10) + (5/13)(13)(4)^2$$
$$= -432 - 8 + 120 + 80 = -240 \text{ ips}^2$$

from which we obtain

$$\mathbf{a}_C = -712\mathbf{i} - 240\mathbf{j} \text{ ips}^2 = 751 \,(-0.947\mathbf{i} - 0.320\mathbf{j}) \text{ ips}^2 \qquad \textit{Ans.}$$

This example illustrates how the rigid-body relative velocity and relative acceleration equations can be applied in sequence to various points in the body as shown in Eqs. (a) and (b).

5.10 Kinematics of Fluid Motion

We could view a system such as a flowing fluid, as being composed of an infinite number of discrete particles and thereby use the various relationships developed in Section 5.2 to describe the motion of any one particle. However, this approach is not satisfactory since a separate relationship would be required for each of the particles. Furthermore, there is no simple equation that relates the motion of one particle of the system to some other particle as for rigid-body motion. To overcome this difficulty the motion is characterized by defining the velocity at all spatial points within the system rather than looking at particular particles. Thus, the velocity $\mathbf{q}$ may be a function of position $\mathbf{r}$ and time t so that

$$\mathbf{q} = f(\mathbf{r}, t) \qquad (5.62)$$

When a particle is located at the position $\mathbf{r}$, it will have a velocity $\mathbf{q}$ at some time t. However, at a later instant of time some other particle will occupy this position $\mathbf{r}$ and have the velocity specified by Eq. (5.62). This method of describing a fluid motion is commonly called the *Eulerian method*. A useful pictorial representation of a flowing fluid can be obtained by drawing lines in the fluid so that the tangents at all points along the line coincide with the velocity vectors at these points (Fig. 5.27); these are called *streamlines*.

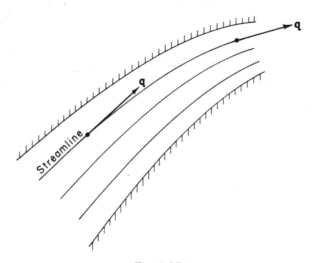

Fig. 5.27

The velocity $\mathbf{q}$ can be expressed in terms of its rectangular components; i.e.,

$$\mathbf{q} = u\mathbf{i} + v\mathbf{j} + w\mathbf{k} \qquad (5.63)$$

where u, v, and w are the components of velocity in the x, y, and z directions respectively. Note that

$$u = u(x, y, z, t), \qquad v = v(x, y, z, t), \qquad w = w(x, y, z, t) \qquad (5.64)$$

As a fluid particle moves from point P to point P' (Fig. 5.28), its

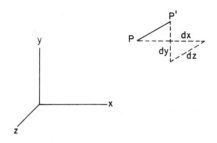

Fig. 5.28

velocity will change due to its change in position and the passage of time. We can express this change in the x component of velocity as

$$du = \frac{\partial u}{\partial x} \, dx + \frac{\partial u}{\partial y} \, dy + \frac{\partial u}{\partial z} \, dz + \frac{\partial u}{\partial t} \, dt$$

If we now divide each term in the equation by dt, we have an expression for the x component of the acceleration of the particle

$$a_x = \frac{du}{dt} = \frac{\partial u}{\partial x} \frac{dx}{dt} + \frac{\partial u}{\partial y} \frac{dy}{dt} + \frac{\partial u}{\partial z} \frac{dz}{dt} + \frac{\partial u}{\partial t} \qquad (5.65)$$

However, the time derivatives of the space variables in this equation represent the components of velocity of the particle; i.e.,

$$u = \frac{dx}{dt}, \qquad v = \frac{dy}{dt}, \qquad w = \frac{dx}{dt}$$

Thus

$$a_x = u \frac{\partial u}{\partial x} + v \frac{\partial u}{\partial y} + w \frac{\partial u}{\partial z} + \frac{\partial u}{\partial t} \qquad (5.66a)$$

Similarly,

$$a_y = u \frac{\partial v}{\partial x} + v \frac{\partial v}{\partial y} + w \frac{\partial v}{\partial z} + \frac{\partial v}{\partial t} \qquad (5.66b)$$

and

$$a_z = u \frac{\partial w}{\partial x} + v \frac{\partial w}{\partial y} + w \frac{\partial w}{\partial z} + \frac{\partial w}{\partial t} \qquad (5.66c)$$

We can conveniently express each of the acceleration components in vector notation as

$$a_x = \mathbf{q} \cdot \nabla u + \frac{\partial u}{\partial t}, \qquad a_y = \mathbf{q} \cdot \nabla v + \frac{\partial v}{\partial t}, \qquad a_z = \mathbf{q} \cdot \nabla w + \frac{\partial w}{\partial t}$$

or

$$\mathbf{a} = (\mathbf{q} \cdot \nabla)\mathbf{q} + \frac{\partial \mathbf{q}}{\partial t} \tag{5.67}$$

where $\mathbf{q} \cdot \nabla$ is the scalar operator

$$\mathbf{q} \cdot \nabla = u \frac{\partial}{\partial x} + v \frac{\partial}{\partial y} + w \frac{\partial}{\partial z}$$

The first term on the right side of Eq. 5.67 is the *convective acceleration* since it is due to velocity changes arising from a change in position of a fluid particle. The last term in Eq. 5.67 is the *local acceleration* and represents the time rate of change of velocity at a point. Note that if the velocity is steady; i.e., does not depend on time, the local acceleration is zero. This method of describing the velocity and acceleration of particles is particularly valuable in the study of fluid mechanics.

The change in any characteristic of the flow field that can be "convected" with the fluid can be expressed in the form of Eq. (5.66a). For example, the rate of change in the temperature T of a fluid particle as it moves with the fluid can be expressed as

$$\frac{dT}{dt} = u \frac{\partial T}{\partial x} + v \frac{\partial T}{\partial y} + w \frac{\partial T}{\partial z} + \frac{\partial T}{\partial t} \tag{5.68}$$

Since this derivative represents the rate of change of a given quantity as it moves with the fluid, it is commonly referred to as the *material derivative* and denoted with the symbol D/Dt; i.e.,

$$\frac{D}{Dt} = u \frac{\partial}{\partial x} + v \frac{\partial}{\partial y} + w \frac{\partial}{\partial z} + \frac{\partial}{\partial t}$$

EXAMPLE 5.23

The velocity in a certain two-dimensional flow field is given by the expression $\mathbf{q} = 2xt\mathbf{i} - 2yt\mathbf{j}$ where the velocity is in fps when x, y, and t are measured in feet and seconds respectively. Determine (a) the local and convective components of acceleration in the x direction and (b) the magnitude and direction of the acceleration at the point $x = y = 1$ ft at the time $t = 0$ s.

SOLUTION

(a) The x component of acceleration is given by the equation

$$a_x = u \frac{\partial u}{\partial x} + v \frac{\partial u}{\partial y} + \frac{\partial u}{\partial t}$$

Thus

$$a_x \text{ (convective)} = 2xt(2t) - 2yt(0) = 4xt^2 \qquad \text{Ans.}$$

and

$$a_x \text{ (local)} = 2x \qquad \text{Ans.}$$

(b) At $t = 0, a_x = 2x$. The y component of the acceleration is

$$a_y = u \frac{\partial v}{\partial x} + v \frac{\partial v}{\partial y} + \frac{\partial v}{\partial t} = 4yt^2 - 2y$$

and at $t = 0, a_y = -2y$. Thus for $x = y = 1$ ft

$$a = 2i - 2j \text{ fps}^2 \qquad \text{Ans.}$$

or

$$a = 2.82 \text{ fps}^2 \quad \text{\raisebox{0pt}{\tiny 1}} \qquad \text{Ans.}$$

PROBLEMS

For all kinematics problems in this chapter the student should assume that all ropes (cords, cables, etc.) are inextensible. In stating answers to the problems, the i and j unit vectors are assumed to the right and upward unless indicated otherwise in the problem figure.

5.1. The path of a particle moving in space is described by

$$r = (5t^2 + 2t + 3)i + (t + 4)j + (2e^t)k$$

where r is in feet when t is in seconds. Determine the position, velocity, and acceleration when $t = 2.0$ s.

Ans. $r = 27i + 6j + 14.8k$ ft, $\qquad v = 22i + 1j + 14.8k$ fps

$a = 10i + 0j + 14.8k$ fps^2

5.2. A particle moving in the xy plane has a velocity of

$$v = 2i + (4t - 1)j$$

where v is in ips when t is in seconds. The particle is located at $r = 4i + 3j$ in. when $t = 1$ s. Determine the equation of the path in terms of x and y.

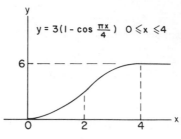

$$y = 3(1 - \cos \tfrac{\pi x}{4}) \quad 0 \leqslant x \leqslant 4$$

Fig. P5.3

5.3. A particle moving in the xy plane is to follow the path $y = 3[1 - \cos(\pi x/4)]$ for $0 \leqslant x \leqslant 4$ and a horizontal path for $x < 0$ and $x > 4$ (Fig. P5.3). The x position of the particle is given by $x = t^3 - 3t$ where x is in centimetres when t is in seconds. Determine the position, velocity, and acceleration of the particle when $t = 2.0$ s.

Ans. $\mathbf{r} = 2\mathbf{i} + 3\mathbf{j}$ cm, $\mathbf{v} = 9\mathbf{i} + 27.2\mathbf{j}$ cm/s, $\mathbf{a} = 12\mathbf{i} + 28.3\mathbf{j}$ cm/s^2

5.4. An inextensible string is connected to the wall at point A, is 11 m long, and 1 m is wrapped around pulley B (Fig. P5.4). The center of the pulley

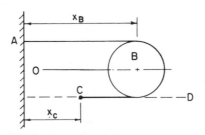

Fig. P5.4

moves along the straight line OB and has its position given by $x_B = (t^3 - 4t^2 + 3)$ where x_B is in metres when t is in seconds. Assuming that the string remains taut, determine the position, velocity, and acceleration of point C which moves rectilinearly along line CD when $t = 4$ s.

5.5. A particle in rectilinear motion has an initial velocity of 90 ips to the right and is given an acceleration of 15 ips^2 to the left for a period of 10 s. Determine: (a) the velocity at time $t = 10$ s, (b) the linear displacement from $t = 4$ s to $t = 10$ s, and (c) the total distance traveled during the 10-s time interval.

Ans. (a) $\mathbf{v} = 60$ ips $\leftarrow$,   (b) $\mathbf{u} = 90$ in. $\leftarrow$,   (c) $S = 390$ in.

5.6. A particle moves along a horizontal axis such that $v = At^2 - 4$, where v is in fps when t is in seconds and A is a constant. When $t = 0$ s, the velocity is 4 fps to the right. When $t = 1$ s, the acceleration is 6 fps^2 to the left and the position of the particle is 5 ft to the right of the origin. Determine the position of the particle when $t = 3$ s.

5.7. The angular velocity of a line rotating in a vertical plane is given by $\omega = 3t^2 - 8t$ where ω is in rad/s when t is in seconds. When $t = 2$ s, the angular velocity is counterclockwise and the angular displacement is 2 rad clockwise. Determine the angular displacement and angular acceleration when $t = 3$ s.

Ans. $\theta = 1.0$ rad $\circlearrowright$, $\alpha = 10.0$ rad/s^2 $\circlearrowright$

5.8. The angular velocity of a line moving in a horizontal plane varies as in Fig. P5.8 where clockwise is positive. Determine the angular displacement and the total angle turned during the first 16 s.

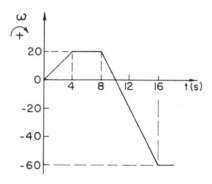

Fig. P5.8

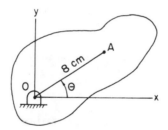

Fig. P5.9

5.9. The rigid body in Fig. P5.9 rotates about an axis through 0 with an angular velocity given by $\omega = 6t - 2t^2$ where ω is in rad/s (counterclockwise positive) when t is in seconds. When $t = 2$ s, $\theta = 30°$. Determine the velocity and acceleration of point A in the body for this position by setting up a position function and differentiating.

Ans. $\mathbf{v} = -16\mathbf{i} + 27.7\mathbf{j}$ cm/s, $\mathbf{a} = -103\mathbf{i} - 77.9\mathbf{j}$ cm/s^2

5.10. The crank arm OA in Fig. P5.10 rotates with a constant angular velocity of 6 rad/s clockwise. Set up an expression for the position of piston B and determine the velocity of piston B when θ is 30° counterclockwise by differentiation of the position function.

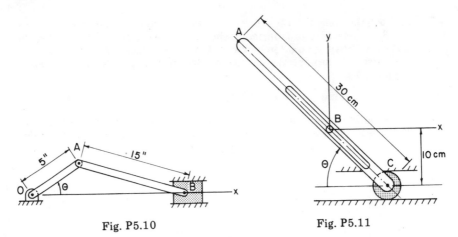

Fig. P5.10   Fig. P5.11

5.11. Bar ABC in Fig. P5.11 moves with end C along a fixed horizontal line, while the slot in ABC slides along the fixed pin at B. When $\theta = 30°$, the velocity of C is $v_c = 120$ cm/s to the left. Determine (a) the angular velocity of ABC and (b) the velocity of end A when in this position by setting up position functions and differentiating.

Ans. (a) $\omega = 3$ rad/s $\circlearrowright$,   (b) $v_A = -75i + 78j$ cm/s

5.12. Show that the radius of curvature ρ for any plane curve $y = y(x)$ as in Fig. P5.12 is given by

$$\rho = \frac{[1 + (dy/dx)^2]^{3/2}}{d^2y/dx^2}$$

which can be evaluated at any point on the curve.

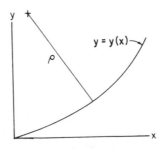

Fig. P5.12

5.13. A particle is traveling on a fixed curve in the xy plane given by $y = 2x^2$ where x and y are in feet. Determine (a) the radius of curvature when $x = 0$, 2, and 4 ft, and (b) the velocity when $t = 2.0$ s if $x = 2$ ft and $v = 10t^2 - 3$ where v is in fps when t is in seconds.

Ans. (a) ρ = 0.25, 131, and 1027 ft. See Problem 5.12

(b) **v** = 37(0.124**i** + 0.992**j**) fps

5.14. Particle P travels in a circular path about point 0 with a radius of 2 ft (Fig. P5.14). The angle θ is given by $\theta = (\pi/12)(t^2 - 2t)$, where θ is in radians when t is in seconds. When $t = 3.0$ s, determine (a) the velocity, (b) the acceleration, (c) and the radius of curvature.

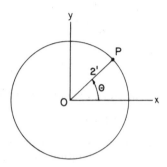

Fig. P5.14

5.15. The position of a particle moving on a circular path is given by $s = 6 + 3t - 9t^2$ where s is measured from a convenient reference, and s is in metres when t is in seconds. When $t = 1$ s, the normal acceleration is 25 m/s^2. Determine the magnitude of the acceleration and the radius of the circle.

Ans. |**a**| = 30.8 m/s^2, ρ = 9.0 m

5.16. The acceleration of a particle moving in the xy plane is given by **a** = $3t^2$**i** + 3t**j** where **a** is in in./s^2 when t is in seconds. When $t = 0$, **v** = 0. When $t = 2$ s, determine (a) the velocity, (b) the tangential acceleration, and (c) the radius of curvature.

5.17. The angular velocity of bar AB in Fig. P5.17 varies according to $\omega = 3t^2 - 8t$ where ω is in rad/s when t is in seconds. The bar is in the position shown and is rotating counterclockwise when $t = 2.0$ s. Determine (a) the acceleration of point B when $t = 2.0$ s and (b) the angular displacement of the bar during the time interval $t = 2$–4 s.

Ans. (a) **a** = -64**i** + 16**j** fps^2, (b) $\theta = 8$ rad $\downarrow$

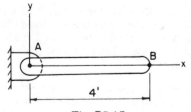

Fig. P5.17

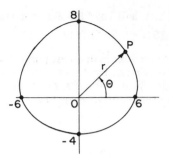

Fig. P5.18

5.18. Particle P moves in a plane along the curve described by

$$r = 6 + 2 \sin \theta, \qquad \theta = 2\pi t^2$$

as in Fig. P5.18 where r is in inches, θ is in radians, and t is in seconds. Determine the velocity and acceleration of P when $t = 1$ s.

5.19. Particle P travels in a circular path of radius 6 m (Fig. P5.19). The angular position of the particle is given by $\theta = 4\pi(e^{-t/2} - 1)$. Determine (a) the position, (b) the velocity, (c) the acceleration of the particle at $t = 2.0$ s.

Ans. (a) $\theta = 7.93$ rad $\big\downarrow$, (b) $\mathbf{v}_p = -13.85\, \mathbf{e}_\theta$ m/s
(c) $\mathbf{a}_P = -32.0\, \mathbf{e}_r + 6.94\, \mathbf{e}_\theta$ m/s^2

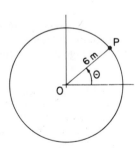

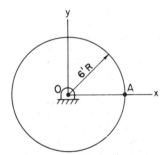

Fig. P5.19   Fig. P5.20

5.20. A child is running radially from the center 0 of a merry-go-round with a constant speed of 2 fps relative to the platform (Fig. P5.20). The merry-go-round has a constant angular velocity of 2 rad/s clockwise. Determine (a) the velocity and (b) the acceleration of the child when located at the edge of the platform at A.

5.21. A particle P is moving in space so that $r = 6t^2$, $\theta = 2t^2 - 3$, and $z = 9t^4$ where r and z are in metres and θ is in radians when t is in seconds. Determine (a) the velocity and (b) the acceleration of P for any time t.

Ans.   (a) $v_P = (12t)e_r + (24t^3)e_\theta + (36t^3)k$ m/s

      (b) $a_P = (12 - 96t^4)e_r + (120t^2)e_\theta + (108t^2)k$ m/s

5.22. The position s of particle P along tube AB in Fig. P5.22 is given by $s = 5t^3 - 10t^2 + 10$ where s is in inches when t is in seconds. The slope of tube AB with respect to the xy plane remains 3 vertical and 4 horizontal as the tube rotates about the z axis through A. The angular position of tube AB is given by $\theta = 2t - 4t^2$ where θ is in radians when t is in seconds. Determine (a) an expression for the position vector in cylindrical coordinates, (b) the velocity of P when $t = 1.0$ s, and (c) the acceleration of P when $t = 1.0$ s.

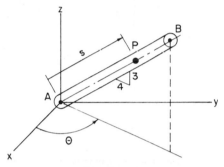

Fig. P5.22

5.23. Car A is approaching a highway intersection with a constant speed of 60 mph going north, while car B is approaching with a constant speed of 45 mph going west. Determine (a) the velocity of car B relative to car A and (b) the displacement of car A relative to car B during a 2.0 s time interval.

Ans.   (a) $v_{B/A} = -66i - 88j = 110 \underset{3}{\measuredangle}^4$ fps,   (b) $r_{A/B} = 220 \underset{3}{\measuredangle}^4$ ft

5.24. A motorboat is to cross the half-mile-wide river from point A to B (Fig. P5.24). The motorboat can maintain a speed of 10 mph relative to the water while the river flows with an average velocity of 2 mph. Determine (a) the direction the boat must head to follow a straight line from A to B, (b) the absolute velocity of the boat, and (c) how long it will take the boat to cross the river.

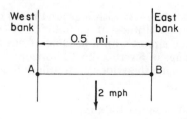

Fig. P5.24

5.25. A defensive end A in Fig. P5.25 intercepts a flat pass and runs straight downfield with a constant speed of 8 yd/s. At the moment shown, the quarterback B takes up pursuit with a speed of 11 yd/s. Determine (a) the optimum angle θ so that the quarterback catches the end in the least amount of time, (b) the least amount of time, and (c) how far the end re- turned the intercepted pass.

Ans. (a) $\theta = 46.6°$, (b) $t = 2.91$ s, (c) $d = 23.3$ yd

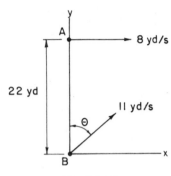

Fig. P5.25

5.26. An air particle is entering a fan blade with a velocity of 400 fps as in Fig. P5.26 where the fan blade has a velocity of 120 fps. The angle β de- fines the tangent line of the blade. Determine (a) the angle β and (b) the velocity of the air particle relative to the blade if the relative velocity of the air is to be tangent to the blade.

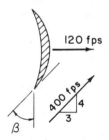

Fig. P5.26

5.27. Jet airliner A is heading east with a ground speed of 550 mph as it passes over an airport at an altitude of 25,000 ft. A second jet airliner B is climbing out of the airport with a heading of due south and a velocity of 260 mph with a slope of 5 vertical to 12 horizontal when it has an altitude of 4000 ft. Determine the velocity of airliner B as observed by the pilot of airliner A.

Ans. $v_{B/A} = -550i - 240j + 100k$ mph

5.28. A lawnmower is operating on a level lawn and is moving in a straight line. Classify the motion of the mower wheels, the mower frame, and the cutting blade if the mower is (a) a rotary type and (b) a reel type.

5.29. The wheel of radius r in Fig. P5.29 rolls without slipping along a straight line on the horizontal surface. Determine the linear displacement of point A in the wheel when the wheel turns through an angle of $270°$ clockwise.

Ans. $u = 5.71ri - rj$

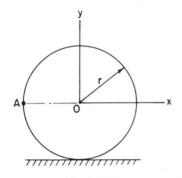

Fig. P5.29

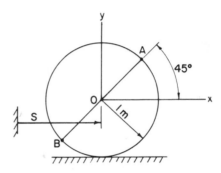

Fig. P5.30

5.30. A 2-m diameter wheel rolls without slipping on the horizontal surface in Fig. P5.30, where the position of the center of the wheel is given by $s = 2t^3 - t^2$ where s is in metres when t is in seconds. When $t = 1.0$ s, point A is located on the rim of the wheel as shown. Determine the velocity of point A and (b) the velocity of point A with respect to B when $t = 1.0$ s.

5.31. The 6-ft diameter wheel A rotates about the fixed z axis through 0 and is connected to body B by an inextensible cord wrapped around body A, which passes over pulley C (Fig. P5.31). Determine (a) the kinematic relationship between the angular displacement of body A and the linear displacement of body B, (b) the angular velocity ω of body A and the linear velocity of body B, and (c) the type of motion bodies A and B have.

Ans.   (a) $y = -3\theta$,   (b) $v = -3\omega$

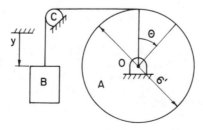

Fig. P5.31

5.32. The 4-ft diameter wheel A rotates about the horizontal axis through 0 and is connected to body B by an inextensible cord passing over pulley C (Fig. P5.32). Body B has a downward velocity of 20 fps at the instant shown. For this instant determine (a) the angular velocity of body A, (b) the velocity of point P in body A, and (c) the velocity of point P with respect to body B.

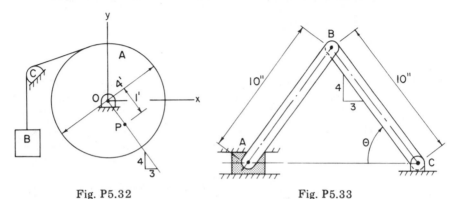

Fig. P5.32   Fig. P5.33

5.33. Body A in Fig. P5.33 is moving to the left with a velocity 192 ips along the horizontal plane when in the position shown. Determine the angular velocity of members AB and BC.

$Ans.$ $\omega_{AB} = 12$ rad/s $\downdownarrows$, $\omega_{BC} = 12$ rad/s $\uparrow$

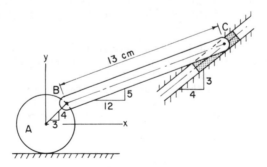

Fig. P5.34

5.34. The 6-cm diameter gear A in Fig. P5.34 rolls without slipping on the horizontal surface and is pinned to bar BC at B. End C of bar BC is pinned to a block at C which slides in the slot. When in the position shown, gear A has an angular velocity of 12 rad/s counterclockwise. Determine the angular velocity of bar BC and the velocity of the block at C.

5.35. Gear D has a diameter of 4 in. and rolls inside the 10-in. diameter ring gear with a constant angular velocity of 6 rad/s (Fig. P5.35). Determine

(a) the angular velocity of arm AB and (b) the velocity of point P when in the position shown.

Ans. (a) $\boldsymbol{\omega}_{AB} = 4$ rad/s $\circlearrowright$,   (b) $\mathbf{v}_P = 12\mathbf{i} - 12\mathbf{j} = 17$ $^1\!\diagdown_1$ ips

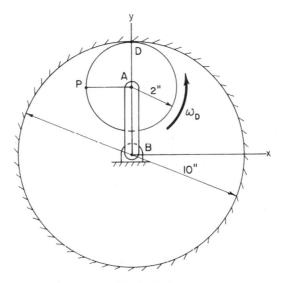

Fig. P5.35

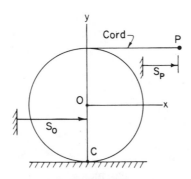

Fig. P5.36

5.36. The 1-m diameter wheel in Fig. P5.36 rolls and slips along the horizontal surface and has an inextensible cord wrapped around it. The position of point P is given by $s_P = t^3 - 8t$, while the position of the center of the wheel 0 is given by $s_0 = 3t^2 - t + 3$, where s_P and s_0 are in metres when t is in seconds. When $t = 1.0$ s, determine (a) the angular velocity of the wheel and (b) the velocity of the point of contact (on wheel) C.

5.37. Two bars AB and DE are pinned togehter at E and move with angular velocities of $\omega_{AB} = 5$ rad/s clockwise and $\omega_{DE} = 3$ rad/s counterclockwise in the vertical plane (Fig. P5.37). Determine (a) the location of the instantaneous center of bar AB and (b) the velocity of point A when in the position shown.

Ans. $v_A = 25 \;\angle\!^4_3$ m/s

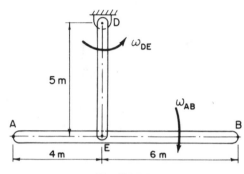

Fig. P5.37

5.38. A mechanism consists of three bars AB, BC, and CD that are pinned together and move in the vertical plane (Fig. P5.38). Bar AB is rotating with a constant angular velocity of 6 rad/s as shown. For the position shown determine (a) the instantaneous centers for bars AB, BC, and CD, (b) the angular velocity of bar BC, (c) the velocity of pin C, and (d) the angular velocity of bar CD.

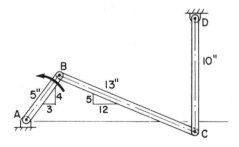

Fig. P5.38

5.39. Solve problem 5.33 by the method of instantaneous centers.

5.40. Solve problem 5.35 by the method of instantaneous centers.

5.41. Solve problem 5.36 by the method of instantaneous centers.

5.42. Two gears A and D are pinned to bar BC (Fig. P5.42). Gear A has a constant clockwise angular velocity of 2 rad/s, while bar BC has a constant clockwise angular velocity of 1 rad/s. For the position shown, determine (a) the location of the instantaneous center of gear D, (b) the angular velocity of gear D, and (c) the velocity of point E on gear D.

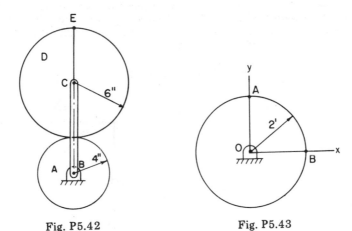

Fig. P5.42   Fig. P5.43

5.43. The 4-ft diameter wheel in Fig. P5.43 rotates about an axis through 0 with an angular velocity of 6 rad/s clockwise and an angular acceleration of 4 rad/s² counterclockwise. For the position shown determine (a) the acceleration of points A and B and (b) the acceleration of point A with respect to point B by two methods.

Ans. (a) $a_A = -8i - 72j$ fps², $a_B = -72i + 8j$ fps²

(b) $a_{A/B} = 64i - 80j$ fps²

5.44. The 10-cm diameter wheel in Fig. P5.44 rolls without slipping along the horizontal surface with the center of the wheel 0 moving according to $s = 3t^3 - 3t^2 - 14t + 19$, where s is in centimetres when t is in seconds. When $t = 2.0$ s, the wheel is in the position shown. For this position determine (a) the angular velocity and acceleration of the wheel, (b) the

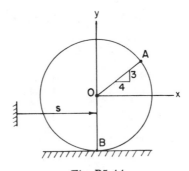

Fig. P5.44

acceleration of point A, and (c) the acceleration of the point of contact B (on wheel) at this instant.

5.45. The 4-ft diameter wheel A in Fig. P5.45 rotates about a fixed axis at 0 and is connected to body B by an inextensible cord passing over C. Body B has a downward velocity of 20 fps and an upward acceleration of 4 fps^2 when in the position shown. For this instant determine the velocity and the acceleration of point P in body A.

Ans. $v_p = 8i + 6j = 10 \underset{4}{\overset{3}{\diagup}}$ fps, $a_p = -61.6i + 78.8j$ fps^2

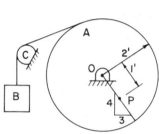 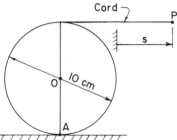

Fig. P5.45   Fig. P5.46

5.46. A 10-cm diameter wheel rolls without slipping along the horizontal surface and has an inextensible cord wrapped around it (Fig. P5.46). The position of point P in the cord is given by $s = 10t^3 - 80t$ where s is in centimetres when t is in seconds. When $t = 2$ s, determine (a) the angular velocity and angular acceleration of the wheel and (b) the velocity and acceleration of the point of contact A (which is on the wheel).

5.47. A mechanism consists of two bars AB and BC pinned together at B (Fig. P5.47). At a given instant the bars are oriented as shown; AB has an angular velocity of 2 rad/s clockwise and an angular acceleration of 6 rad/s^2 counterclockwise. For this instant determine the combination or combinations of angular velocity and angular acceleration of bar BC so that the acceleration of point C is zero.

Ans. $\omega_{BC} = 3$ rad/s $\circlearrowright$ or $\circlearrowleft$, $\alpha_{BC} = 6$ rad/s^2 $\circlearrowright$

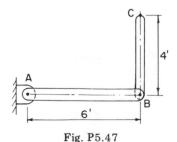

Fig. P5.47

5.48. The 20-ft ladder AB in Fig. P5.48 moves with its ends in contact with the vertical and horizontal surfaces. The velocity of point A is 6 fps downward, and the acceleration of point B is 12 fps^2 to the right when in the position shown. For the position shown determine (a) the angular velocity and acceleration of the ladder, (b) the velocity of point C, and (c) the acceleration of point C.

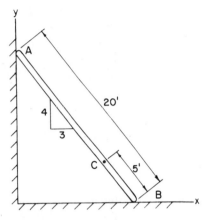

Fig. P5.48

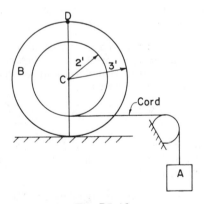

Fig. P5.49

5.49. Body B rolls without slipping and is connected to body A by a cord passing over the smooth pin in Fig. P5.49. The velocity of body A is 5 fps downward while the acceleration of point C in body B is 21 fps^2 to the left. Determine the acceleration of body A with respect to point D in body B for the instant shown.

Ans. $\mathbf{a}_{A/D} = 42\mathbf{i} + 82\mathbf{j}$ fps^2

5.50. Gear D has a diameter of 4 ft and rolls without slipping with a constant

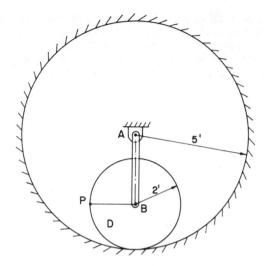

Fig. P5.50

angular velocity of 6 rad/s counterclockwise inside a 10-ft diameter fixed-ring gear (Fig. P5.50). For the position shown determine (a) the angular velocity of arm AB, (b) the acceleration of point P, and (c) the angular acceleration of arm AB and the acceleration of point P if gear D has an angular acceleration of 9 rad/s^2 clockwise at this instant and the angular velocity is still 6 rad/s counterclockwise.

5.51. Gear A in Problem 5.34 has an angular acceleration of 6 rad/s^2 clockwise when in the position shown as well as the angular velocity of 12 rad/s counterclockwise. For this instant determine the angular acceleration of bar BC and the acceleration of block C.

Ans. α_{BC} = 6.74 rad/s^2 ↻ , a_C = 623 cm/s^2 ◁ $^3_4$

5.52. For the wheel described in Problem 5.36, determine (a) the angular acceleration of the wheel and (b) the acceleration of point C (the point of contact on the wheel) when t = 3.0 s.

5.53. A fluid particle moves along the straight path of Fig. P5.53 between points A and B, and the magnitude of the velocity is given by the equation $q = Bx + V_0$ where x is the position coordinate shown in the figure and B

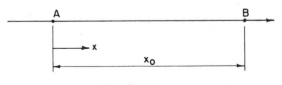

Fig. P5.53

and V_0 are constants. Determine the local, convective, and total acceleration of the particle in the x direction when it is at point B.

Ans.   Local acc. = 0
         Convective acc. = total acc. = $B(Bx + V_0)i$

5.54. The velocity in a certain flow field is given by the equation

$$q = 3yz^2 i + xzj + yk$$

Determine an expression for the acceleration in the x direction.

5.55. The velocity components in a two-dimensional flow field are given by the
equations

$$u = 3(x^2 - y^2), \qquad v = -6xy$$

where the velocity is in fps when x and y are in ft. Determine the velocity
and acceleration of a fluid particle at the point $x = y = 1$ ft.

Ans. $q = 6.00$ fps ↓, $a = 50.9$ fps² ◢

chapter 6

Deformation and Strain

6.1 Introduction

The state of stress that develops at an arbitrary point within a body as a result of surface or body-force loadings was discussed in Chapter 3. The relationships obtained were based on equilibrium considerations, and since no assumptions were made regarding deformations or physical properties of the body, the results are valid for the idealized "rigid" body or for a "real" deformable body. Since our ultimate concern is with real bodies, this chapter will emphasize the nature of the deformations associated with a real body under an arbitrary system of loads.

6.2 Displacement and Strain

When a body of arbitrary shape is subjected to a system of loads, individual points of the body generally move. This movement of a point with respect to some convenient reference system of axes is a vector quantity known as a *displacement*. Since different points in the body can undergo different movements as in Fig. 6.1 for points A, B, C, and N, a different displacement vector must be associated with each point of the body. For example, the displacement **u** is associated with point N in Fig. 6.1. Displacements may be expressed in mathematical form as functions of the coordinates of the points with which they are associated. In cases where a rectangular coordinate system is used, the displacement vector may be written as

$$\mathbf{u} = u_x\,\mathbf{i} + u_y\,\mathbf{j} + u_z\,\mathbf{k} \tag{6.1}$$

where u_x, u_y, and u_z are components of the displacement in the x, y, and z directions respectively. In this case **u** and its components u_x, u_y, and u_z could all be functions of x, y, and z.

The displacement associated with an individual point can result

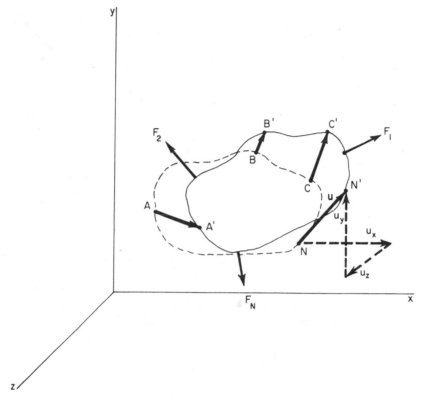

Fig. 6.1

from either or both of the following effects: (1) a translation and/or rotation of the body as a whole and (2) movement of points of the body relative to each other. The translation and/or rotation of the body as a whole is known as *rigid-body motion.* This type of motion is applicable to either the idealized rigid body or the real deformable body. The movement of points of the body relative to each other is known as a *deformation* and is associated only with real bodies. Rigid-body motion can be large or small. Deformations associated with engineering materials generally are small.

Strain is a geometric quantity related to the deformation displacements.   In Chapter 3 two types of stresses were discussed, normal and shear.   This same classification will be used for strains. *Normal strains* will be used to provide a measure of the elongation or contraction of an arbitrary line segment in the body during deformation. *Shear strains* will be used to provide a measure of the angular distortions associated with the deformation.

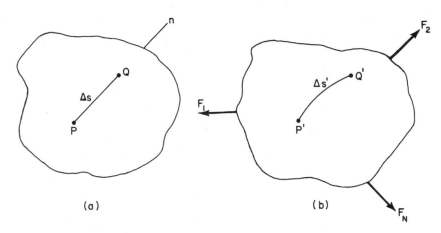

Fig. 6.2

Consider a body that is initially undeformed and unloaded as in Fig. 6.2a. When external loads are applied to the body, it will move as a rigid body and will also deform (Fig. 6.2b). As a result of the deformation an arbitrary straight line PQ of length Δs in the n direction in the undeformed body becomes the curve $P'Q'$ of length $\Delta s'$ in the deformed body. The *normal* strain at point P in the n direction $\epsilon_n(P)$ is defined as

$$\epsilon_n(P) = \lim_{\Delta s \to 0} \left(\frac{\Delta s' - \Delta s}{\Delta s} \right) \tag{6.2}$$

In a similar manner the shear strain is obtained by considering two straight lines PQ and PR in the mutually perpendicular n and t directions (Fig. 6.3a). When the body is loaded, these lines become

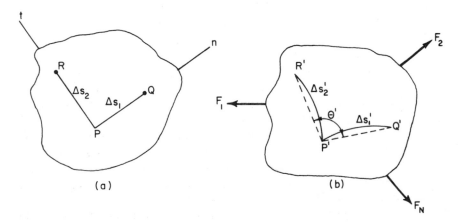

Fig. 6.3

curves $P'Q'$ and $P'R'$ (Fig. 6.3b). If the angle between the chords $P'Q'$ and $P'R'$ is taken as θ', the *shear strain* at point P associated with the n and t directions $\gamma_{nt}(P)$ is defined as

$$\gamma_{nt}(P) = \lim_{\substack{\Delta s_1 \to 0 \\ \Delta s_2 \to 0}} [(\pi/2) - \theta'] \tag{6.3}$$

From the definition of normal strain given by Eq. (6.2), it is evident that normal strain is positive when the line elongates and negative when it contracts. From Eq. (6.3) we observe that shear strains are positive when the angle between the positive n and t directions decreases. If the angle increases, the shear strain is negative.

6.3   Strain-Displacement Relationships

Relationships between the six Cartesian components of strain (ϵ_x, ϵ_y, ϵ_z, γ_{xy}, γ_{yz}, and γ_{zx}) and the three Cartesian components of displacement (u_x, u_y, and u_z) at an arbitrary point P in a body under load can be determined by considering the deformation associated with a small volume of material surrounding the point. For convenience consider the volume to have the shape of a rectangular parallelepiped with its sides oriented perpendicular to the reference x, y, and z axes in the undeformed state (Fig. 6.4a). If we further assume that the element of volume is extremely small, we can consider the strains to be uniform (constant along any arbitrary straight line) and independent of position (equal along parallel straight lines) within the volume. Thus planes will remain plane and straight lines will remain straight in the deformed element (Fig. 6.4b).

In Fig. 6.4 it can be seen that the general point P is displaced through a distance u_x in the x direction, u_y in the y direction, and u_z in the z direction. The other corners of the parallelepiped are also displaced and generally will be displaced by amounts differing from those at point P. For example, the displacements associated with point Q may be obtained from a Taylor series expansion of displacement about point P as $u_x + (\partial u_x / \partial x)\, dx$ in the x direction, $u_y + (\partial u_y / \partial x)\, dx$ in the y direction, and $u_z + (\partial u_z / \partial x)\, dx$ in the z direction. These are the only significant terms appearing in the expansion since y and z are constant along PQ and the element is sufficiently small for higher order terms to be neglected.

The normal strain along line PQ may be computed by using the definition for normal strain expressed in Eq. (6.2). Since the element is sufficiently small for the strain to be assumed uniform, Eq. (6.2) may be written

$$\epsilon_x = \frac{dx' - dx}{dx} \tag{6.4}$$

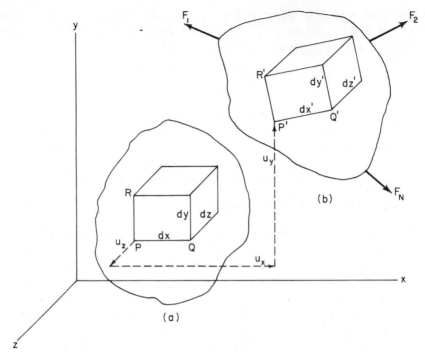

Fig. 6.4

which is equivalent to

$$dx' = (1 + \epsilon_x)\, dx \tag{6.5}$$

As in Fig. 6.5, the deformed length dx' may be expressed in terms of the displacement gradients as

$$(dx')^2 = \left[\left(1 + \frac{\partial u_x}{\partial x}\right) dx\right]^2 + \left(\frac{\partial u_y}{\partial x}\, dx\right)^2 + \left(\frac{\partial u_z}{\partial x}\, dx\right)^2 \tag{a}$$

Squaring Eq. (6.5) and substituting Eq. (a) yields

$$(1 + 2\epsilon_x + \epsilon_x^2)\,(dx)^2$$

$$= \left[1 + 2\frac{\partial u_x}{\partial x} + \left(\frac{\partial u_x}{\partial x}\right)^2 + \left(\frac{\partial u_y}{\partial x}\right)^2 + \left(\frac{\partial u_z}{\partial x}\right)^2\right](dx)^2 \tag{b}$$

For cases where strains and derivatives of displacements are sufficiently small that higher order terms can be neglected, Eq. (b) reduces to

$$\epsilon_x = \frac{\partial u_x}{\partial x} \tag{6.6a}$$

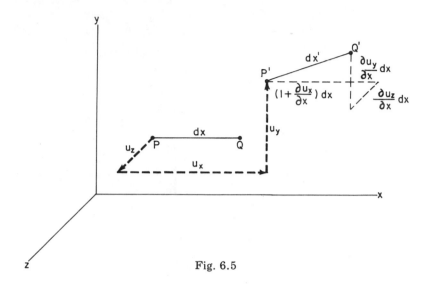

Fig. 6.5

In a similar manner

$$\epsilon_y = \frac{\partial u_y}{\partial y} , \qquad \epsilon_z = \frac{\partial u_z}{\partial z} \tag{6.6b}$$

The shear strain components are related to the displacements by considering the angular distortions associated with the edges of the parallelepiped. For example, consider lines PQ and PR in Fig. 6.6. The angle θ' between $P'Q'$ and $P'R'$ in the deformed state can be determined since it can be shown that the cosine of the angle between any two intersecting lines in space is the sum of the paired products of the direction cosines of the lines with respect to any reference coordinate system. Thus

$$\cos \theta' = \left[\left(1 + \frac{\partial u_x}{\partial x}\right) \frac{dx}{dx'}\right] \left(\frac{\partial u_x}{\partial y} \frac{dy}{dy'}\right)$$

$$+ \left(\frac{\partial u_y}{\partial x} \frac{dx}{dx'}\right) \left[\left(1 + \frac{\partial u_y}{\partial y}\right) \frac{dy}{dy'}\right] + \left(\frac{\partial u_z}{\partial x} \frac{dx}{dx'}\right) \left(\frac{\partial u_z}{\partial y} \frac{dy}{dy'}\right) \tag{a}$$

From the definition of shear strain as indicated in Eq. (6.3) $\gamma_{xy} = [(\pi/2) - \theta']$. For the case of small strains

$$\gamma_{xy} = \sin \gamma_{xy} = \sin [(\pi/2) - \theta'] = \cos \theta' \tag{b}$$

Substituting Eq. (a) into Eq. (b) and simplifying yields

$$\gamma_{xy} dx' dy' = \left[\left(1 + \frac{\partial u_x}{\partial x}\right) \frac{\partial u_x}{\partial y} + \left(1 + \frac{\partial u_y}{\partial y}\right) \frac{\partial u_y}{\partial x} + \frac{\partial u_z}{\partial x} \frac{\partial u_z}{\partial y}\right] dx \, dy$$

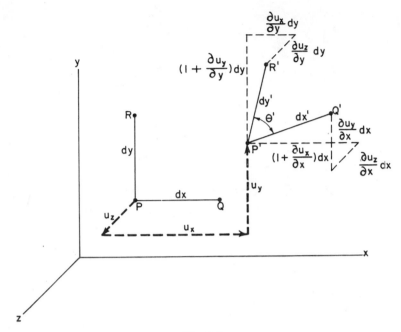

Fig. 6.6

From Eq. (6.5)

$$dx' = (1 + \epsilon_x)\, dx \quad dy \quad dy' = (1 + \epsilon_y)\, dy$$

therefore

$$\gamma_{xy}\,(1 + \epsilon_x + \epsilon_y + \epsilon_x \epsilon_y)\, dx\, dy$$

$$= \frac{\partial u_x}{\partial y} + \frac{\partial u_y}{\partial x} + \frac{\partial u_x}{\partial x}\frac{\partial u_x}{\partial y} + \frac{\partial u_y}{\partial x}\frac{\partial u_y}{\partial y} + \frac{\partial u_z}{\partial x}\frac{\partial u_z}{\partial y}$$

Again for cases where strains and derivatives of displacements are sufficiently small for higher order terms to be neglected, the previous expression reduces to

$$\gamma_{xy} = \frac{\partial u_x}{\partial y} + \frac{\partial u_y}{\partial x} \tag{6.6c}$$

and in a similar manner

$$\gamma_{yz} = \frac{\partial u_y}{\partial z} + \frac{\partial u_z}{\partial y}\,, \qquad \gamma_{zx} = \frac{\partial u_z}{\partial x} + \frac{\partial u_x}{\partial z} \tag{6.6d}$$

Eqs. (6.6) are known as the strain-displacement equations and in this reduced form are widely used for the analysis of engineering structures.   When materials such as aluminum and steel are used in their elastic ranges, the strains and displacements generally are so small that no significant error is introduced by neglecting the higher order terms in the general expressions.

EXAMPLE 6.1

A rectangle $ABCD$ having the dimensions in Fig. E6.1a is drawn on a thin sheet of rubber. When loads are applied to the sheet, the rectangle is distorted into the shape $A'B'C'D'$ (Fig. E6.1b). Determine ϵ_x, ϵ_y, and γ_{xy}. Assume that the rubber distorts uniformly.

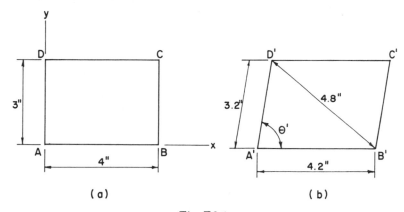

Fig. E6.1

SOLUTION

From the definition of normal strain given by Eq. (6.4)

$$\epsilon_x = \frac{dx' - dx}{dx} = \frac{4.2 - 4.0}{4.0} = 0.05 \text{ in./in.} \qquad \textit{Ans.}$$

$$\epsilon_y = \frac{dy' - dy}{dy} = \frac{3.2 - 3.0}{3.0} = 0.067 \text{ in./in.} \qquad \textit{Ans.}$$

The angle θ' between the sides $A'B'$ and $A'D'$ of the distorted rectangle can be computed by using the law of cosines. Thus

$$\cos \theta' = -\frac{(B'D')^2 - (A'B')^2 - (A'D')^2}{2(A'B')(A'D')} = -\frac{(4.8)^2 - (4.2)^2 - (3.2)^2}{2(4.2)(3.2)} = 0.180$$

therefore $\theta' = 76.6°$.

From the definition of shear strain given by Eq. (6.3)

$$\gamma_{xy} = (\pi/2) - \theta' = 90° - 76.6° = 10.4° = 0.181 \text{ rad} \qquad \textit{Ans.}$$

6.4 Two-dimensional or Plane Strain

The method of relating the six Cartesian components of strain associated with our reference xyz coordinate system to the normal and shear strains associated with other orthogonal directions will be illustrated by considering the two-dimensional or *plane strain* case. If we consider the xy plane as the reference plane, then for the special case under consideration $\epsilon_z = \gamma_{zx} = \gamma_{zy} = 0$. The more general three-dimensional problem can be approached in an identical manner, but since we will need only the plane strain results in future work, our analysis will be limited to this special case. The problem to be considered is illustrated in Fig. 6.7.

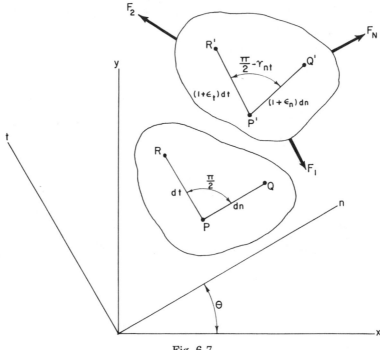

Fig. 6.7

Let two lines PQ and PR be oriented in the n and t directions respectively in an undeformed body. The n and t directions are perpendicular, lie in the xy plane, and are oriented at an angle θ with respect to the reference x and y axes. Under the action of a system of loads the body displaces and deforms so that the lines PQ and PR become $P'Q'$ and $P'R'$. Since the problem we are considering is one of plane strain, both lines remain in the xy plane. As a result of the loading, point P is displaced through a distance u_n in the n direction,

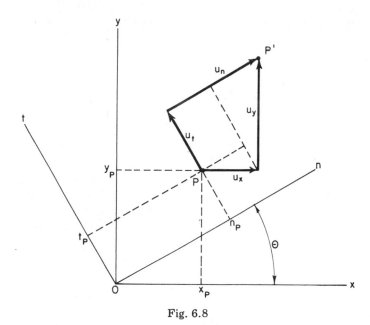

Fig. 6.8

and u_t in the t direction. By using Eqs. (6.6) the strains ϵ_n, ϵ_t, and γ_{nt} can be expressed in terms of these displacements as

$$\epsilon_n = \frac{\partial u_n}{\partial n}, \qquad \epsilon_t = \frac{\partial u_t}{\partial t}, \qquad \gamma_{nt} = \frac{\partial u_n}{\partial t} + \frac{\partial u_t}{\partial n} \tag{a}$$

The displacements u_n and u_t are related to the displacements u_x and u_y and the angle θ (Fig. 6.8). From the figure we see that

$$u_n = u_x \cos\theta + u_y \sin\theta, \qquad u_t = u_y \cos\theta - u_x \sin\theta \tag{b}$$

The initial position of P can be specified in terms of either xy coordinates or nt coordinates. The two systems are related by the following expressions (as in Fig. 6.8):

$$x = n \cos\theta - t \sin\theta, \qquad y = n \sin\theta + t \cos\theta \tag{c}$$

Finally, the strains associated with the two coordinate systems can be related by employing the chain rule for partial derivatives. Thus

$$\epsilon_n = \frac{\partial u_n}{\partial n} = \frac{\partial u_n}{\partial x}\frac{\partial x}{\partial n} + \frac{\partial u_n}{\partial y}\frac{\partial y}{\partial n}$$

$$\epsilon_t = \frac{\partial u_t}{\partial t} = \frac{\partial u_t}{\partial x}\frac{\partial x}{\partial t} + \frac{\partial u_t}{\partial y}\frac{\partial y}{\partial t} \tag{d}$$

$$\gamma_{nt} = \frac{\partial u_t}{\partial n} + \frac{\partial u_n}{\partial t} = \frac{\partial u_t}{\partial x}\frac{\partial x}{\partial n} + \frac{\partial u_t}{\partial y}\frac{\partial y}{\partial n} + \frac{\partial u_n}{\partial x}\frac{\partial x}{\partial t} + \frac{\partial u_n}{\partial y}\frac{\partial y}{\partial t}$$

Substitution of Eqs. (b) and (c) into (d) yields

$$\epsilon_n = \frac{\partial u_x}{\partial x} \cos^2 \theta + \frac{\partial u_y}{\partial y} \sin^2 \theta + \left(\frac{\partial u_y}{\partial x} + \frac{\partial u_x}{\partial y}\right) \sin \theta \cos \theta$$

$$\epsilon_t = \frac{\partial u_x}{\partial x} \sin^2 \theta + \frac{\partial u_y}{\partial y} \cos^2 \theta - \left(\frac{\partial u_y}{\partial x} + \frac{\partial u_x}{\partial y}\right) \sin \theta \cos \theta \qquad (e)$$

$$\gamma_{nt} = -2 \left(\frac{\partial u_x}{\partial x} - \frac{\partial u_y}{\partial y}\right) \sin \theta \cos \theta + \left(\frac{\partial u_y}{\partial x} + \frac{\partial u_x}{\partial y}\right) (\cos^2 \theta - \sin^2 \theta)$$

An examination of Eqs. (e) indicates that the expression for ϵ_t can be obtained by substituting $(\pi/2) + \theta$ for θ in the expression for ϵ_n. Thus only two of the relationships will be needed for future work, one for normal strain determinations and the other for shear strain determinations. Substitution of Eqs. (6.6) into Eqs. (e) yields transformation equations in terms of the Cartesian components of strain rather than the displacements. Thus

$$\epsilon_n = \epsilon_x \cos^2 \theta + \epsilon_y \sin^2 \theta + \gamma_{xy} \sin \theta \cos \theta \qquad (6.7a)$$

$$\gamma_{nt} = -2(\epsilon_x - \epsilon_y) \sin \theta \cos \theta + \gamma_{xy}(\cos^2 \theta - \sin^2 \theta) \qquad (6.8a)$$

or in double-angle form

$$\epsilon_n = (1/2)(\epsilon_x + \epsilon_y) + (1/2)(\epsilon_x - \epsilon_y) \cos 2\theta + (1/2)(\gamma_{xy}) \sin 2\theta \quad (6.7b)$$

$$\gamma_{nt}/2 = -(1/2)(\epsilon_x - \epsilon_y) \sin 2\theta + (1/2)(\gamma_{xy}) \cos 2\theta \qquad (6.8b)$$

A comparison of Eqs. (6.7) and (6.8) with Eqs. (3.8) and (3.9) indicates that the strain transformation equations can be obtained from the stress transformation equations by substituting ϵ_x for σ_x, ϵ_y for σ_y, and $\gamma_{xy}/2$ for τ_{xy}. Thus all the relationships developed in Chapter 3 for the plane state of stress can be duplicated for the plane state of strain by an appropriate change of variables in the equations. For example, from Eqs. (3.10) and (3.11) we can obtain the orientations and magnitudes of the extreme values of in-plane normal strains as

$$\tan 2\theta_{x\epsilon} = \frac{\gamma_{xy}}{\epsilon_x - \epsilon_y} \qquad (6.9)$$

$$\epsilon_{P1} = \frac{\epsilon_x + \epsilon_y}{2} + \sqrt{\left(\frac{\epsilon_x - \epsilon_y}{2}\right)^2 + \left(\frac{\gamma_{xy}}{2}\right)^2} \qquad (6.10a)$$

$$\epsilon_{P2} = \frac{\epsilon_x + \epsilon_y}{2} - \sqrt{\left(\frac{\epsilon_x - \epsilon_y}{2}\right)^2 + \left(\frac{\gamma_{xy}}{2}\right)^2} \qquad (6.10b)$$

The extreme values of in-plane normal strain together with $\epsilon_z = 0$

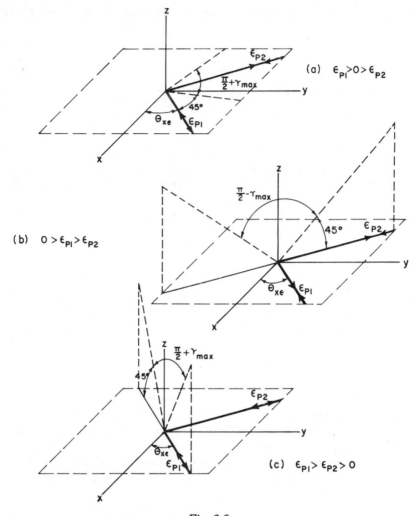

Fig. 6.9

are known as *principal strains*. In a similar manner the extreme values of the in-plane shear strains and the directions with which they are associated are obtained from Eqs. (3.12) and (3.13) as

$$\tan 2\theta_{x\gamma} = -\frac{\epsilon_x - \epsilon_y}{\gamma_{xy}} \tag{6.11}$$

$$\frac{\gamma_P}{2} = \pm \sqrt{\left(\frac{\epsilon_x - \epsilon_y}{2}\right)^2 + \left(\frac{\gamma_{xy}}{2}\right)^2} \tag{6.12}$$

Finally, the principal strains can be ordered and denoted as ϵ_1, ϵ_2, and ϵ_3 following the procedure used for principal stresses. Eq. (3.16) then yields

$$\gamma_{\max}/2 = (\epsilon_1 - \epsilon_3)/2 \qquad (6.13)$$

Depending on the magnitudes of the principal strains, the maximum shear strain may be associated with two mutually perpendicular lines that lie originally in the xy plane (when ϵ_{P1} and ϵ_{P2} have opposite signs) or with two mutually perpendicular lines that are inclined 45° with respect to the reference xy plane (when ϵ_{P1} and ϵ_{P2} have the same sign). The various cases are illustrated in Fig. 6.9.

The pictorial or graphical representation of Eqs. (3.8b) and (3.9b), known as Mohr's circle for stress, can be used with Eqs. (6.7b) and (6.8b) to yield a Mohr's type of circle for strain. The equation for the circle obtained from Eqs. (6.7b) and (6.8b) or by change of variables from Eq. (3.17) is

$$\left(\epsilon_n - \frac{\epsilon_x + \epsilon_y}{2}\right)^2 + \left(\frac{\gamma_{nt}}{2}\right)^2 = \left(\frac{\epsilon_x - \epsilon_y}{2}\right)^2 + \left(\frac{\gamma_{xy}}{2}\right)^2 \qquad (6.14)$$

Equation (6.14) is the equation of a circle in terms of the variables ϵ_n and $\gamma_{nt}/2$. The circle is centered on the ϵ axis at a distance $a = (\epsilon_x + \epsilon_y)/2$ from the origin, and the circle has a radius given by

$$r = \sqrt{\left(\frac{\epsilon_x - \epsilon_y}{2}\right)^2 + \left(\frac{\gamma_{xy}}{2}\right)^2}$$

A typical Mohr's circle for strain is shown in Fig. 6.10. We note on the figure that normal strains are plotted horizontally following the procedure developed previously for stresses. Thus positive or tensile strains are plotted to the right of the origin, while negative or compressive strains are plotted to the left. When plotting shear strains, however, only *half* their values are plotted on the vertical axis. Positive values of shear strain $\gamma_{xy}/2$ are plotted below the abscissa with ϵ_x and above with ϵ_y. Negative values of shear strain $\gamma_{xy}/2$ are plotted above the abscissa with ϵ_x and below with ϵ_y. When this convention is followed, relative angular positions on the real body are in the same direction as on Mohr's circle. Recall, however, that values of the angle are doubled on the Mohr's circle representation.

Thus far we have been dealing with the plane state of strain in which ϵ_z, γ_{zx}, and γ_{zy} were all equal to zero. In many practical problems situations will be encountered where $\gamma_{zx} = \gamma_{zy} = 0$, but $\epsilon_z \neq 0$. This situation occurs on the free surfaces of machine components and structural members where strain measurements are often made. Under this condition the effects of the u_z component of displacement should be noted. From development of expressions

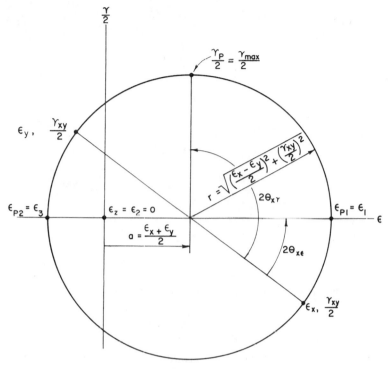

Fig. 6.10

for ϵ_x, ϵ_y, and γ_{xy} in terms of the displacements u_x, u_y, and u_z as given by Eqs. (6.6), we note that u_z enters only as a higher order term. Since the x and y directions were chosen arbitrarily in the xy plane, the displacement u_z will have no effect on any strain component in the plane. Thus Eqs. (6.7) and (6.8) and their Mohr's circle representation are valid not only for the plane strain case but also for the more general case involving out of plane deformation as long as we deal only with lines that lie originally in the xy plane in the undeformed state.

EXAMPLE 6.2

The strain components associated with a state of plane strain are

$$\epsilon_x = 1000 \ \mu\text{in./in.}, \quad \epsilon_y = 400 \ \mu\text{in./in.}, \quad \gamma_{xy} = 800 \ \mu\text{rad}$$

Determine the two in-plane principal strains and the maximum in-plane shear strain. Indicate on a sketch the principal strain directions and the two lines that suffer the greatest angular distortion.

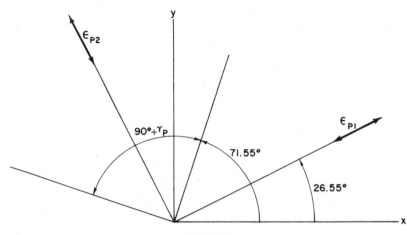

Fig. E6.2

SOLUTION

The two in-plane principal strains are found by using Eq. (6.10). Thus

$$\epsilon_{P1} = \frac{\epsilon_x + \epsilon_y}{2} + \sqrt{\left(\frac{\epsilon_x - \epsilon_y}{2}\right)^2 + \left(\frac{\gamma_{xy}}{2}\right)^2}$$

$$= \frac{1000 + 400}{2} + \sqrt{\left(\frac{1000 - 400}{2}\right)^2 + \left(\frac{800}{2}\right)^2}$$

$$= 700 + 500 = 1200 \; \mu\text{in./in.}$$

$$\epsilon_{P2} = 700 - 500 = 200 \; \mu\text{in./in.} \qquad\qquad Ans.$$

From Eq. (6.12)

$$\frac{\gamma_P}{2} = \sqrt{\left(\frac{\epsilon_x - \epsilon_y}{2}\right)^2 + \left(\frac{\gamma_{xy}}{2}\right)^2} = 500 \; \mu\text{rad}$$

therefore

$$\gamma_P = 1000 \; \mu\text{rad} \qquad\qquad Ans.$$

Once the magnitudes of the principal strains are known, their directions are found by using Eq. (6.9). Thus

$$\tan 2\theta_{x\epsilon} = \gamma_{xy}/(\epsilon_x - \epsilon_y) = 800/(1000 - 400) = 1.333$$

from which

$$2\theta_{x\epsilon} = 53.1°, 233.1°$$

or

$$\theta_{x\epsilon} = 26.55°, 116.55° \qquad\qquad Ans.$$

In a similar manner the two lines that suffer the greatest angular distortions are located using Eq. (6.11).

$$\tan 2\theta_{x\gamma} = -(\epsilon_x - \epsilon_y)/\gamma_{xy} = (1000 - 400)/800 = -0.750$$

from which

$$2\theta_{x\gamma} = 143.1°, 323.1°$$

or

$$\theta_{x\gamma} = 71.55°, 161.55° \qquad Ans.$$

Fig. E6.2 shows the principal strain directions and the lines associated with the maximum shearing strain.

EXAMPLE 6.3

The strain components associated with a state of plane strain are

$$\epsilon_x = 400 \,\mu in./in., \qquad \epsilon_y = -200 \,\mu in./in., \qquad \gamma_{xy} = 800 \,\mu rad$$

Use Mohr's strain circle to determine the two in-plane principal strains and the maximum in-plane shear strain.

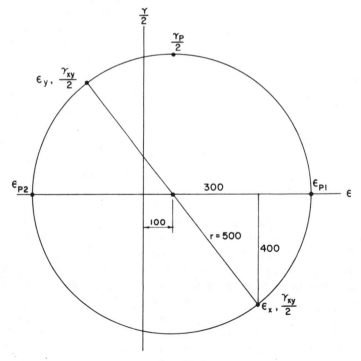

Fig. E6.3

SOLUTION

Mohr's circle for the given state of plane strain is shown in Fig. E6.3. From the sketch of the circle we see that the circle is centered at $a = 100$ μin./in. and has a radius $r = \sqrt{300^2 + 400^2} = 500$ μin./in. Thus

$$\epsilon_{P1} = a + r = 100 + 500 = 600 \ \mu\text{in./in.} \qquad\qquad Ans.$$

$$\epsilon_{P2} = a - r = 100 - 500 = -400 \ \mu\text{in./in.} \qquad\qquad Ans.$$

$$\gamma_P = 2r = 1000 \ \mu\text{in./in.} = 0.001 \text{ rad} \qquad\qquad Ans.$$

We also note from the circle that ϵ_{P1} is associated with a line that is located

$$\theta_{x\epsilon} = (1/2) \tan^{-1} (400/300) = 26.55°$$

counterclockwise from the x axis.

6.5   Measurement of Strain

Individual normal and shear strain components have been expressed in terms of displacements; therefore, if the displacements are known, the strains can be determined. For the most part, strain measurements are confined to the free surfaces of machine components and structural members. The strain components of interest are then ϵ_x, ϵ_y, and γ_{xy} if the z direction is taken normal to the surface. The u_x and u_y displacements needed to determine these strains are generally extremely small in engineering materials (steel, aluminum, etc.). Thus they are difficult to determine with any reasonable degree of accuracy over significant portions of the surface of the body. To overcome this difficulty, strain-measuring transducers such as the electrical resistance strain gage have been developed. These instruments measure differences in displacement (change in length) along some finite straight-line segment. Strain measured in this manner is not exact since the deformation is associated with some finite length (gage length), not a point as the definition for normal strain requires. As long as the gage length is kept small, however, the errors associated with such measurements can be kept within acceptable limits.

While electrical resistance strain gages and other measuring transducers have been developed to measure average normal strains, no transducer has been developed to measure accurately the small angle changes associated with the shear strains. To overcome this difficulty, three normal strain measurements in different directions are usually made and Eq. (6.7a) is used to compute γ_{xy} from the normal strain data. For example, consider the most general case of three arbitrary normal strain measurements shown in Fig. 6.11. From

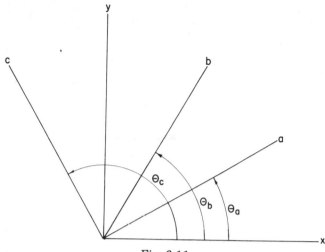

Fig. 6.11

Eq. (6.7a) we may write

$$\epsilon_a = \epsilon_x \cos^2 \theta_a + \epsilon_y \sin^2 \theta_a + \gamma_{xy} \sin \theta_a \cos \theta_a \qquad (6.15a)$$

$$\epsilon_b = \epsilon_x \cos^2 \theta_b + \epsilon_y \sin^2 \theta_b + \gamma_{xy} \sin \theta_b \cos \theta_b \qquad (6.15b)$$

$$\epsilon_c = \epsilon_x \cos^2 \theta_c + \epsilon_y \sin^2 \theta_c + \gamma_{xy} \sin \theta_c \cos \theta_c \qquad (6.15c)$$

From the measured values of ϵ_a, ϵ_b, and ϵ_c and a knowledge of the gage orientations θ_a, θ_b, and θ_c with respect to the reference x axis, the values of ϵ_x, ϵ_y, and γ_{xy} can be determined by simultaneous solution of the three equations. In practice the angles θ_a, θ_b, and θ_c are selected to simplify the calculations. The following two examples illustrate the procedure with commercially available multiple-element strain gages (rosettes) manufactured for this type of determination.

EXAMPLE 6.4

The three-element rectangular rosette strain gage in Fig. E6.4a has been bonded to the surface of a machine part. When the part is loaded, normal strains are as follows:

$$\epsilon_a = 1200 \ \mu in./in., \qquad \epsilon_b = 100 \ \mu in./in., \qquad \epsilon_c = -400 \ \mu in./in.$$

Determine the in-plane principal strains and their orientations with respect to the elements of the rosette.

SOLUTION

For convenience we select the reference x axis to coincide with one of the elements of the rosette (say the a axis). The angles in Eqs. (6.15) become

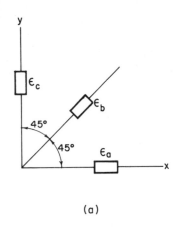

(a)

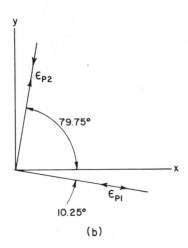

(b)

Fig. E6.4

$\theta_a = 0°$, $\theta_b = 45°$ and $\theta_c = 90°$ with this selection of reference axes. Substitution of these values of the angles in Eqs. (6.15) yields

$$\epsilon_a = \epsilon_x, \qquad \epsilon_b = (1/2)(\epsilon_x + \epsilon_y + \gamma_{xy}), \qquad \epsilon_c = \epsilon_y$$

Thus

$$\epsilon_x = 1200 \ \mu in./in., \qquad \epsilon_y = -400 \ \mu in./in.$$

$$\gamma_{xy} = 2\epsilon_b - \epsilon_a - \epsilon_c = 2(100) - 1200 + 400 = -600 \ \mu in./in. = -0.0006 \ rad$$

The in-plane principal strains are now determined by using Eqs. (6.10) as

$$\epsilon_{P1} = \frac{\epsilon_x + \epsilon_y}{2} + \sqrt{\left(\frac{\epsilon_x - \epsilon_y}{2}\right)^2 + \left(\frac{\gamma_{xy}}{2}\right)^2}$$

$$= \frac{1200 - 400}{2} + \sqrt{\left(\frac{1200 + 400}{2}\right)^2 + \left(\frac{-600}{2}\right)^2} = 1254 \ \mu in./in.$$

$$\epsilon_{P2} = \frac{\epsilon_x + \epsilon_y}{2} - \sqrt{\left(\frac{\epsilon_x - \epsilon_y}{2}\right)^2 + \left(\frac{\gamma_{xy}}{2}\right)^2} = -454 \ \mu in./in. \qquad \qquad Ans.$$

The orientations of the in-plane principal strains with respect to the x axis are given by Eq. (6.9) as

$$\tan 2\theta_{x\epsilon} = \gamma_{xy}/(\epsilon_x - \epsilon_y) = -600/(1200 + 400) = -0.375$$

from which

$$2\theta_{x\epsilon} = -20.5°, 159.5°, \qquad \theta_{x\epsilon} = -10.25°, 79.75° \qquad \qquad Ans.$$

The orientations of the principal strains with respect to the gages of the rosette are shown in Fig. E6.4b.

EXAMPLE 6.5

The three-element delta rosette strain gage in Fig. E6.5a has been bonded to the surface of a structural member. When the member is loaded, normal strains are as follows:

$$\epsilon_a = 1000 \ \mu\text{in./in.}, \quad \epsilon_b = 500 \ \mu\text{in./in.}, \quad \epsilon_c = -500 \ \mu\text{in./in.}$$

Determine the in-plane principal strains and their orientations with respect to the elements of the rosette.

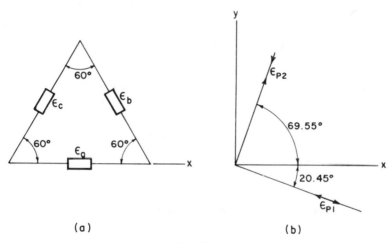

(a)   (b)

Fig. E6.5

SOLUTION

For convenience select the reference x axis to coincide with the horizontal element (a axis) of the rosette. With this selection of orientation for the reference axes, the angles in Eqs. (6.15) become

$$\theta_a = 0°, \quad \theta_b = -60°, \quad \theta_c = 60°$$

Substitution of these values in Eqs. (6.15) yields

$$\epsilon_a = \epsilon_x, \quad \epsilon_b = (1/4)\epsilon_x + (3/4)\epsilon_y - (\sqrt{3}/4)\gamma_{xy}$$
$$\epsilon_c = (1/4)\epsilon_x + (3/4)\epsilon_y + (\sqrt{3}/4)\gamma_{xy} \tag{a}$$

We now solve Eqs. (a) simultaneously for ϵ_x, ϵ_y, and γ_{xy} in terms of ϵ_a, ϵ_b, and ϵ_c to obtain

$$\epsilon_x = \epsilon_a, \quad \epsilon_y = (2/3)(\epsilon_b + \epsilon_c) - (1/3)\epsilon_a, \quad \gamma_{xy} = (2\sqrt{3}/3)(\epsilon_c - \epsilon_b) \tag{b}$$

Substitution of the measured values of strain into Eqs. (b) yields

$$\epsilon_x = 1000 \ \mu\text{in./in.}, \qquad \epsilon_y = -333 \ \mu\text{in./in.}$$

$$\gamma_{xy} = -1155 \ \mu\text{in./in.} = -0.001155 \ \text{rad}$$

The in-plane principal strains are determined using Eqs. (6.10). Thus

$$\epsilon_{P1} = \frac{\epsilon_x + \epsilon_y}{2} + \sqrt{\left(\frac{\epsilon_x - \epsilon_y}{2}\right)^2 + \left(\frac{\gamma_{xy}}{2}\right)^2}$$

$$= \frac{1000 - 333}{2} + \sqrt{\left(\frac{1000 + 333}{2}\right)^2 + \left(\frac{-1155}{2}\right)^2}$$

$$= 333 + 880 = 1213 \ \mu\text{in./in.}$$

$$\epsilon_{P2} = 333 - 880 = -547 \ \mu\text{in./in.} \qquad\qquad\qquad Ans.$$

The orientations of the principal strain directions with respect to the x axis are given by Eq. (6.9) as

$$\tan 2\theta_{x\epsilon} = \gamma_{xy}/(\epsilon_x - \epsilon_y) = -1155/1333 = -0.866$$

therefore

$$2\theta_{x\epsilon} = -40.9°, 139.1°$$

or

$$\theta_{x\epsilon} = -20.45°, 69.55°$$

The orientations of the principal strain directions with respect to the elements of the rosette are shown in Fig. E6.5b.

PROBLEMS

6.1. A rubber band can be placed on a 5-in. diameter cylinder without stretching. Determine the average normal strain in the band when it is stretched to fit over a 6-in. diameter cylinder.

   Ans. $\epsilon = 0.20$ in./in.

6.2. The three steel wires in Fig. P6.2 are initially unstrained. When a load is applied, point A moves vertically downward to point A'. Determine the average normal strain in each of the wires after loading.

6.3. A rectangular frame is composed of four rigid bars (Fig. P6.3). A wire is fastened diagonally across the frame from A to C. If point B is displaced 5 in. to the right, determine the average normal strain in the wire.

   Ans. $\epsilon_{AC} = 0.0955$ in./in.

6.4. A triangular plate of rubber is uniformly deformed as in Fig. P6.4. Determine the average normal and shear strains ϵ_x, ϵ_y, and γ_{xy}.

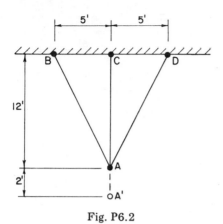

Fig. P6.2

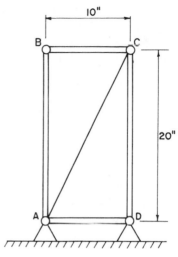

Fig. P6.3

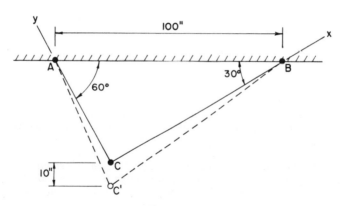

Fig. P6.4

6.5. A thin wire 50 in. long is deformed by heating in such a way that the extensional (normal) strain can be represented by the expression $\epsilon = Kx^2$. If the strain at the midlength of the wire is 1250 μin./in., determine the increase in length of the wire.

Ans. $\Delta L = 0.083$ in.

6.6. If a thin rubber sheet is fastened to the frame in Fig. P6.3 in the undeformed position, determine the average shear strain γ_{xy} in the rubber in the deformed position.

6.7. The rectangular plate in Fig. 6.7, having a length of 50 in. and a width of 24 in., is uniformly deformed in such a way that the length increases by

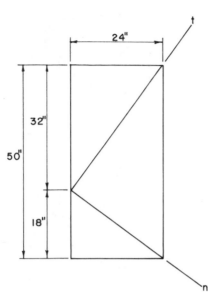

Fig. P6.7

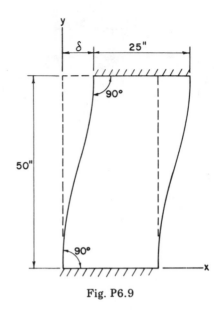

Fig. P6.9

1 in. while the width remains unchanged. Determine the average shear strain γ_{nt}.

Ans. $\gamma_{nt} = -0.0189$ rad

6.8. A rectangular plate $ABCD$ is uniformly deformed into the shape $A'B'C'D'$ shown in Fig. P6.8. Determine the average normal and shear strains ϵ_x, ϵ_y, and γ_{xy}.

6.9. A rectangular plate of rubber (25 in. by 50 in.) is deformed by shifting the upper edge with respect to the lower edge by an amount δ (Fig. P6.9). The displacement of an arbitrary point in the plate can be expressed as

$$u_x = ay^3 + by^2, \qquad u_y = 0$$

Determine the displacement δ required to produce a maximum shear strain $\gamma_{xy} = 0.003$ rad.

Ans. $\delta = 0.1$ in.

6.10. At a point in a structural member the following state of plane strain exists:

$$\epsilon_x = -3000\ \mu\text{in./in.}, \qquad \epsilon_y = 1800\ \mu\text{in./in.}, \qquad \gamma_{xy} = -1400\ \mu\text{rad}$$

Determine the principal strains and the maximum shearing strain.

6.11. At a point in a body the following state of plane strain exists:

$$\epsilon_x = 3000\ \mu\text{in./in.}, \qquad \epsilon_y = 600\ \mu\text{in./in.}, \qquad \gamma_{xy} = 1000\ \mu\text{rad}$$

Determine the principal strains and the maximum shearing strain. Show

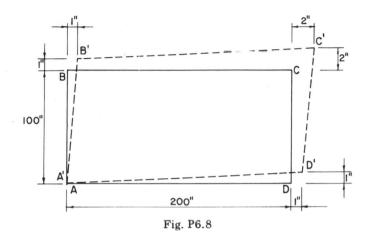

Fig. P6.8

in a suitable sketch the two lines associated with the maximum shearing strain.

Ans. $\epsilon_1 = 3100 \, \mu\text{in./in.}, \quad \epsilon_2 = 500 \, \mu\text{in./in.}, \quad \epsilon_3 = 0$

$\gamma_{\text{max}} = 3100 \, \mu\text{rad}$

6.12. The strain components associated with a state of plane strain are:

$$\epsilon_x = 1500 \, \mu\text{in./in.}, \quad \epsilon_y = -500 \, \mu\text{in./in.}, \quad \gamma_{xy} = 2600 \, \mu\text{rad}$$

Use Mohr's strain circle to determine the in-plane principal strains, the in-plane principal strain directions, and the maximum in-plane shear strain.

6.13. The strain components associated with a state of plane strain are:

$$\epsilon_x = -600 \, \mu\text{in./in.}, \quad \epsilon_y = 1800 \, \mu\text{in./in.}, \quad \gamma_{xy} = -1000 \, \mu\text{rad}$$

Use Mohr's strain circle to determine the in-plane principal strains and their directions.

Ans. $\epsilon_{P1} = 1900 \, \mu\text{in./in.}, \quad \epsilon_{P2} = -700 \, \mu\text{in./in.}, \quad \theta_{x\epsilon} = 11.3°, 101.3°$

6.14. The strain components associated with a plane state of strain are:

$$\epsilon_x = 400 \, \mu\text{in./in.}, \quad \epsilon_y = 2800 \, \mu\text{in./in.}, \quad \gamma_{xy} = -1000 \, \mu\text{rad}$$

Use Mohr's strain circle to determine the principal strains and the maximum shear strain at the point.

6.15. At a point on the surface of a machine component a rectangular rosette strain gage has been used to measure the following strains (see Fig. P6.15):

$$\epsilon_a = 600 \, \mu\text{in./in.}, \quad \epsilon_b = 500 \, \mu\text{in./in.}, \quad \epsilon_c = -200 \, \mu\text{in./in.}$$

Determine the in-plane principal strains and their directions.

Ans. $\epsilon_{p1} = 700 \, \mu\text{in./in.}, \quad \epsilon_{p2} = -300 \, \mu\text{in./in.}, \quad \theta_{x\epsilon} = 18.5°, 108.5°$

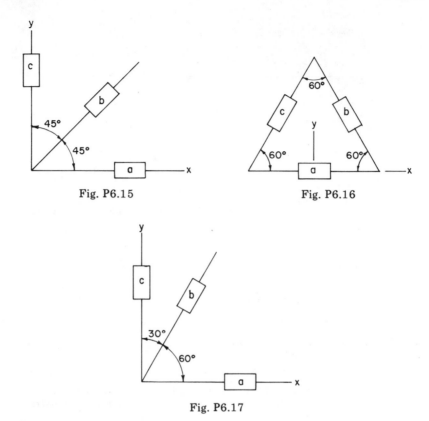

Fig. P6.15   Fig. P6.16

Fig. P6.17

6.16. At a point on the surface of a structural member a delta rosette strain gage has been used to measure the following strains (see Fig. P6.16):

$$\epsilon_a = 900\ \mu in./in., \qquad \epsilon_b = 600\ \mu in./in., \qquad \epsilon_c = 1200\ \mu in./in.$$

Determine the in-plane principal strains and their directions.

6.17. At a point on the surface of a machine component three strain gages were used to obtain the following strains (see Fig. P6.17):

$$\epsilon_a = 1000\ \mu in./in., \qquad \epsilon_b = 800\ \mu in./in., \qquad \epsilon_c = 400\ \mu in./in.$$

Determine the shearing strain γ_{xy} and the maximum in-plane shearing strain.

Ans. $\gamma_{xy} = 578\ \mu rad, \qquad \gamma_p = 833\ \mu rad$

part IV

RIGID-BODY DYNAMICS AND VIBRATIONS

chapter 7

Rigid-Body Dynamics

7.1 Introduction

Newton's laws of motion apply only to a single particle; therefore, they are inadequate to describe the behavior of rigid bodies. In this chapter Newton's laws of motion are extended to systems of particles and symmetrical rigid bodies in plane motion. These extended laws give the differential equations of motion that relate the linear and angular accelerated motion of the body to the forces and moments acting on it. From these equations we can solve for either (1) the instantaneous accelerations due to known forces and moments or (2) the forces and moments necessary to produce a prescribed motion. The study of forces and moments that either cause the motion or are the result of a prescribed motion is called *kinetics*.

Before proceeding, the student should review Chapter 5. In the study of kinetics, it is important to recognize the type of motion the body can have so that the proper kinematic relationships are used to describe the motion.

The basic procedure for solving kinetics problems is to:

1. Draw a complete free-body diagram of all the forces acting on the particle (or body).
2. Select a convenient coordinate system to describe the motion.
3. Apply the equations of motion that relate the forces (and moments) to the motion of the particle (or body).
4. Establish the kinematic relationships the motion must satisfy.
5. Solve the resulting equations of motion for the desired result, which may include integrating the equation of motion to obtain both velocity and displacement.

7.2   Principle of Motion of the Mass Center

Here we shall develop the principle of motion of the mass center of a system of particles that can be a rigid body (or bodies), an elastic body, a fluid, or simply a can and the air inside it. We shall do this by differentiating the equation that locates the mass center twice with respect to time, recognizing that Newton's second law for a single particle is contained in the terms, and then examining the meaning of the resulting force summation.

Consider a system of n particles (Fig. 7.1) where three are shown.

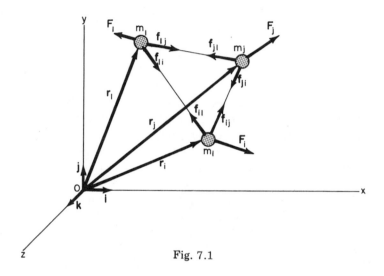

Fig. 7.1

We can label the particles in any order from 1 thru n so that m_1 is the mass of the first particle, m_i of the ith particle, and m_j of the jth particle. Similarly, r_1, r_i, and r_j are the position vectors locating each particle, while the *resultant external force* acting on the first particle is given by F_1, the ith particle by F_i, and the jth particle by F_j, etc.

In dealing with a system of particles, we must allow for the existence of *internal forces* f_{ij}, where f_{ij} is the force exerted on the ith particle by the jth particle. Thus according to Newton's third law,

$$f_{ij} + f_{ji} = 0 \qquad\qquad (a)$$

since f_{ji} is equal in magnitude but oppositely directed to f_{ij} as shown.

Now consider the location of the mass center of the system of particles obtained from

$$mr_c = \sum_{i=1}^{n} m_i r_i \qquad (b)$$

where m is the total mass of the system given by $m = \sum_{i=1}^{n} m_i$, which is constant. Differentiation of Eq. (b) with respect to time gives

$$G_c = m\dot{r}_c = \sum_{i=1}^{n} m_i \dot{r}_i = \sum_{i=1}^{n} G_i \qquad (7.1)$$

The *linear momentum* of a particle is defined as the product of the mass times the velocity; i.e., $G = mv$. Thus Eq. (7.1) states that the linear momentum of a system of particles G_c is equal to the sum of the linear momentum G_i of each particle in the system. The fact that we can relate the momentum of the mass center to the momentum of each particle by such a simple means as vector addition is an important concept.

Differentiation of Eq. (7.1) with respect to time gives

$$\dot{G}_c = m\ddot{r}_c = \sum_{i=1}^{n} m_i \ddot{r}_i = \sum_{i=1}^{n} \dot{G}_i \qquad (c)$$

which states that the time rate of change of the linear momentum of the mass center is equal to the sum of the time rate of change of the momentum of each particle within the system.

Newton's second law can be applied to any of the n particles in the system, so for the ith particle

$$F_i + \sum_{j=1}^{n} f_{ij} = m_i \ddot{r}_i = \dot{G}_i \qquad (d)$$

where $\sum_{j=1}^{n} f_{ij}$ is the resultant of all the internal forces acting on the ith particle. From the definition of f_{ij}, $f_{ii} = 0$ since the particle cannot exert an internal force on itself. However, note that both internal and external forces can be due to the same sources; i.e., (1) particle to particle contact, (2) forces of gravitational attraction, (3) electric and magnetic fields, etc. The classification of these forces as internal or external depends upon whether the particle (or body) causing the force is within the system of particles (internal force) or from outside the system (external force).

The right-hand side of Eq. (d) is the time rate of change of the linear momentum of the ith particle so that substitution of Eq. (d) into Eq. (c) gives

$$\dot{\mathbf{G}}_c = m\ddot{\mathbf{r}}_c = \sum_{i=1}^{n} \left(\mathbf{F}_i + \sum_{j=1}^{n} \mathbf{f}_{ij} \right) = \sum_{i=1}^{n} \mathbf{F}_i + \sum_{i=1}^{n} \sum_{j=1}^{n} \mathbf{f}_{ij} \qquad (e)$$

where

$$\sum_{i=1}^{n} \sum_{j=1}^{n} \mathbf{f}_{ij} = \mathbf{f}_{12} + \mathbf{f}_{21} + \cdots + \mathbf{f}_{ij} + \mathbf{f}_{ji} + \cdots + \mathbf{f}_{1n} + \mathbf{f}_{n1} + \cdots = 0$$

by virtue of Eq. (a); i.e., the resultant of all internal forces is zero since they occur in equal and opposite pairs. If $\mathbf{F} = \Sigma_{i=1}^{n} \mathbf{F}_i$ is the resultant of all the external forces acting on the system of particles, Eq. (e) becomes

$$\mathbf{F} = \dot{\mathbf{G}}_c = m\ddot{\mathbf{r}}_c = \sum_{i=1}^{n} m_i \ddot{\mathbf{r}}_i \qquad (7.2)$$

which is called the *principle of motion of the mass center* and applies only to systems of constant mass. We can express Eq. (7.2) in any coordinate system; e.g., in rectangular coordinates

$$\mathbf{i}\colon\ F_x = \dot{G}_{cx} = m\ddot{x}_c, \qquad \mathbf{j}\colon\ F_y = \dot{G}_{cy} = m\ddot{y}_c, \qquad \mathbf{k}\colon\ F_z = \dot{G}_{cz} = m\ddot{z}_c \qquad (7.3)$$

where subscript c is used to emphasize that the equations refer to the mass center.

Equation (7.2) states that the resultant external force acting on a system of particles is equal to the time rate of change of the linear momentum of the mass center (or the mass of the system times the acceleration of the mass center); i.e., we can obtain the motion of the mass center of any arbitrary system of particles as though the system were a single particle of mass m.

Two observations in the development of Eq. (7.2) merit emphasis. First, the equation applies to any conceivable constant mass system of particles since no kinematic equations relating the motion of one particle to that of any other are needed in the development; therefore the system can be composed of any combination of particles and rigid bodies that may or may not posses kinematic equations relating their motions. Second, no information is provided as to the location of the line of action of the resultant force since we simply summed forces in the development. Therefore, all we can learn from the principle of motion of the mass center is how the mass center moves; i.e., the mass center moves "as though" the system were a single particle of mass m with the resultant force passing through the mass center (see Section 7.4).

EXAMPLE 7.1

Body A in Fig. E7.1a weighs 20 lb and body B weighs 60 lb. The two bodies are initially at rest on the smooth horizontal plane when the 50-lb force is applied. Determine the minimum coefficient of friction between A and B so that they move together.

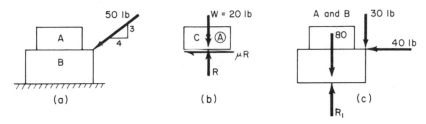

Fig. E7.1

SOLUTION

A free-body diagram of body A is shown in Fig. E7.1b from which we obtain

$$\left(\uparrow\right) \Sigma F_y = m_A (a_c)_{Ay} = 0, \quad R - 20 = 0$$

or

$$R = 20 \text{ lb}$$

and

$$\left(\longleftrightarrow\right) \Sigma F_x = m_A (a_c)_{Ax}, \quad -\mu R = (20/32.2)a_A$$

or

$$\mu = -a_A/32.2 \tag{a}$$

where a_A is assumed to be positive to the right.

From the free-body diagram of bodies A and B combined as in Fig. E7.1c,

$$\left(\longleftrightarrow\right) \Sigma F_x = m_{(A\ \&\ B)}(a_c)_x, \quad 40 = (80/32.2)a_c$$

or

$$a_c = 16.1 \text{ fps}^2 \tag{b}$$

for the combined system. The acceleration of the mass center and each of the bodies is the same since they are to move together. Hence

$$a_A = -a_c = -16.1 \qquad (c)$$

since to the right is positive in formulating Eq. (a) and to the left is positive in formulating Eq. (b). Substitution of Eq. (c) into Eq. (a) gives

$$\mu = -(-16.1)/32.2 = 0.50 \qquad \qquad Ans.$$

Consistency in signs is important in solving kinetics problems.

EXAMPLE 7.2

Body A in Fig. E7.2a weighs 32.2 lb and is sliding up the plane. The coefficient of friction between A and the plane is 0.50. Determine the weight of body B attached to body A with a flexible cable of negligible mass so that body A has an acceleration of 16.1 fps^2 up the plane.

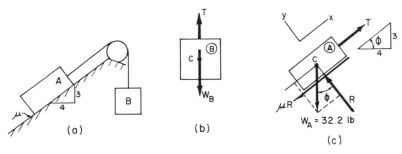

Fig. E7.2

SOLUTION

From the free-body diagram of body B (Fig. E7.2b) we have

$$\textcircled{\uparrow}\, \Sigma F_y = m_B \ddot{y}, \qquad T - W_B = (W_B/32.2)(-16.1) = -W_B/2$$

or

$$W_B = 2T \qquad (a)$$

since the acceleration of bodies A and B are the same as long as there is tension in the cable. The tension can be obtained from the free-body diagram of body A (Fig. E7.2c) giving

$$\textcircled{\nwarrow}\, \Sigma F_y = m_A a_{Ay} = 0, \qquad R - 32.2(4/5) = 0$$

or

$$R = 32.2(0.80) = 25.8 \text{ lb}$$

and

$$\left(\nearrow\right)\Sigma F_x = m_A a_{Ax}, \quad T - \mu R - (3/5)W_A = (32.2/32.2)(16.1)$$

or

$$T = 0.5(25.8) + (0.60)(32.2) + 16.1 = 48.3 \text{ lb}$$

Thus from Eq. (a),

$$W_B = 2T = 96.6 \text{ lb} \qquad\qquad Ans.$$

EXAMPLE 7.3

An external force P is applied to mass m_1 (90 kg) of the idealized mechanism in Fig. E7.3a until the 1000 N/m spring attached to mass m_2 (60 kg) is stretched a distance of 0.8 m from its equilibrium position. The masses are connected by an ideal cable of negligible mass that passes over two massless pulleys and moves in horizontal slots. Determine the acceleration of both masses and the mass center of the system at the instant force P is removed if the system is frictionless.

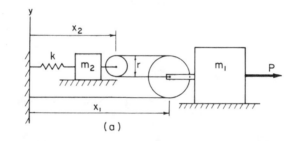

(a)

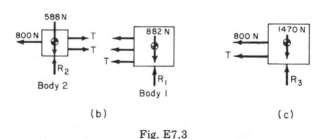

Body 2

Body I

(b)

(c)

Fig. E7.3

SOLUTION

Let x_1 locate m_1 and x_2 locate m_2 (Fig. E7.3a) and note that y_1, y_2, z_1, and z_2 are constant since the masses move in horizontal slots. The free-body diagrams of each mass are shown in Fig. E7.3b where T is the unknown cable tension and the spring force is 800 N to the left since the spring was stretched. Application of Eq. (7.2) to the horizontal direction for each mass gives

Body 1	Body 2

$\overset{\longleftrightarrow}{\bigcirc} F_x = m_1 \ddot{x}_{c1}$ $\overset{\longleftrightarrow}{\bigcirc} F_x = m_2 \ddot{x}_{c2}$

$-3T = 90\,\ddot{x}_1$ (a) $2T - 800 = 60\,\ddot{x}_2$ (b)

Equations (a) and (b) contain three unknowns; i.e., T, $\ddot{x}_1$, and $\ddot{x}_2$.

We can obtain a third equation relating x_1 and x_2 by noting that the length of the cable ℓ is constant. From Fig. E7.3a

$$\ell = \ell_{p1} + \ell_{p2} + x_1 + 2(x_1 - x_2)$$

where ℓ_{p1} and ℓ_{p2} are the constant lengths around pulleys 1 and 2 respectively. Differentiating this equation twice with respect to time gives

$$3\ddot{x}_1 = 2\ddot{x}_2 \qquad\qquad (c)$$

as the kinematic relationship between accelerations. Solving for the tension T in Eqs. (a) and (b) gives

$$T = 400 + 30\ddot{x}_2 = -30\ddot{x}_1 \qquad\qquad (d)$$

which, on substitution of $\ddot{x}_1$ from Eq. (c), gives

$$50\ddot{x}_2 = -400$$

or

$$\ddot{x}_2 = -8 \text{ m/s}^2 \text{ or } 8 \text{ m/s}^2 \leftarrow \qquad\qquad Ans.$$

where the minus sign simply tells us the acceleration is in the negative x_2 direction. From Eq. (c) we find that

$$\ddot{x}_1 = (2/3)\ddot{x}_2 = (2/3)(-8) = -5.33 \text{ m/s}^2 \text{ or } 5.33 \text{ m/s}^2 \leftarrow \qquad Ans.$$

We can obtain the acceleration of the mass center from Eq. (7.2)

$$\dot{\mathbf{G}}_c = m\ddot{\mathbf{r}}_c = \sum_{i=1}^{n} m_i \ddot{\mathbf{r}}_i$$

which becomes for the x direction

$$m\ddot{x}_c = \sum_{i=1}^{2} m_i \ddot{x}_i = m_1 \ddot{x}_1 + m_2 \ddot{x}_2 \qquad\qquad (e)$$

Substitution of numerical values gives

$$\ddot{x}_c = [90(-5.33) + 60(-8)]/150 = -960/150 = -6.4 \text{ m/s}^2$$

or

$$\ddot{x}_c = 6.4 \text{ m/s}^2 \leftarrow \qquad\qquad Ans.$$

We can check the validity of the results by combining both masses and using the free-body diagram of Fig. E7.3c. Here application of Eq. (7.2) gives

$$(\longleftrightarrow) F_x = m\ddot{x}_c, \qquad -800 - T = 150\,\ddot{x}_c \tag{f}$$

where T can be found by combining Eqs. (a), (c), and (e) to give $T = -25\,\ddot{x}_c$ so that Eq. (f) gives

$$\ddot{x}_c = -800/125 = -6.4 \ \text{m/s}^2$$

which is the same as before. We could have solved for the acceleration of the mass center by this method but still needed two free-body diagrams (body 1 and the combined system), the kinematic relationship of Eq. (c), and the relationship between the motion of the mass center and the individual masses, Eq. (e). Hence one method is as suitable as the other.

7.3 Properties of Mass Moments and Products of Inertia

The inertia of a single particle is totally described by its mass. The inertia of a system of particles is also described by its mass as we work with the motion of the mass center. However, when the system of particles is a rigid body, additional inertia properties are needed to relate the moments of the external forces to the angular motion of the body. These inertia properties are called *mass moments of inertia* and *mass products of inertia*.

Let the *xyz* coordinate system be attached to the rigid body in Fig. 7.2 with the origin at 0. The moment of inertia of mass with re-

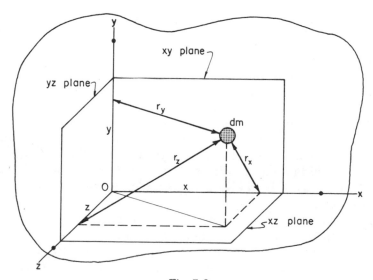

Fig. 7.2

spect to the z axis is defined by the integral

$$I_z = \int_m r_z^2 \, dm = \int_m (x^2 + y^2) \, dm \qquad (7.4)$$

Similarly, for the x and y axes the body has moments of inertia given by

$$I_x = \int_m r_x^2 \, dm = \int_m (y^2 + z^2) \, dm$$

and

$$I_y = \int_m r_y^2 \, dm = \int_m (x^2 + z^2) \, dm$$

The product of inertia of mass with respect to the xy and yz planes is defined by the integral

$$I_{xz} = \int_m xz \, dm \qquad (7.5)$$

Similarly, with respect to the yx and xz planes

$$I_{yz} = \int_m yz \, dm \qquad (7.6)$$

and with respect to the xz and zy planes

$$I_{xy} = \int_m xy \, dm \qquad (7.7)$$

Note that the xy and yz planes, the yx and xz planes, and the xz and zy planes intersect at the y, x, and z axes respectively. Remember that I_{xz} is related to the xy and yz planes by the fact that the planes to which the product of inertia is referred are those whose intersection forms the axis missing in the subscript. Hence we can split the subscript and add the missing orthogonal coordinate twice to obtain the reference planes; i.e., xz becomes xy and yz. The definitions of the products and moments of masses are similar to those for volumes where dm is replaced by dV, and care should be exercised not to confuse these quantities.

Equations (7.4), (7.5), (7.6), and (7.7) show that the moments and products of inertia have the same dimensions, either ML^2 or FLT^2. This means the units will be slug-ft^2 or ft-lb-s^2 (also in.-lb-s^2) in the gravitational English system and kg-m^2 or N-m-s^2 (also N-cm-s^2) in the SI system.

The product of inertia is zero when one of its planes passes through the mass center and the body is symmetrical with respect to that plane. For example, if the xy plane in Fig. 7.2 passes through the mass center, the body is symmetrical with respect to the xy plane if for every x and y there is as much mass for each positive z as there is for each negative z. As a consequence of this symmetry, both I_{xz} and I_{yz} would be zero, but I_{xy} may or may not be zero. When the z axis is one of symmetry, both I_{xz} and I_{yz} will be zero since for every value of z there is an equal amount of mass at each plus and minus x and y due to symmetry with respect to the z axis. Thus the products of inertia I_{xz} and I_{yz} for a rigid body are zero if either (1) the xy plane passing through the mass center is a plane of symmetry or (2) the body is symmetrical with respect to the axis of rotation passing through the mass center. These statements are illustrated in Fig. 7.3 for a rigid body composed of four equal masses. The mass center C is at the origin of the coordinate system, and the products and moments of inertia are shown.

The products of inertia I_{xz} and I_{yz} will be zero when the z axis is any axis parallel to the axis of symmetry from the parallel plane

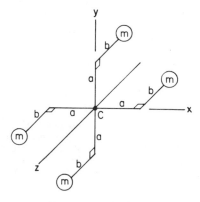

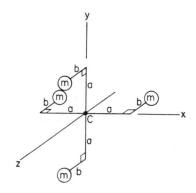

Nonsymmetrical

$I_x = 2mb^2 + 2m(b^2 + a^2)$

$I_y = 2mb^2 + 2m(b^2 + a^2)$

$I_z = 4ma^2$

$I_{xy} = 0$

$I_{xz} = m(a)(-b) + m(-a)(b) = -2mba$

$I_{yz} = m(a)(-b) + m(-a)(b) = -2mba$

Symmetrical

$I_x = 2mb^2 + 2m(b^2 + a^2)$

$I_y = 2mb^2 + 2m(b^2 + a^2)$

$I_z = 4ma^2$

$I_{xy} = 0$

$I_{xz} = m(a)(-b) + m(-a)(-b) = 0$

$I_{yz} = m(a)(b) + m(-a)(b) = 0$

Fig. 7.3

Table 7.1. Mass moment and products of inertia of homogeneous bodies with respect to the mass center

	Body	Moments and products	Remarks
Sphere		$I_x = I_y = I_z = 2mr^2/5$ $I_{xy} = I_{xz} = I_{yz} = 0$	Note absolute symmetry for any three orthogonal axes
Solid circular cylinder		$I_x = I_y = m(3r^2 + \ell^2)/12$ $I_z = mr^2/2$ $I_{xy} = I_{xz} = I_{yz} = 0$	Two limiting cases: (a) slender rod $\ell \gg r$ $\quad I_x = I_y = m\ell^2/12$ (b) thin disk $r \gg \ell$ $\quad I_x = I_y = mr^2/4$
Thin circular ring		$I_x = I_y = m(6r^2 + \ell^2)/12$ $I_z = mr^2$ $I_{xy} = I_{xz} = I_{yz} = 0$	Two limiting cases: (a) long thin ring $\ell \gg r$ $\quad I_x = I_y = m\ell^2/12$ (b) short thin ring $r \gg \ell$ $\quad I_x = I_y = mr^2/2$ r is average radius in this case
Rectangular prism		$I_x = m(b^2 + c^2)/12$ $I_y = m(a^2 + c^2)/12$ $I_z = m(a^2 + b^2)/12$ $I_{xy} = I_{xz} = I_{yz} = 0$	Note that a thin plate occurs when one dimension is small compared to the other dimensions
Cone		$I_x = I_y = 3m(4r^2 + \ell^2)/80$ $I_z = 3mr^2/10$ $I_{xy} = I_{yz} = I_{xz} = 0$	

theorem. See Example 7.4. The moments and products of inertia as defined by Eqs. (7.4), (7.5), (7.6), and (7.7) for many common engineering shapes are tabulated in Table 7.1.

Parallel Axis and Parallel Plane Theorems—We often know the products and moments of inertia of a rigid body with respect to its mass center since the body is one of standard shape. Frequently, we must know the moments and products of inertia with respect to a parallel set of axes x, y, z (Fig. 7.4) where the x', y', z' axes pass through the mass center C.

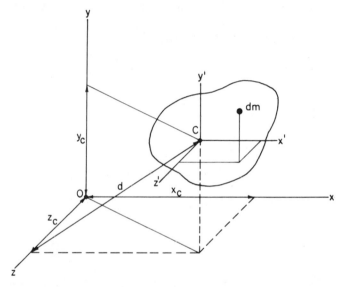

Fig. 7.4

The moment of inertia of the mass element dm with respect to the z axis is

$$dI_z = (x^2 + y^2)\, dm = [(x_c + x')^2 + (y_c + y')^2\; dm]$$
$$= (x_c^2 + y_c^2)\, dm + (x'^2 + y'^2)\, dm + 2x_c x'\, dm + 2y_c y'\, dm$$

so that integration over the entire mass gives

$$I_z = (x_c^2 + y_c^2) \int_m dm + \int_m (x'^2 + y'^2)\, dm + 2x_c \int_m x'\, dm + 2y_c \int_m y'\, dm$$

We recognize from Fig. 7.4 that $x_c^2 + y_c^2 = d^2$ (square of the distance between the parallel z and z' axes), from Eq. (7.4) that

$$I_{z_c} = \int_m (x'^2 + y'^2)\, dm$$

and from the fact that x' and y' are measured from the mass center that $\int_m x' \, dm = \int_m y' \, dm = 0$. Thus

$$I_z = I_{z_c} + md^2 = I_c + md^2 \qquad (7.8)$$

where I_c is the moment of inertia of mass of the body with respect to an axis parallel to the z axis and passing through the mass center. We call Eq. (7.8) the *parallel axis theorem for rigid bodies*, which states that the moment of inertia of mass of a body with respect to any axis is equal to the moment of inertia of mass with respect to a parallel axis passing through the mass center I_c plus the product of the mass and the distance squared between the two parallel axes md^2.

We can develop a similar relationship for products of inertia. The product of inertia of the mass dm with respect to the xy and yz planes is given by

$$dI_{xz} = xz \, dm = (x_c + x')(z_c + z') \, dm$$
$$= x_c z_c \, dm + z_c x' \, dm + x_c z' \, dm + x'z' \, dm$$

which on integration over the mass m becomes

$$I_{xz} = x_c z_c \int_m dm + z_c \int_m x' \, dm + x_c \int_m z' \, dm + \int_m x'z' \, dm$$
$$I_{xz} = x_c z_c m + I_{xz_c} \qquad (7.9)$$

since $\int_m x' \, dm = \int_m z' \, dm = 0$. We call Eq. (7.9) the *parallel plane theorem for rigid bodies* since I_{xz} is the product of inertia of the body with respect to xy and yz planes that are parallel to the $x'y'$ and $y'z'$ planes. Similarly,

$$I_{yz} = y_c z_c m + I_{yz_c} \qquad (7.10)$$

If we let the xy plane and the $x'y'$ plane be the same, then $z_c = 0$, and $I_{xz} = I_{xz_c}$ and $I_{yz} = I_{yz_c}$. Thus when the xy plane passes through the mass center, the products of inertia I_{xz} and I_{yz} are invariant, which holds regardless of the location of the origin in the xy plane. This is not true for I_{xy}, however.

Radius of Gyration—The concept of radius of gyration of mass is a useful means for expressing the moment of inertia of mass of a body with respect to a given axis since its dimensions are proportional to the product of the mass of the body and a length squared. Hence we define the radius of gyration of mass of a rigid body with respect to an axis as the length that must be squared and multiplied by the mass of the body to give the moment of its inertia with respect to that axis; in equation form

$$I_z = k_z^2 m \tag{7.11}$$

An interpretation of the radius of gyration is that it is the distance from the axis where we can imagine the entire mass of the body to be concentrated as a thin circular ring with the same moment of inertia of mass. The radius of gyration does not locate any particular point in the body such as the mass center, but it is useful when working with machine parts that have the same geometric shape but are made from different homogeneous materials. Finally, from the parallel axis theorem we can show that the radius of gyration for an axis 2 ft from another parallel axis passing through the mass center does not differ by 2 ft:

$$I_z = I_c + md^2, \qquad mk_z^2 = mk_c^2 + md^2, \qquad k_z = (k_c^2 + 4)^{1/2}$$

unless, of course, k_c^2 is very small compared to 4 in this example.

Moments of Inertia of Composite Bodies—In engineering practice we often encounter rigid bodies made up of a number of simpler shapes (or bodies) such as rods, cylinders, disks, spheres, and cones. We can use the parallel axis (or parallel plane) theorem to obtain the moment (or product) of inertia of the entire body with respect to a given axis (or planes) provided we can determine the moment (or product) of inertia for each of the simpler bodies with respect to a parallel axis (or planes) passing through the mass center of that body. The moment (or product) of inertia of the entire body is simply the algebraic sum of the moments (or products) of inertia of each of the simpler bodies with respect to the same axis (or planes). See Table 7.1. The following example will illustrate the bookkeeping procedure used.

EXAMPLE 7.4

Body 1 in Fig. E7.4 is a nonhomogeneous cylinder weighing 483 lb with its mass center at C; body 2 is a slender homogeneous bar weighing 64.4 lb; and body 3 is a homogeneous sphere weighing 966 lb. The radius of gyration of mass of body 1 with respect to an axis through A perpendicular to the page is 2.0 ft, and all three bodies are symmetrical with respect to the plane of the page. Determine (a) the mass moment of inertia of the entire assembly with respect to an axis perpendicular to the page passing through the mass center of the sphere (body 3) and (b) the mass product of inertia I_{xz} for the coordinates shown.

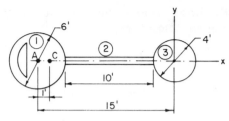

Fig. E7.4

SOLUTION

(a) The mass moment of inertia of the composite body can be obtained from the parallel axis theorem given by Eq. (7.8), which is modified to be

$$I_z = \sum_{i=1}^{3} I_{zi} = \sum_{i=1}^{3} (I_{ci} + m_i d_i^2) \tag{a}$$

for this case. The simplest way to use Eq. (a) is to construct the following table:

(a) Body i	(b) Mass m_i (slugs)	(c) d_i (ft)	(d) $m_i d_i^2$ (slug-ft^2)	(e) I_{ci} (slug-ft^2)	(f) I_{zi} (slug-ft^2)
1	15.0	14.0	2940	45.0	2985.0
2	2.0	7.0	98	16.7	114.7
3	30.0	0	0	48.0	48.0
					3147.7

The masses of the various bodies are obtained from $m_i = W_i/32.2$, while the transfer distances from the mass center of each body to the mass center of body 3 are obtained from the geometry of Fig. E7.4 and entered in column (c). The computation for column (d) is straightforward.

The values entered in column (e) represent the mass moment of inertia for each body with respect to an axis parallel to the z axis passing thru the mass center. From the data for body 1, $k_A = 2.0$ ft so that $I_A = mk_A^2$ and

$$I_{c1} = I_A - ma^2 = 15(2)^2 - 15(1)^2 = 45.0 \text{ slug-ft}^2$$

from Eq. (7.8) where $a = 1.0$ ft is the transfer distance between points A and C. Body 2 is recognized to be a slender homogeneous bar, so from Table 7.1

$$I_{c2} = m\ell^2/12 = 2(10)^2/12 = 16.7 \text{ slug-ft}^2$$

and body 3 is a sphere, so from Table 7.1

$$I_{c3} = (2/5)mr^2 = (2/5)(30)(2)^2 = 48.0 \text{ slug-ft}^2$$

The values entered in column (f) are the sum of columns (d) and (e) for each body. The mass moment of inertia is the sum of values in column (f) which has a total of

$$I_z = 3150 \text{ slug-ft}^2 \qquad \qquad Ans.$$

for slide rule accuracy.

(b) The product of inertia for the composite body can be obtained from the parallel plane theorem [see Eq. (7.9)], which is modified to be

$$I_{xz} = \sum_{i=1}^{3} I_{xz_i} = \sum_{i=1}^{3} I_{xz_{c_i}} + m_i x_i z_{c_i} \qquad \text{(b)}$$

for this case. The problem states that each body is symmetrical with respect to the plane of the page, so

$$I_{xz_{c_1}} = I_{xz_{c_2}} = I_{xz_{c_3}} = 0$$

and $z_{c_1} = z_{c_2} = z_{c_3} = 0$. Thus Eq. (b) gives

$$I_{xz} = 0 \qquad \qquad Ans.$$

This result will always occur when (a) each body making up the composite body is itself symmetrical with respect to the xy plane and (b) the mass center of each body lies in the xy plane. Hence we can ignore products of inertia as long as these two requirements are satisfied.

7.4   Equations of Motion for a Rigid Body in Plane Motion

We can use Eq. (7.2) to obtain the motion of the mass center of a rigid body in plane motion. Plane motion of a rigid body is the superposition of translation and rotation about an axis in such a manner that the angular velocity and acceleration vectors remain parallel. The principle of motion of the mass center provides no information about either the angular quantities describing the angular motion of the body or the location of the line of action of the resultant external force.

Let point A be any point in the plane of motion passing through the mass center C of the rigid body in plane motion (Fig. 7.5). The XYZ coordinate system is fixed in space, while the xyz coordinate system is attached to and rotates with the rigid body and has unit vectors i, j, k as shown. Let dm be the mass of a small volume of the body which is located with respect to A by r_2, the mass center by r_1, and the fixed point 0 by R. The mass center C is located with respect to A by r_{AC} and point A is located with respect to 0 by R_A. Note that r_{AC} lies in the plane of motion (which passes through the

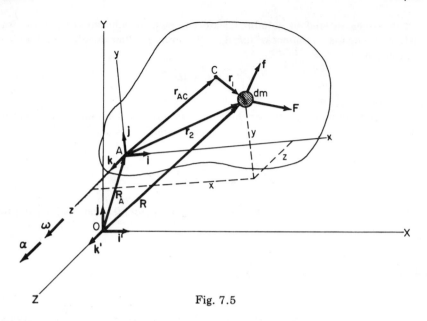

Fig. 7.5

mass center and point A). The resultant external and internal forces acting on dm are **F** and **f** respectively. The angular velocity and acceleration of the body are assumed to be positive counterclockwise.

The moment of forces **F** and **f** (which act on the differential mass dm) with respect to point A is

$$d\mathbf{M}_A = \mathbf{r}_2 \times (\mathbf{F} + \mathbf{f}) \tag{a}$$

but from Newton's second law for a particle we have

$$\mathbf{F} + \mathbf{f} = dm\ \mathbf{a} = dm\ \ddot{\mathbf{R}} \tag{b}$$

so that Eq. (a) can be written as

$$d\mathbf{M}_A = \mathbf{r}_2 \times (\mathbf{F} + \mathbf{f}) = \mathbf{r}_2 \times \ddot{\mathbf{R}}\ dm \tag{c}$$

which is a vector moment equation. Integration of Eq. (c) over the entire mass of the body gives the moment of all external forces with respect to point A since the resultant moment of all internal forces is zero (recall that internal forces always occur in equal and opposite pairs and hence cancel out). Thus

$$\mathbf{M}_A = \int_m \mathbf{r}_2 \times (\mathbf{F} + \mathbf{f}) = \int_m \mathbf{r}_2 \times \ddot{\mathbf{R}}\ dm \tag{d}$$

as a formal form of the desired relationship. It is instructive, however, to evaluate this integral in terms of its scalar components
M_{Ax}, M_{Ay}, M_{Az}.

First, the acceleration of the differential mass dm in the rigid
body can be written as

$$\ddot{R} = a_A + a_{dm/A} = a_A + r\alpha e_t + r\omega^2 e_n$$

as in Fig. 7.6a where

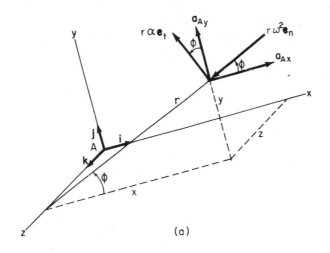

(a)

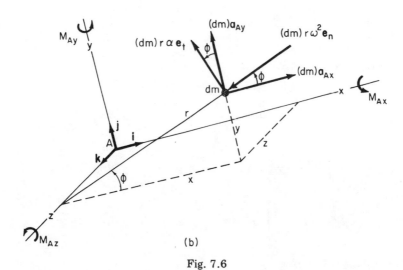

(b)

Fig. 7.6

$$\mathbf{e}_t = -(y/r)\mathbf{i} + (x/r)\mathbf{j} = -\sin\phi\,\mathbf{i} + \cos\phi\,\mathbf{j}$$

$$\mathbf{e}_n = -(x/r)\mathbf{i} - (y/r)\mathbf{j} = -\cos\phi\,\mathbf{i} - \sin\phi\,\mathbf{j}$$

$$\mathbf{a}_A = a_{Ax}\mathbf{i} + a_{Ay}\mathbf{j}$$

and r is the distance from the z axis to the mass dm. Second, according to Eq. (b) the forces $\mathbf{F} + \mathbf{f}$ shown in Fig. 7.5 as acting on mass dm can be replaced by $dm\,\ddot{\mathbf{R}}$ as in Fig. 7.6b. Third, according to Eq. (c) the moment of the forces $\mathbf{F} + \mathbf{f}$ is the same as the moment of $dm\,\ddot{\mathbf{R}}$. Thus, the three scalar moment components can be written down for mass dm from Fig. 7.6b,

$$dM_{Ax} = -\alpha(xz)\,dm + \omega^2(yz)\,dm - a_{Ay}\,z\,dm$$

$$dM_{Ay} = -\alpha(yz)\,dm - \omega^2(xz)\,dm + a_{Ax}\,z\,dm$$

$$dM_{Az} = \alpha r^2\,dm + a_{Ay}x\,dm - a_{Ax}\,y\,dm$$

On integration over mass m these equations become

$$M_{Ax} = -\alpha \int_m xz\,dm + \omega^2 \int_m yz\,dm - a_{Ay}\int_m z\,dm$$

$$M_{Ay} = -\alpha \int_m yz\,dm - \omega^2 \int_m xz\,dm + a_{Ax}\int_m z\,dm \qquad (e)$$

$$M_{Az} = \alpha \int_m r^2\,dm + a_{Ay}\int_m x\,dm - a_{Ax}\int_m y\,dm$$

Fourth, we recognize that

$$I_{xz} = \int_m xz\,dm \qquad\qquad mx_c = \int_m x\,dm$$

$$I_{yz} = \int_m yz\,dm \qquad\qquad my_c = \int_m y\,dm$$

$$I_{zA} = I_A = \int_m r^2\,dm \qquad mz_c = \int_m z\,dm = 0$$

so that Eqs. (e) become

$$M_{Ax} = -I_{xz}\alpha + I_{yz}\omega^2$$

$$M_{Ay} = -I_{yz}\alpha - I_{xz}\omega^2 \qquad (7.12)$$

$$M_{Az} = I_A\alpha + x_c ma_{Ay} - y_c ma_{Ax}$$

which relate the moments of the external forces about the x, y, and z axes to the angular motion of the rigid body at a given instant. Equations (7.12) can be written in vector form similar to Eq. (d) giving

$$\mathsf{M}_A = \mathsf{r}_{AC} \times ma_A + (-I_{xz}\alpha + I_{yz}\omega^2)\mathsf{i}$$
$$- (I_{yz}\alpha + I_{xz}\omega^2)\mathsf{j} + I_A\alpha\mathsf{k} \qquad (7.13)$$

where $\mathsf{r}_{AC} = x_c\mathsf{i} + y_c\mathsf{j}$ (see Fig. 7.5). Equation (7.13) relates the moments of the external forces with respect to point A to the angular motion of the rigid body at a given instant. The student should verify that Eqs. (7.12) and (7.13) are exactly the same.

The product of inertia terms I_{xz} and I_{yz} are a measure of the lack of symmetry of the body with respect to the xy plane. Hence the body tries to rotate about an axis other than the z axis so that the external force system must produce moments about the x and y-axes such that these terms are opposed. These product of inertia terms are zero when the body is symmetrical with respect to the plane of motion. In all problems in this chapter, it is assumed that the body is symmetrical with respect to the plane of motion. Hence the moment equation reduces to

$$\mathsf{M}_A = M_{Az}\mathsf{k} = \mathsf{r}_{AC} \times ma_A + I_A\alpha\mathsf{k} \qquad (7.14)$$

Equation (7.14) can be reduced to a simpler form by elimination of the cross-product term. This occurs when (1) point A is the mass center C since $\mathsf{r}_{AC} = 0$ and is always a valid point to use; (2) the acceleration of point A is zero ($a_A = 0$), which implies point A must be a point on a fixed axis of rotation (pure rotation); (3) the acceleration of point A and r_{AC} are parallel vectors, which occurs so infrequently that we shall ignore this possibility. Thus Eq. (7.14) can be written as

$$\mathsf{M}_A = I_A\,\alpha\,\mathsf{k} \qquad (7.15)$$

when A is either (1) the mass center (always valid) or (2) the fixed axis of rotation for pure rotation problems where I_A is the moment of inertia of mass about the z axis, which is perpendicular to the plane of motion and passing through point A.

Equation (7.15) has the three scalar equations

$$M_{Ax} = 0 \tag{7.16a}$$

$$M_{Ay} = 0 \tag{7.16b}$$

$$M_{Az} = I_A \alpha \tag{7.16c}$$

which state that the moments of the external forces about the x and y axes must be zero, while the moment of the external forces about the z axis is related to the angular acceleration of the rigid body.

The moment of inertia of mass I_A in Eq. 7.16c is a measure of the *angular inertia* of the body, just as mass m in Eq. (7.2) is a measure of the *linear inertia*. This angular inertia depends not only on how large the body mass is but also on how this mass is distributed relative to the axis of rotation thru point A.

We can conclude that the equations of motion for a symmetrical rigid body in plane motion, which are always valid, consist of the principle of motion of the mass center given by $\mathbf{F} = m\ddot{\mathbf{r}}_c = m\mathbf{a}_c$ and the moment equation given by

$$\mathbf{M}_A = I_A \, \alpha \, \mathbf{k} \tag{7.15}$$

where A is either (1) the mass center (always valid) or (2) fixed axis of rotation for pure rotation problems (as a special case).

Since plane motion is the superposition of translation and pure rotation (rotation about a fixed axis), these equations apply to these rigid-body motions as well and are summarized in Table 7.2. The most basic approach to solving rigid-body dynamics problems (which include plane motion, pure rotation, and translation) is to work from the mass center with one exception, that of pure rotation where we have a choice of applying Eq. (7.15) to either the axis of rotation or the mass center.

Table 7.2.   Rigid body equations of motion

Type of motion	Force equation	Moment equation
Translation*	$\mathbf{F} = m\ddot{\mathbf{r}}_c$	$\mathbf{M}_c = 0$ since $\boldsymbol{\omega} = \boldsymbol{\alpha} = 0$
Pure rotation†	$\mathbf{F} = m\ddot{\mathbf{r}}_c$	$\mathbf{M}_A = I_A \, \alpha \, \mathbf{k}$
Plane motion*	$\mathbf{F} = m\ddot{\mathbf{r}}_c$	$\mathbf{M}_c = I_c \, \alpha \, \mathbf{k}$

Note: The products of inertia are zero when either the xy plane is a plane of symmetry or the axis of rotation thru the mass center is an axis of symmetry.

*We can always apply the moment equation to the mass center without fear of neglecting $\mathbf{r}_{AC} \times (m\mathbf{a}_A)$ since $\mathbf{r}_{AC}$ is zero for this case. See Eq. (7.14).

†A is either the mass center or a fixed ($\mathbf{R}_A = \mathbf{v}_A = \mathbf{a}_A = 0$) axis of rotation. Note that the value of I_A depends on whether A is the mass center or the axis of rotation.

A basic procedure is used in solving dynamics problems where a mathematical model of physical reality is being formulated so that the physical behavior of the system can be predicted or explained. There are five steps to follow:

1. Study the problem carefully to ascertain what information is available, what additional information may be needed, what is to be found, and what approach may be required to solve for the unknown information.
2. Draw a complete free-body diagram of the body (or bodies) showing all the forces acting on each. Are any assumptions as to the direction, magnitude, or source of one of the forces or moments being made? This is particularly important when dealing with forces involving friction.
3. Decide what kinematic description (plane motion, pure rotation, or translation) is required to describe the motion of each body. Are there any kinematic equations relating the motions (linear and/or angular) of each body?
4. Apply the equations of motion as given in Table 7.2 for a rigid body to each free-body diagram. Recall that each of these two vector equations has three scalar equations, some of which may be identically satisfied due to zero terms on both sides.
5. Substitute the data from the problem and solve for the desired unknowns. Keep in mind that a consistent system of units and a consistent sign convention must be employed in all the equations.

The following examples illustrate the use of the equations of motion for problems in translation, pure rotation, and plane motion.

EXAMPLE 7.5

The hangar deck of an aircraft carrier can be quickly divided into smaller areas by fire doors to localize a fire. These doors are $24' \times 40' \times 1'$ thick, weigh 1 ton, are supported by two rollers on a track, and are opened or closed by a cable (Fig. E7.5a). Determine the maximum acceleration of a typical fire door if neither roller is to leave (lift-off) the track.

SOLUTION

Draw a free-body diagram of the door (Fig. E7.5b) where we have neglected frictional forces at rollers A and B, replaced the cable by an unknown force F, and calculated the perpendicular distances from the mass center to the lines of action of each force. We recognize that the motion is one of *translation* and the motion of the door will be horizontal. Thus we establish the xyz coordinate system (z is perpendicular to the page) as shown and note that the kinematic conditions are $\omega = \alpha = y_c = z_c = 0$. Application of equations of motion for

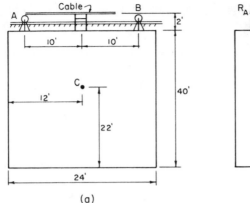

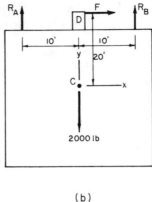

(a)   (b)

Fig. E7.5

translation to find $\ddot{x}_c$ maximum gives

$$\overset{\longleftrightarrow}{+}\, F_x = m\ddot{x}_c, \qquad F = m\ddot{x}_c = (2000/32.2)\ddot{x}_c \quad \text{(contains answer)} \qquad \text{(a)}$$

$$\overset{\uparrow}{+}\, F_y = m\ddot{y}_c = 0, \qquad R_A + R_B - 2000 = 0 \qquad \text{(b)}$$

$$\overset{\curvearrowright}{+}\, M_{cz} = 0, \qquad -10R_A - 20F + 10R_B = 0$$

or

$$R_B - R_A = 2F \qquad \text{(c)}$$

Both Eqs. (b) and (c) give F as a function of R_A and R_B. Subtracting Eqs. (b) from Eqs. (c),

$$F = 1000 - R_A \qquad \text{(d)}$$

which is a maximum when $R_A = 0$ since R_A cannot be negative due to the roller lifting off the track. Hence we conclude from Eq. (d) that $F_{max} = 1000$ lb so that Eqs. (a) give

$$\ddot{x}_{c\,max} = F_{max}/m = (1000/2000)(32.2) = 16.1 \text{ fps}^2 \rightarrow \qquad Ans.$$

The free-body diagram and the recognition of the type of motion of the rigid body are vital in problem solving. We can approach the problem most efficiently when we look for the equation of motion that contains the answer and then ask which of the remaining equations will provide the necessary additional information to obtain the desired result.

EXAMPLE 7.6

The odd-shaped 64.4-lb body in Fig. E7.6a is symmetrical with respect to the plane of motion and is rotating about the axle at 0 with an angular velocity

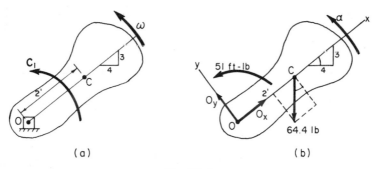

Fig. E7.6

of 4 rad/s counterclockwise when in the position shown. The radius of gyration of mass with respect to the mass center is 3 ft, and a 51 ft-lb couple C_1 is applied to the body. Determine (a) the angular acceleration of the body and (b) the bearing reactions acting on the body at this instant. Neglect bearing friction.

SOLUTION

(a) The motion of the body is pure rotation about the axis through 0. For this type of problem, it is convenient to attach the xy coordinate system in the normal and tangential directions (Fig. E7.6b) since the acceleration of the mass center is easily computed. Hence the pin reactions are usually assumed to be in the normal and tangential directions as well.

Since the body is in pure rotation, we can sum moments about either the mass center or the fixed axis of rotation. Usually, summing moments about the fixed axis of rotation is advantageous since the bearing reactions 0_x and 0_y are eliminated. Thus from Fig. E7.6b, on taking α positive counterclockwise,

$$\overset{+}{\curvearrowleft}\Sigma M_0 = I_0\,\alpha = (I_c + md^2)\alpha, \quad 51 - 2(4/5)(64.4) = [2(3)^2 + 2(2)^2]\alpha = 26\alpha$$

or

$$\alpha = (51 - 103)/26 = 2.0 \text{ rad/s}^2 \qquad\qquad Ans.$$

i.e., α is clockwise rather than counterclockwise as assumed initially.

(b) The bearing reactions can be obtained from applying the principle of motion of the mass center to the free-body diagram of Fig. E7.6b, giving

$$\overset{\nearrow}{}\Sigma F_x = ma_{cx}, \quad 0_x - (3/5)(64.4) = 2A_{cx} = -2(2)(4)^2 = -64$$

or

$$0_x = 38.6 - 64 = -25.4 = 25.4 \text{ lb} \quad \overset{3}{\diagup_4} \qquad\qquad Ans.$$

for the normal direction and

$$\overset{\searrow}{}\Sigma F_y = ma_{cy}, \quad 0_y - (4/5)(64.4) = 2a_{cy} = 2(2)(-2) = -8$$

or

$$0_y = 51.5 - 8 = 43.5 \text{ lb}$$ Ans.

for the tangential direction where it is noted that a_{cy} is negative for this problem since α is clockwise.

EXAMPLE 7.7

The proposed design of the impact pendulum in Fig. E7.7a is to consist of a slender rigid bar AB of length ℓ and mass m. The pendulum is released from rest and strikes the anvil at D. Determine the location of the anvil so that the horizontal pin reaction at A is zero throughout the impact.

SOLUTION

The slender bar has pure rotation; hence summation of moments about the axis of rotation at A in Fig. E7.7b gives

$$\circlearrowright \Sigma M_A = I_A \alpha, \quad aF = I_A \alpha = [(m\ell^2/12) + (m\ell^2/4)]\alpha = (m\ell^2/3)\alpha \quad \text{(a)}$$

where $I_c = m\ell^2/12$ for a slender bar. We can obtain an expression for F in terms of the angular acceleration by applying the principle of motion of the mass center in the horizontal direction to obtain

$$\longleftrightarrow \Sigma F_x = m\ddot{x}_c, \quad F = m(\ell/2)\alpha \quad \text{(b)}$$

since $R_{Ax} = 0$. Substitution of Eq. (b) into Eq. (a) gives

$$a = I_A \alpha / [m(\ell/2)\alpha] = 2m\ell^2/3m\ell = (2/3)\ell \quad \text{Ans. (c)}$$

where the location of point D (anvil) is called the *center of percussion*.

We can obtain the same results working from the mass center:

$$\circlearrowright \Sigma M_{cz} = I_c \alpha, \quad F[a - (\ell/2)] = I_c \alpha = (m\ell^2/12)\alpha$$

or

$$aF = (\ell/2)F + (m\ell^2/12)\alpha$$

where substitution of Eq. (b) gives

$$aF = am\ell\alpha/2 = [(m\ell^2/12) + (m\ell^2/4)]\alpha = (m\ell^2/3)\alpha$$

as before, where we recognize that $I_A = m\ell^2/3$. Hence regardless of whether we work from the mass center or the fixed axis of rotation, we obtain the same result.

The idea of the center of percussion is important in the design of shock-loading type of machinery so that the horizontal component of the bearing reaction will be zero. Note from Eq. (c) that the location of the center of percussion is independent of both the mass of the body m and the angular ac-

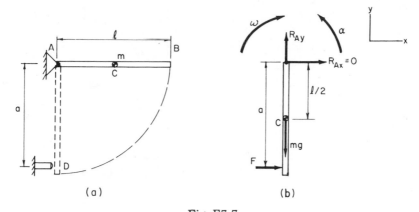

Fig. E7.7

celeration α. We can generalize Eq. (c) by noting that $I_A/m = k_A^2$ is the square of the radius of gyration of mass with respect to the axis of rotation and that $\ell/2 = d$ is the distance from the mass center to the axis of rotation. Thus Eq. (c) can be written as

$$a = (I_A/m)(1/d) = k_A^2/d = k_c^2/d + d \qquad (d)$$

where k_c is the radius of gyration of mass with respect to the mass center. Hence the location of the center of percussion with respect to the mass center $(a - d)$ is solely a function of the radius of gyration of mass with respect to the mass center and the distance d. The greater the distance d, the closer are the center of percussion and the mass center.

To obtain a feel for the importance of the center of percussion, we need only to recall how much a baseball bat stings our hands when its end hits the ball. Striking the ball at or near the center of percussion leaves no lasting impression.

EXAMPLE 7.8

The winch in Fig. E7.8a is attached to a heavy container by a cable that wraps around the winch. The winch weighs 12 kg and has a radius of gyration with respect to the mass center of 0.5 m. The container weighs 100 kg and the coefficient of friction between the container and the horizontal surface is 0.3. Determine the maximum couple **C** that can be applied to the winch and still have the container slide on the horizontal surface.

SOLUTION

Since we are looking for the maximum couple **C**, we turn our attention to the winch which is a rigid body in pure rotation about its mass center. From the free body of Fig. E7.8c,

$$\underset{\curvearrowright}{\textstyle\bigoplus}\Sigma M_{cz} = I_c\alpha, \qquad C - 0.5T = 12(0.5)^2\alpha$$

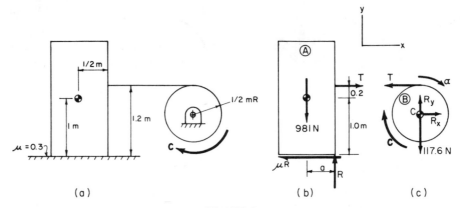

Fig. E7.8

or

$$C = 0.5T + 3\alpha \qquad\qquad (a)$$

To solve for C, we must know $T_{\max}$ and α, so we look at the container (body A), which is translating. Kinematically, the motions of bodies A and B are related by $x_c = r\theta$ or

$$\ddot{x}_c = r\alpha = 0.5\alpha \qquad\qquad (b)$$

where x_c is positive to the right and α is positive clockwise.

The tension T can be obtained from the free body in Fig. E7.8b,

$$(\longleftrightarrow)\ \Sigma F_x = m_A \ddot{x}_c, \qquad T - \mu R = 100\ddot{x}_c$$

or

$$T - 0.3R = 100\ddot{x}_c \qquad\qquad (c)$$

where R can be obtained from

$$(\updownarrow)\ \Sigma F_y = m_A \ddot{y}_c = 0, \qquad R - 981 = 0$$

or $R = 981\ N$.

We still do not know T or $\ddot{x}_c$, so we sum moments about the mass center of body A:

$$(\overset{+}{\curvearrowright})\ \Sigma M_c = I_{c_A}\, \alpha = 0, \qquad aR - (1)(0.3)(R) - 0.2T = 0$$

or

$$T = [(a - 0.3)R]/0.2 \qquad\qquad (d)$$

where a is the distance from the mass center to where R is acting. When $a =$

0.5 m, T is a maximum since a value of T greater than this will cause body A to rotate. Hence setting $a = 0.5$ in Eq. (d) gives

$$T_m = [(0.5 - 0.3)/0.2] R = R = 981 \, N$$

so that from Eq. (c) the maximum acceleration of body A is

$$\ddot{x}_c = (T - 0.3R)/m = [(1 - 0.3)981]/100 = 6.87 \text{ m/s}^2$$

and from Eq. (b) the maximum angular acceleration of body B is

$$\alpha = \ddot{x}_c/r = 6.87/0.5 = 13.74 \text{ rad/s}^2$$

Substitution of these values for T and α into Eq. (a) gives

$$\mathbf{C} = 0.5T + 3\alpha = 0.5(981) + 3(13.74) = 532 \text{ N-m} \qquad\qquad Ans.$$

EXAMPLE 7.9

Body A in Fig. E7.9a weighs 32.2 lb, slides in a smooth horizontal slot, and is connected to a uniform 96.6-lb bar BD by a smooth pin at B. Body A is acted on by 6-lb force $\mathbf{P}$ while bar BD is acted on by 12 ft-lb couple $\mathbf{C}_1$. Determine the linear acceleration of body A and the angular acceleration of bar BD when in the position shown.

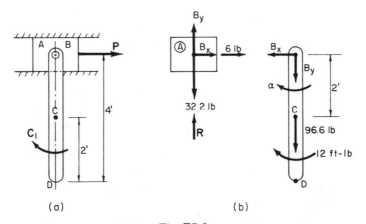

(a)   (b)

Fig. E7.9

SOLUTION

The motion of body A is translation, while that of bar BD is plane motion. Free-body diagrams are shown in Fig. E7.9b from which we have

Body A

$$\left(\overset{\curvearrowright}{\leftrightarrow}\right) \Sigma F_x = m_A a_{Ax}, \qquad 6 + B_x = a_{Ax} \qquad\qquad\qquad (a)$$

Body BD

$$\overset{\longleftrightarrow}{}\ \Sigma F_x = m_B a_{cx}, \qquad -B_x = 3a_{cx} \tag{b}$$

$$\overset{+}{\curvearrowright}\ \Sigma M_c = I_c \alpha, \qquad 12 - 2B_x = \frac{3(4)^2}{12} \alpha = 4\alpha \tag{c}$$

The accelerations are related kinematically by

$$\overset{\longleftrightarrow}{}\ a_{cx} = a_{Ax} - 2\alpha \tag{d}$$

where a_{cx} and a_{Ax} are positive to the right and α is positive clockwise. Combining Eqs. (a), (b), and (d) gives

$$6 = 4a_{Ax} - 6\alpha \tag{e}$$

while combining Eqs. (b), (c), and (d) gives

$$12 = -6a_{Ax} + 16\alpha \tag{f}$$

from which simultaneous solution of Eqs. (e) and (f) gives

$$\alpha = 3 \text{ rad/s}^2 \,\big)\ , \qquad a_A = 6 \text{ fps}^2 \rightarrow \qquad\qquad Ans.$$

7.5   Engineering Problems in Kinetics—Integrated Response

The equations developed in Section 7.4 are used to solve for the instantaneous relationships between forces and moments and the linear and angular motion of the body. It is often necessary to determine not only these instantaneous relationships between forces (and moments) and the accelerated motion of the body but also the linear and angular velocity as a function of time or position. Hence it becomes necessary to integrate the differential equations of motion, which can be done in three basic ways: (1) the formal methods of calculus and differential equations; (2) numerical integration, using a digital computer; or (3) analog integration, using an analog computer.

Consider the rigid body, which is symmetrical with respect to the plane of motion, (Fig. 7.7a) to be in plane motion with n external forces (F_1, F_2, ... F_p, ... F_n) and a resultant couple C acting upon it. We can replace each of the external forces by an equal force acting at the mass center and a couple about the mass center. This gives the equivalent free-body diagram (Fig. 7.7b) where

$$F = \sum_{p=1}^{n} F_p \tag{a}$$

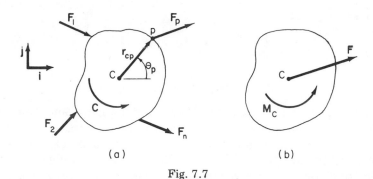

Fig. 7.7

and

$$M_c = C + \sum_{p=1}^{n} r_{cp} \times F_p \qquad (b)$$

are the resultant force and couple acting on the body. In general, any of the external forces F_p can be dependent upon time, position of the point of application (spring), or velocity of the point of application (fluid resistance). Thus

$$F_p = F_p (r_p, v_p, t) \qquad (c)$$

From plane motion kinematics and Fig. 7.7a we can write the general functional position and velocity relationships for point P as

$$r_p = r_c + r_{cp} (\cos \theta_p i + \sin \theta_p j) = r_p (r_c, r_{cp}, \theta, \theta_{0p}) \qquad (d)$$

and

$$v_p = v_c + r_{cp} \omega(-\sin \theta_p i + \cos \theta_p j) = v_p (v_c, r_{cp}, \theta_{0p}, \theta, \omega) \qquad (e)$$

where r_c is the position and v_c is the velocity of the mass center. The angle θ_p is related to the rigid-body angular position θ by $\theta_p = \theta_{0p} + \theta$, where θ_{0p} is a constant angular difference so that the angular velocity of line CP is the same as the angular velocity of the rigid body. Equations (d) and (e) show that the position and velocity of P can be expressed in terms of the position and velocity of the mass center C, the location of P relative to C, and the angular position and angular velocity of the rigid body. Hence Eq. (c) can be written as

$$F_p = F_p (r_c, v_c, r_{cp}, \theta, \omega, t) \qquad (f)$$

where r_{cp} takes into account both r_{cp} and θ_{0p}. In view of Eq. (f) the resultant force and couple given in Eqs. (a) and (b) can be written as

$$F = F(r_c, v_c, r_{cp}, \theta, \omega, t) \tag{g}$$

and

$$M_c = M_c(r_c, v_c, r_{cp}, \theta, \omega, t) \tag{h}$$

Equations (g) and (h) show that the resultant force and couple are dependent on the position and velocity of the mass center, the angular position and angular velocity of the rigid body, the passage of time, and the location of force application to the body.

We conclude that the differential equations of motion for a rigid in-plane motion have a general functional form of

$$[F(r_c, v_c, r_{cp}, \theta, \omega, t)]_x = m\ddot{x}_c = ma_{cx}$$
$$[F(r_c, v_c, r_{cp}, \theta, \omega, t)]_y = m\ddot{y}_c = ma_{cy} \tag{7.17}$$

$$M_c(r_c, v_c, r_{cp}, \theta, \omega, t) = I_c\alpha k \tag{7.18}$$

We cannot state a priori that

$$[F(r_c, v_c, r_{cp}, \theta, \omega, t)]_x = F_x(x_c, v_{cx}, r_{cp}, \theta, \omega, t)$$

until we have analyzed the nature of the external forces that contribute to the resultant force in the x direction because F_x is dependent on y_c and v_{cy} unless proven otherwise.

In many problems an examination of kinematics of body motion and the nature of the forces and moments acting on the body will allow Eqs. (7.17) to reduce to

$$F(x_c, v_{cx}, r_{cp}, \theta, \omega, t) = m\ddot{x}_c \tag{7.19}$$

$$F(y_c, v_{cy}, r_{cp}, \theta, \omega, t) = m\ddot{y}_c \tag{7.20}$$

As written, these equations are dependent (coupled) differential equations of motion since the solution to one is dependent on the solution of the other. These equations can be linear or nonlinear functions of the variables. There are situations where the kinematic conditions of a given problem will uncouple and even possibly reduce these equations to a single differential equation of motion; e.g., see Examples 7.10 and 7.12. In other situations the simplest means of obtaining a solution is to program the differential equations of motion on an analog or digital computer.

In many problems the kinematics and the force system are such that Eqs. (7.18), (7.19), and (7.20) take on special forms that can be solved by direct integration. These are:

1. Forces and couples are functions of position only; i.e.,

$$\left. \begin{array}{l} F_x(x_c) = m\ddot{x}_c \\ F_y(y_c) = m\ddot{y}_c \end{array} \right\} \quad \mathbf{F}(\mathbf{r}_c) = m\ddot{\mathbf{r}}_c \tag{7.21}$$

$$\mathbf{M}_c(\theta) = I_c \alpha\mathbf{k} = I_c\,\omega\frac{d\omega}{d\theta}\,\mathbf{k} \tag{7.22}$$

2. Forces and couples are functions of time only; i.e.,

$$\left. \begin{array}{l} F_x(t) = m\ddot{x}_c \\ F_y(t) = m\ddot{y}_c \end{array} \right\} \quad \mathbf{F}(t) = m\,\ddot{\mathbf{r}}_c = m\frac{d\mathbf{v}_c}{dt} \tag{7.23}$$

$$\mathbf{M}_c(t) = I_c\alpha\mathbf{k} = I_c\frac{d\omega}{dt}\,\mathbf{k} \tag{7.24}$$

3. Forces and couples are functions of velocity only; i.e.,

$$\left. \begin{array}{l} F_x(v_{cx}) = m\ddot{x}_c \\ F_y(v_{cy}) = m\ddot{y}_c \end{array} \right\} \quad \mathbf{F}(\mathbf{v}_c) = m\ddot{\mathbf{r}}_c = m\frac{d\mathbf{v}_c}{dt} \tag{7.25}$$

$$\mathbf{M}_c(\omega) = I_c\alpha\mathbf{k} = I_c\frac{d\omega}{dt}\,\mathbf{k} \tag{7.26}$$

where $\mathbf{F}$ = constant and $\mathbf{M}_c$ = constant are a special case of each of the three forms. There is no need to treat constant forces and moments as a special case since all the above forms can be integrated with constant values on the left-hand side.

7.6 Principle of Work and Energy

Equations (7.21) and (7.22) can be integrated into two basic forms of the principle of work and energy, which are useful in obtaining velocity displacement relationships, particularly in problems involving spring elements and constant forces.

Equations (7.21) and (7.22) are written here for convenience:

$$\mathbf{F}(\mathbf{r}_c) = m\ddot{\mathbf{r}}_c = m\frac{d\mathbf{v}_c}{dt}\,, \qquad \mathbf{M}_c(\theta) = I_c\alpha\mathbf{k} = I_c\,\omega\frac{d\omega}{d\theta}\,\mathbf{k}$$

Recall that Eq. (7.21) describes the motion of the mass center "as though" the rigid body were a particle and Eq. (7.22) describes the angular motion of the rigid body with respect to the mass center.

We can integrate Eq. (7.21) by dotting the equation with the differential displacement $d\mathbf{r}_c$ to obtain

$$\mathbf{F}(\mathbf{r}_c) \cdot d\mathbf{r}_c = m\left(\frac{d\mathbf{v}_c}{dt}\right) \cdot d\mathbf{r}_c \tag{a}$$

where the terms $F(r_c) \cdot dr_c$ represent the differential work done by the resultant force. Only the component of the resultant force in the direction of dr_c can do work and this differential work is positive or negative depending on whether the component of $F(r_c)$ is in the direction of or opposed to dr_c respectively.

To integrate Eq. (a), we must recognize that $dr_c = v_c\, dt$, so that the right-hand side of Eq. (a) becomes

$$\frac{dv_c}{dt} \cdot dr_c = dv_c \cdot v_c = \frac{d(v_c \cdot v_c)}{2} = \frac{d(v_c^2)}{2}$$

since $dv_c = (dv_c/dt)\, dt$. Thus Eq. (a) becomes

$$dWk_{\text{forces}} = F(r_c) \cdot dr_c = (m/2)\, d(v_c^2)$$

which is a scalar equation. Integration gives

$$Wk_{\text{forces}} = \int_{r_{ci}}^{r_{cf}} F(r_c) \cdot dr_c = \int_{v_{ci}^2}^{v_{cf}^2} \frac{m}{2}\, d(v_c)^2 = \frac{m}{2} v_{cf}^2 - \frac{m}{2} v_{ci}^2$$

or

$$Wk_{\text{forces}} = (m/2)v_{cf}^2 - (m/2)v_{ci}^2 = \Delta(\text{K.E.})_{\text{translation}} \qquad (7.27)$$

where the subscripts i and f refer to the initial and final linear positions and velocities respectively. Equation (7.27) states that the work done by the external forces due to the translation of the mass center (as though the body did not rotate) is equal to the change in the kinetic energy (K.E.) of the mass center of the body.

Equation (7.22) can be rewritten as

$$M_c(\theta)\, d\theta = I_c\, \omega\, d\omega \qquad (b)$$

because the moments about the x and y axes are zero for a rigid body in plane motion. Hence these moments will do no work. The left-hand side of Eq. (b) represents the differential work done by the external couples with respect to the mass center. We see that the variables are separated in Eq. (b), so that direct integration gives

$$Wk_{\text{couples}} = \int_{\theta_i}^{\theta_f} M_c(\theta)\, d\theta = \int_{\omega_i}^{\omega_f} I_c\, \omega\, d\omega = \frac{I_c}{2} \omega_f^2 - \frac{I_c}{2} \omega_i^2$$

or

$$Wk_{\text{couples}} = (I_c/2)\omega_f^2 - (I_c/2)\omega_i^2 = \Delta(\text{K.E.})_{\text{rotation}} \qquad (7.28)$$

where the subscripts i and f refer to the initial and final angular positions and velocities respectively. Equation (7.28) shows that work done by the external couples with respect to the mass center is equal to the change in the rotational kinetic energy of the rigid body due to rotation about the mass center. The rotational kinetic energy of a rigid body in plane motion (due to rotation about the mass center) is given by $(1/2)I_c\,\omega^2$.

Equations (7.27) and (7.28) are independent scalar equations as written, which may or may not be coupled kinematically; and as such they constitute one important form of the principle of work and energy. The second important form can be obtained by adding Eqs. (7.27) and (7.28) (since they are scalar equations) to obtain

$$Wk_{\text{forces}} + Wk_{\text{couples}} = (1/2)mv_{cf}^2 + (1/2)I_c\,\omega_f^2$$
$$- (1/2)mv_{ci}^2 - (1/2)I_c\,\omega_i^2$$

or

$$Wk = \Delta(\text{K.E.}) = \Delta[(1/2)mv_c^2 + (1/2)I_c\,\omega^2] \qquad (7.29)$$

where Wk is the total work done by the external forces and couples. The total kinetic energy of the body is the sum of the translational kinetic energy of the mass center plus the rotational kinetic energy with respect to the mass center. We can show that the total kinetic energy of a rigid body in plane motion is given as

$$(1/2)mv_c^2 + (1/2)I_c\,\omega^2$$

by integrating the kinetic energy expression for a differential mass dm over the entire mass of the body to obtain

$$T = \int_m \frac{(\mathbf{v}\cdot\mathbf{v})}{2}\,dm$$

where $\mathbf{v} = \mathbf{v}_c + r\omega\mathbf{e}_t$ and r is measured from the mass center. Equation (7.29) tells us that the work done on the rigid body by the external forces and couples is equal to the change in the kinetic energy of the rigid body.

The units of work are ft-lb or, N-m while those of kinetic energy are slug-ft^2/s^2 (which reduce to ft-lb) or kg-m^2/s^2 (which reduce to N-m).

Since translation and pure rotation are special cases of plane motion, Eq. (7.29) reduces to

$$Wk = \Delta[(1/2)mv_c^2] \qquad \text{(translation)}$$

which is the same as Eq. (7.27) and

$$Wk = \Delta[(1/2)mv_c^2 + (1/2)I_c\,\omega^2] = \Delta[(1/2)I_A\,\omega^2]$$

$$\text{(pure rotation)} \qquad (7.30)$$

since $v_c = r\omega$ for a rigid body in pure rotation, where r is the distance from the axis of rotation at A to the mass center at C.

When working with multibody problems, we can use Eq. (7.29) for the system of bodies if we interpret the right-hand side as the change in kinetic energy of the system. The kinetic energy for a system of connected bodies is the sum of the kinetic energies of the individual bodies; i.e.,

$$T_{\text{system}} = \sum_{i=1}^{n} \left(\frac{1}{2} m_i v_{ic}^2 + \frac{1}{2} I_{ci} \omega_i^2 \right) \tag{7.31}$$

The kinetic energy equations are summarized in Table 7.3. In multibody problems there are usually several kinematic relationships so that Eqs. (7.29) and (7.31) can be written in terms of a single coordinate. In addition, note that the internal forces do no net work on the system of bodies because the internal forces always occur in equal and opposite pairs while the displacement associated with the internal force is the same for both bodies.

Table 7.3.   Kinetic energy equations for rigid bodies

Type of motion	Single body	n Bodies
Translation	$(1/2)mv_c^2$	$\displaystyle\sum_{i=1}^{n} \frac{1}{2} m_i v_{ci}^2$
Pure rotation*	$(1/2)mv_c^2 + (1/2)I_c\omega^2 = (1/2)I_A\omega^2$	$\displaystyle\sum_{i=1}^{n} \frac{1}{2} I_{Ai}\omega_i^2$
Plane motion†	$(1/2)mv_c^2 + (1/2)I_c\omega^2$	$\displaystyle\sum_{i=1}^{n} \left(\frac{1}{2} m_i v_{ci}^2 + \frac{1}{2} I_{ci}\omega_i^2 \right)$

*$v_c = r\omega$ for body in pure rotation.
†Most general form with translation and pure rotation as special cases.

Work. Two force systems occur so often that the work expression merits special discussion. These are the work done by a constant force (i.e., weight of the body) and that done by spring elements.

When a constant force $\mathbf{F}$ (magnitude and direction) moves from A to B (Fig. 7.8), the work done is given by

$$Wk = \int_{r_i}^{r_f} \mathbf{F} \cdot d\mathbf{r} = \mathbf{F} \cdot \int_{r_i}^{r_f} d\mathbf{r} = \mathbf{F} \cdot (\mathbf{r}_f - \mathbf{r}_i)$$

$$= \mathbf{F} \cdot \mathbf{u} = |\mathbf{F}||\mathbf{u}| \cos\theta \tag{c}$$

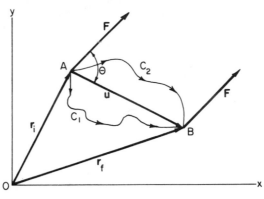

Fig. 7.8

which we see is independent of the path from A to B; i.e., the work done is dependent only on the end points and is independent of the path the point of application follows. Thus whether the point of application follows curve C_1 or C_2 or the straight line, the work done is the same as given by Eq. (c).

A spring is a body that deforms under the action of a force (or couple). The relationship between the force (or couple) and the deformation depends on the material used, the shape, and the dimensions of the spring body as described in Chapter 10. Usually most springs exhibit linear behavior between load and deformation so that

$$F = ks \qquad (7.32)$$

where k = the *spring modulus* with units of lb/ft or N/m
       s = the deformation from the unstretched length with units of ft or m

and

$$M = k_t\theta \qquad (7.33)$$

where k_t = the *torsional spring modulus* with units of ft-lb/rad or N-m/rad
       θ = the angular deformation from the unstretched position with units of rad

The work done on the spring of free length ℓ (Fig. 7.9a) in stretching from s_2 to s_3 (Fig. 7.9b) is

$$Wk_{2 \to 3 \text{ spring}} = \int_{r_2}^{r_3} \mathbf{F} \cdot d\mathbf{r} = k \int_{s_2}^{s_3} s\, ds = \frac{ks_3^2}{2} - \frac{ks_2^2}{2} \qquad (d)$$

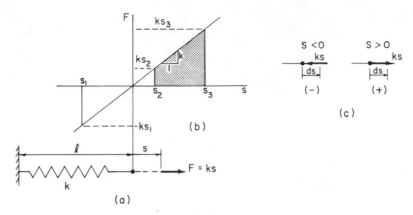

Fig. 7.9

which is proportional to the shaded area since the area is given by

$$Wk_{2\to3 \text{ spring}} = F_{\text{ave}} \, \Delta s = \frac{k(s_3 + s_2)}{2} (s_3 - s_2) = \frac{k}{2} s_3^2 - \frac{k}{2} s_1^2 \qquad (e)$$

The work done on the spring in moving from a compressed condition at 1 to an extended position at 3 becomes

$$Wk_{1\to3 \text{ spring}} = \int_{s_1}^{s_3} ks \, ds = \frac{k}{2} s_3^2 - \frac{k}{2} s_1^2 \qquad (f)$$

where negative work is done on the spring when $s < 0$ since ds is to the right and ks is to the left (Fig. 7.9c). Thus the spring equation takes into account whether the work is positive or negative.

The work done on the body by the spring is simply the negative of the work done on the spring by the body. Hence we frequently compute the work done on a spring separately and then take the negative of the spring work as the work done on the rigid body since the work on the spring can be computed by means of Eqs. (d), (e), or (f). Similar equations can be developed for the torsional spring where k is replaced by k_t and s by θ.

Power. Power is defined as the time rate of doing work. Thus

$$P = \frac{dWk}{dt} = \mathbf{F} \cdot \frac{d\mathbf{r}}{dt} = \mathbf{F} \cdot \mathbf{v} \qquad (7.34)$$

which is a scalar quantity. The units of power are ft-lb/s in the English gravitational system of units and N-m/s in the SI system of units. A commonly used power unit is *horsepower* (hp) defined as

$$1 \text{ hp} = 33{,}000 \text{ ft-lb/min} = 550 \text{ ft-lb/s} = 746 \text{ N-m/s}$$

The following examples illustrate the use of the work and energy approach in solving dynamics problems.

EXAMPLE 7.10

A 32.2-lb body A is connected to a 96.6-lb body B by an inextensible cord passing over a frictionless pulley at C (Fig. E7.10a). The spring has a modulus of 16 lb/ft and is initially stretched 1.0 ft. Determine (a) the velocity of body B as a function of its vertical displacement y, (b) the velocity of body B when it has moved 2.0 ft downward, and (c) the displacement y of body B when the two bodies come to rest. Both bodies are initially at rest and the coefficient of friction between body A and the plane is 0.50.

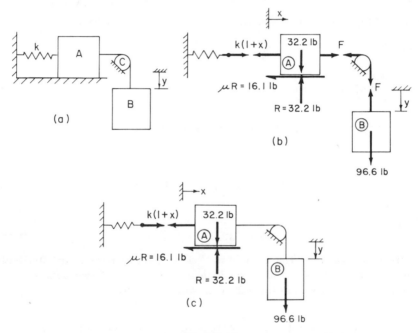

Fig. E7.10

SOLUTION

(a) A free-body diagram of the two bodies is shown in Fig. E7.10b where the spring force is given by $k(1 + x)$ since the spring is initially stretched 1.0 ft. The motion of body A is described in terms of coordinate x, while body B is described in terms of coordinate y. Since the cord is inextensible, the position, velocity, and accelerations are related by

$$x = y, \qquad v_A = v_B = v, \qquad a_A = a_B = a = \frac{dv}{dt} = v\frac{dv}{dy} \qquad \text{(a)}$$

By summing forces in the vertical direction on body A, we find the normal re-action $R = 32.2$ lb. The friction force is 16.1 lb to the left since body A will initially move to the right. At this point we can solve the problem either by integrating the equations of motion or by employing the principle of work and energy. Both methods are illustrated as follows.

Differential equations. From the free-body diagrams we have

Body A $\Sigma F = ma$ Body B

$\overset{\longleftrightarrow}{\bigcirc} -16(1 + x) - 16.1 + F = \ddot{x}$ $\overset{\updownarrow}{\bigcirc} 96.6 - F = 3\ddot{y}$

which can be combined to eliminate the cord force F to give

$$96.6 - 32.1 - 16x = \ddot{x} + 3\ddot{y} \tag{b}$$

In view of Eqs. (a) we can rewrite Eq. (b) in terms of v and y to have

$$64.5 - 16y = 4v \frac{dv}{dy} \tag{c}$$

Integration of Eq. (c) gives

$$\int_0^y (64.5 - 16y)\, dy = 4 \int_0^v v\, dv$$

or

$$64.5y - 8y^2 = 2v^2$$

or

$$v = \sqrt{32.25y - 4y^2} \qquad\qquad Ans.\ (d)$$

Work and energy. The cord force F is an internal force in the free-body diagram in Fig. E7.10c. The work done on the system of two bodies by the external forces is the sum of the individual works given by

$$Wk_{\text{spring}} = -\int_0^y k(1 + y)\, dy = -ky - \frac{ky^2}{2} = -16y - 8y^2$$

$$Wk_{\text{friction}} = -16.1x = -16.1y$$

and

$$Wk_{w_B} = 96.6y$$

or

$$Wk = 64.5y - 8y^2 \tag{e}$$

The kinetic energy of the system is given by

$$(1/2)m_A v_A^2 + (1/2)m_B v_B^2 = 2v^2$$

so that the principle of work and energy gives

$$64.5y - 8y^2 = 2v^2 - 2(0)^2$$

or

$$v = \sqrt{32.25y - 4y^2} \qquad\qquad Ans. \quad (f)$$

as before.

(b) The velocity of body B when $y = 2.0$ ft becomes

$$\mathbf{v} = \sqrt{32.25(2) - 4(4)} = \sqrt{48.5} = 6.96 \text{ fps } \downarrow \qquad\qquad Ans.$$

(c) From either Eq. (e) or (f) we see the velocity is zero when the total work done is zero. Thus from Eq. (f),

$$y^2 - 8.06y = y(y - 8.06) = 0$$

or

$$y = 0$$

and

$$y = 8.06 \text{ ft} \qquad\qquad Ans.$$

The zero root corresponds to the initial position.

EXAMPLE 7.11

The 981-N body 1 in Fig. E7.11a has a diameter of 4 m, rotates about the axis at 0, and is connected to the 294.3-N body 2 by a cord wrapped around body 1. The radius of gyration of mass with respect to the mass center of body 1 is 1.2 m. When in the position shown, the mass center of body 1 is 1 m vertically above the axis of rotation, the angular velocity of body 1 is 2 rad/s clockwise, and the 1000 N/m spring is compressed 0.5 m. The spring is attached to body 1 at point A by a pin. Determine the velocity of body 2 when body 1 has rotated 180° clockwise.

SOLUTION

From the free-body diagram (Fig. E7.11b) we see that the forces 0_x and 0_y will do no work, while the weights W_1 and W_2 do positive work on the system of bodies and the spring does negative work. Also, W_1 and W_2 are constant forces, and the motion of body 2 is related to the motion of body 1 by $y = r\theta$ so that body 2 moves downward a distance of 2π m. Thus, the work done on the system by bodies 1 and 2 by the weight forces becomes

$$Wk_{1,2} = 981(2) + 294.3(2\pi) = 3811 \text{ N-m} \qquad\qquad (a)$$

To determine the work done by the spring, consider the movement of the

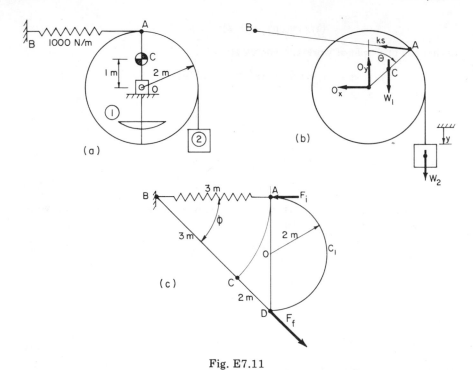

Fig. E7.11

spring itself (Fig. E7.11c). The point of application of the spring force actually follows arc C_1 in moving from its initial position A to its final position D. However, we can also find an equivalent path by allowing the spring to rotate through an angle ϕ so that the initial compressive force F_i is always normal to the arc AC and hence does no work. Then we allow the spring to be elongated along the straight path CD to a final length of 5 m. Since the spring is initially compressed 0.5 m, we conclude its unstretched length ℓ_0 is 3.5 m so that $s_i = -0.5$ m and $s_f = 1.5$ m where the relationship $\ell_s = \ell_0 + s$ must be satisfied. Hence the work done on the spring becomes

$$Wk_{\text{on spring}} = k \int_{s_i}^{s_f} s \, ds = \frac{k}{2}(s_f^2 - s_i^2) = \frac{1000}{2}(2.25 - 0.25) = 1000 \text{ N-m} \quad \text{(b)}$$

which is the work done on the spring in moving from C to D along the straight line. The work done on body 1 (and hence the system of body 1 and 2) is the negative of the value obtained in Eq. (b).

The kinetic energy of the system of body 1 and body 2 is

$$T = (1/2)m_1 v_1^2 + (1/2)I_{c1}\omega_1^2 + (1/2)m_2 v_2^2 + (1/2)I_{c2}\omega_2^2 \quad \text{(c)}$$

but we have from kinematics that

$$v_1 = r\omega = \omega, \qquad \omega_1 = \omega$$
$$v_2 = r\omega = 2\omega, \qquad \omega_2 = 0$$

so that Eq. (c) becomes

$$T = (1/2)[100(1)^2 + 100(1.2)^2 + 30(2)^2]\omega^2 = 182\omega^2 \tag{d}$$

Thus from the principle of work and energy,

$$Wk = \Delta T$$

$$3811 - 1000 = 182(\omega_f^2 - \omega_i^2) = 182[\omega_f^2 - (2)^2] \tag{e}$$

or

$$\omega_f = \sqrt{3540/182} = \sqrt{19.45} = 4.41 \text{ rad/s}$$

so that the final velocity of body 2 becomes

$$v_f = 2\omega_f = 8.82 \text{ m/s} \downarrow \qquad\qquad Ans.$$

Equation (e) can be set up in terms of the initial and final velocities of body 2 just as well as the angular velocity of body 1.

EXAMPLE 7.12

The homogeneous 64.4-pound 4-ft diameter cylinder in Fig. E7.12a has an initial velocity of 5 fps to the right when the cable connected to a spring with a modulus of 96.6 lb/ft becomes taut. The coefficient of friction between the cylinder and the surface is 0.5. Determine the distance the cylinder moves in coming to rest.

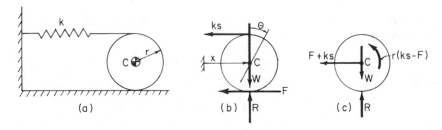

Fig. E7.12

SOLUTION

In the free-body diagram of the cylinder (Fig. E7.12b) F is the friction force between the cylinder and the surface, x is the location of the mass center, and θ is the angular position of the cylinder. All coordinates are referenced to the spring being unstretched so that $s = x + r\theta$. An equivalent force system is shown in Fig. E7.12c, from which

$$(\uparrow)\ \Sigma F_y = ma_{cy} = 0 \qquad \text{or} \qquad R = W$$

$$(\leftrightarrow)\ \Sigma F_x = m\ddot{x} \qquad \text{or} \qquad -F - k(x + r\theta) = m\ddot{x} \tag{a}$$

$$(\curvearrowright)\ \Sigma M_c = I_c\alpha \qquad \text{or} \qquad rF - rk(x + r\theta) = I_c\alpha \tag{b}$$

where the resultant force and couple are functions of both x and θ. Equations (a) and (b) are examples of Eqs. (7.21) and (7.22) in which the desired form has not yet been achieved.

As long as $F \leqslant \mu w$; the coordinates x and θ are related kinematically by

$$x = r\theta \tag{c}$$

$$\dot{x} = v = r\omega \tag{d}$$

$$\ddot{x} = a = r\alpha \tag{e}$$

and Eqs. (a) and (b) reduce to

$$-F - 2kx = m\ddot{x} = mv \frac{dv}{dx} \tag{f}$$

$$rF - 2r^2 k\theta = I_c \alpha = I_c \, \omega \frac{d\omega}{d\theta} \tag{g}$$

which are still coupled. Solving for F in Eq. (f) and making use of Eqs. (c) and (e), Eq. (g) becomes

$$-4kr^2 \theta = (I_c + mr^2)\alpha = (I_c + mr^2)\omega \frac{d\omega}{d\theta} \tag{h}$$

or

$$-128.8\theta \; d\theta = \omega \, d\omega$$

on substitution of $m = 2$ slugs, $I_c = 4$ slug-ft^2, $k = 96.6$ lb/ft, and $r = 2$ ft. Integration gives

$$-128.8\theta^2 = \omega_f^2 - \omega_i^2 \tag{i}$$

where $\omega_f = 0$ and $\omega_i = 2.5$ rad/s so that

$$\theta = \sqrt{6.25/128.8} = \sqrt{0.0485} = 0.220 \text{ rad} \, \circlearrowright$$

or

$$x = 2\theta = 0.440 \text{ ft} \rightarrow \qquad\qquad Ans.$$

This answer is dependent on Eq. (h) being valid, which in turn is dependent on the assumption that $F \leqslant \mu w$. To check the validity of the answer, return to Eq. (h) and solve for α,

$$\ddot{x} = r\alpha = -\frac{4r^2 k(r\theta)}{I_c + mr^2} = -\frac{4r^2 kx}{I_c + mr^2} \tag{j}$$

by virtue of Eqs. (c) and (e). Then substitute Eq. (j) into Eq. (f),

$$F = -2kx + \frac{4mr^2 kx}{I_c + mr^2} = 2kx \left(\frac{mr^2 - I_c}{mr^2 + I_c} \right) \leqslant \mu w$$

Therefore, the values of x for which Eq. (h) is valid give

$$x \leqslant \left(\frac{\mu w}{2k}\right)\left(\frac{mr^2 + I_c}{mr^2 - I_c}\right) = 3\left(\frac{\mu w}{2k}\right) = 3\frac{(0.5)(64.4)}{(2)(96.6)} - 0.50 \text{ ft}$$

which is slightly greater than the answer. The coefficient of friction was used to show that friction can be a real limiting factor in this type of problem.

We can obtain Eq. (i) directly from the principle of work and energy. The solution becomes

$$T = (1/2)mv_c^2 + (1/2)I_c\omega^2 = (1/2)(I_c + mr^2)\omega^2$$

for kinetic energy of a wheel that rolls without slipping;

$$Wk_{\text{spring}} = \int_0^s ks\ ds = \frac{ks^2}{2} = \frac{k(x + r\theta)^2}{2} = \frac{4kr^2\theta^2}{2}$$

as the work done on the spring; so that Eq. (7.29) becomes

$$- 4kr^2\theta^2/2 = (1/2)(I_c + mr^2)(\omega_f^2 - \omega_i^2)$$

or

$$-128.8\theta^2 = \omega_f^2 - \omega_i^2 \qquad\qquad\qquad (k)$$

as before. The kinematic assumption that $x = r\theta$ (no slipping) places the same restriction on Eq. (k) that was on Eq. (i). Note that the force F does no net work when the cylinder rolls without slipping. The friction force μw does do work when the cylinder rolls and slips.

7.7 Principles of Linear and Angular Impulse and Momentum

The principle of *conservation* of *linear* or *angular momentum* follows naturally from the first time integrals of Eqs. (7.23) and (7.24). These equations give us the linear and angular velocities as functions of time so that the linear and angular displacements as functions of time can be obtained by performing a second time integration.

Motion of Mass Center—The principle of motion of the mass center (which applies to systems of particles of constant mass as well as the motion of the mass center of a rigid body) when the resultant external force is only time dependent can be written from Eq. (7.23) as

$$\mathbf{F}(t) = m\frac{d\mathbf{r}_c}{dt} = m\frac{d\mathbf{v}_c}{dt}$$

where we see that the variables are separable. Integration gives

$$\int_0^t \mathbf{F}\,(t)\ dt = m \int_{v_{ci}}^{v_{cf}} d\mathbf{v}_c = m\mathbf{v}_{cf} - m\mathbf{v}_{ci} \qquad (7.35)$$

or $\hat{\mathbf{F}} = \Delta(\mathbf{LM}) = \Delta(m\mathbf{v}_c)$ where $\mathbf{v}_{ci}$ is the initial velocity when $t = 0$ and $\mathbf{v}_{cf}$ is the final velocity at time t. Equation (7.35) tells us that the *linear impulse* $\hat{\mathbf{F}}$ is equal to the change in the *linear momentum* ($\mathbf{LM} = m\mathbf{v}_c$) of the mass center. Equation (7.35) is a vector equation and as such, has three scalar equations, each of which can be integrated separately. The units for $\hat{\mathbf{F}}$ are either lb-s or N-s while those of momentum are slug-ft/s (which are the same as lb-s) or kg-m/s (which are the same as N-s).

The momentum of the mass center of a system of particles (or rigid bodies) can also be written as

$$\mathbf{G}_c = m\mathbf{v}_c = \sum_{i=1}^{m} m_i \mathbf{v}_i \qquad (7.36)$$

Often when dealing with multibody problems we find that $\mathbf{F}(t) = 0$ during the time interval of interest so that the linear impulse $\hat{\mathbf{F}}$ is zero. From Eq. (7.35) we see that the linear momentum of the mass center is conserved. However, as we can see from Eq. (7.36), this does not mean the momentum of each particle will remain unchanged. The principle of conservation of linear momentum can be applied in any direction in which the resultant external force is zero during the time interval of interest.

We can solve for the position vector locating the mass center by solving for $\mathbf{v}_{cf}$ in Eq. (7.35) and integrating; i.e.,

$$\mathbf{v}_{cf} = \frac{d\mathbf{r}_c}{dt} = \mathbf{v}_{ci} + \frac{1}{m} \int_0^t \mathbf{F}(t)\ dt$$

so that

$$\mathbf{r}_c = \mathbf{r}_{ci} + \mathbf{v}_{ci}t + \frac{1}{m} \int_0^t \left[\int_0^t \mathbf{F}(\tau)d\tau \right] dt \qquad (7.37)$$

where $\mathbf{r}_{ci}$ is the initial position of the mass center and τ is a dummy variable for the first integration contained in the brackets. Hence when the external forces are known functions of time, it is possible to completely determine the velocity and position of the mass center by integration. The following examples illustrate the use of this integration procedure.

EXAMPLE 7.13

A 193.2-pound fisherman realized when he arrived at the stern of his 161-pound boat that he had forgotten his worms on the dock (Fig. E7.13a). If the fisherman and boat are stationary and in the position shown when he begins to walk to the bow, determine if he can reach his can of worms. Neglect friction between boat and water.

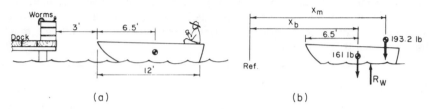

Fig. E7.13

SOLUTION

We observe from the free-body diagram (Fig. E7.13b) that the force of the water on the boat R_w is vertical. For this system of bodies (boat and man),

$$\overset{(\leftrightarrow)}{} \Sigma F_x = M\ddot{x}_c = m_b\ddot{x}_b + m_m\ddot{x}_m = 0$$

$$\overset{(\uparrow)}{} \Sigma F_y = M\ddot{y}_c = R_w - 193.2 - 161 = 0 \tag{a}$$

where $\ddot{y}_c = 0$ since there is no significant vertical motion of the mass center.

From Eq. (a) integration gives

$$M\dot{x}_c = m_m\dot{x}_m + m_b\dot{x}_b = \text{constant} \tag{b}$$

which states that the momentum of the mass center must remain constant, but at time $t = 0$ we are told that the system is at rest. Thus Eq. (b) reduces to

$$M\dot{x}_c = 0 = m_m\dot{x}_m + m_b\dot{x}_b \tag{c}$$

which tells us that the momentum of the system will remain zero regardless of what the fisherman does since he can only exert internal forces on the rest of the system. Hence integration of Eq. (c) gives

$$Mx_c = m_m x_m + m_b x_b = \text{constant} \tag{d}$$

The constant of integration can be evaluated from the initial positions:

$$t = 0, x_b = 3 + 6.5 = 9.5 \text{ ft}, x_m = 15 \text{ ft}, m_m = 6 \text{ slugs, and } m_b = 5 \text{ slugs}$$

so that Eq. (d) becomes

$$Mx_c = 6(15) + (5)(9.5) = 137.5 \text{ slug-ft}$$

When the man reaches the front of the boat, his position is x_m and that of the boat becomes $x_b = x_m + 6.5$. Thus Eq. (d) becomes

$$6x_m + 5(x_m + 6.5) = 137.5$$

or

$$x_m = 105/11 = 9.55 \text{ ft} \qquad\qquad Ans.$$

and the fisherman is unable to reach his can of worms. However, if we had allowed for the existence of friction and had taken into account the motion of the water, the fisherman would have come closer to the dock but not much closer than 9.5 ft.

EXAMPLE 7.14

A 0.1-lb projectile has an initial velocity of 1000 fps to the right before impacting a 10-lb block of wood (Fig. E7.14a). Determine the velocity of departure of the projectile and block if the projectile is embedded in the block. Neglect friction between m_2 and the plane.

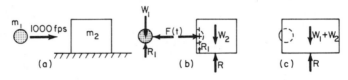

Fig. E7.14

SOLUTION

Two methods can be employed to solve this problem. The first is to consider the free-body diagram of each mass (Fig. E7.14b). Observe that the vertical reaction R_1 (= W_1 since $m_1\ddot{y}_1$ is assumed to be zero during the embedding process) acts on both bodies. The vertical reaction R on body 2 increases from W_2 to $W_1 + W_2$ as the embedding process takes place. In the horizontal direction

$$\underline{\text{Body 1}} \qquad \overset{(\;+\!\!\rightarrow\;)}{} \Sigma F_x = m\frac{dv}{dt} \qquad \underline{\text{Body 2}}$$

$$-F(t) = m_1 \frac{dv_1}{dt} \qquad\qquad F(t) = m_2 \frac{dv_2}{dt}$$

Integration gives

$$m_1 \int_{1000}^{v_1} dv_1 = -\int_0^t F(t)\, dt, \qquad m_2 \int_0^{v_2} dv_2 = \int_0^t F(t)\, dt$$

or

$$m_1 v_1 - 1000 m_1 = - \int_0^t F(t)\, dt = -m_2 v_2 + 0 \qquad \text{(a)}$$

where we recognize that $\int_0^t F(t)\, dt$ is the common impulse experienced by both masses. In addition, we recognize that v_1 and v_2 are equal once the projectile is firmly embedded in the block. Thus Eq. (a) becomes

$$[(0.1/g) + (10/g)]v = 1000\,(0.1/g) \qquad \text{(b)}$$

or

$$\mathbf{v} = 100/10.1 = 9.9 \text{ fps} \rightarrow \qquad\qquad Ans.$$

The second method is to consider the momentum of the system of block and projectile in the horizontal direction. From a free-body diagram of the system (Fig. 7.14c) we note that the resultant force in the horizontal direction is zero. Thus

$$\overset{(\longleftrightarrow)}{} \Sigma F = M\ddot{x}_c = M\frac{dv_c}{dt} = 0$$

from which integration gives

$$M\int_{v_{ci}}^{v_{cf}} dv_c = 0 \qquad \text{(c)}$$

or

$$Mv_{cf} = Mv_{ci}$$

which states that the momentum of the mass center is conserved during impact. We can write Eq. (c) as

$$m_1 v_1 + m_2 v_2 = m_1 v_{1i} + m_2 v_{2i} \qquad \text{(d)}$$

from the principle of motion of the mass center. Since $v_1 = v_2$ after the projectile is fully embedded in the block and $v_{1i} = 1000$ fps $\rightarrow$ and $v_{2i} = 0$, Eq. (d) becomes

$$[(0.1/g) + (10/g)]v = (0.1/g)\,1000 + (10/g)(0)$$

which is exactly the same as Eq. (b). Thus either method gives the same answer, which they should since the principle of motion of the mass center allows us to choose the system in any convenient fashion.

In the first method the impulse $\int F(t)\, dt$ appeared in both integrated equations and hence could be eliminated by combining the two. In the second method, the linear momentum of the mass center of the system was conserved since $F(t)$ becomes an internal force for the system. However, the linear momentum of the projectile and the block changes; therefore they are each subjected to the linear impulse $\int F(t)\, dt$, to the left on body 1 and to the right on body 2.

Angular Motion—We can solve for the angular motion of a rigid body in plane motion when the resultant external couple (moment of the external forces with respect to the mass center) is known as a function of time by integrating Eq. (7.24) to obtain

$$\hat{M}_c = \int_0^t M_c(t)\, dt = kI_c \int_{\omega_i}^{\omega_f} d\omega = (I_c\,\omega_f - I_c\,\omega_i)k$$

or

$$\hat{M}_c = \int_0^t M_c(t)\, dt = I_c\,\omega_f - I_c\,\omega_i = \Delta(AM)_c \qquad (7.38)$$

since $M_{cx} = M_{cy} = 0$ for symmetrical rigid bodies. Equation (7.38) states that the *angular impulse* $\hat{M}_c$ with respect to the z axis thru the mass center is equal to the change in *angular momentum* $(AM = I_c\,\omega)$ of the rigid body with respect to the z axis passing thru the mass center. The angular momentum of a rigid body with respect to the z axis passing thru the mass center is given by $I_c\,\omega$. The sense of the angular momentum is the same as ω. The units of angular impulse are ft-lb-s or N-m-s while those of angular momentum are slug-ft^2/s (which reduce to ft-lb-s) or kg-m^2/s (which reduce to N-m-s).

The angular momentum of a single particle with respect to a given axis is defined as the moment of the linear momentum of the particle with respect to that axis. For the particle of mass m and velocity v in the plane of the page (Fig. 7.10) the linear momentum

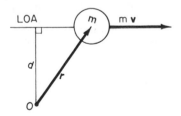

Fig. 7.10

is mv and the angular momentum with respect to the axis perpendicular to the page and thru point 0 is

$$(AM)_0 = (mv)d\,\big\rangle = r \times mv \qquad (7.39)$$

where d is the perpendicular distance from 0 to the line of action (LOA) of the momentum vector mv and r is the position vector locating m with respect to 0. The sense of the angular momentum is clockwise.

The angular momentum of a rigid body in plane motion with respect to the mass center can be obtained by applying Eq. (7.39) to mass dm in Fig. 7.5 and then integrating over the mass of the body. The result is

$$AM_c = \int_m \mathbf{r}_1 \times (dm\ \mathbf{v}) = \left[\int_m \mathbf{r}_1\ dm \right] \times \mathbf{v}_c$$

$$+ \int_m \mathbf{r}_1 \times (r\omega \mathbf{e}_t)\ dm = I_c\,\omega \mathbf{k} \qquad (a)$$

where $\mathbf{r}_1$ is measured from the mass center to mass dm and

$$\mathbf{v} = \mathbf{v}_c + r\omega \mathbf{e}_t$$

is the velocity of mass dm. The student should prove that Eq. (a) is indeed valid for symmetrical rigid bodies.

We see from Eq. (7.38) that the angular momentum is conserved if $M_c(t)$ is zero for the time interval of interest. This conservation often occurs in connected multibody problems, as shown in the examples.

The angular position of the rigid body can be obtained by solving for the angular velocity in Eq. (7.38) and integrating; i.e.,

$$\omega = \frac{d\theta}{dt} = \omega_i + \frac{1}{I_c} \int_0^t M_c(\tau)\ d\tau \qquad (b)$$

where the dummy time variable τ has been introduced into the first integral. Integration of Eq. (b) gives

$$\theta = \theta_i + \omega_i t + \frac{1}{I_c} \int_0^t \left[\int_0^t M_c(\tau)\ d\tau \right] dt \qquad (7.40)$$

where θ_i is the initial angular position of the rigid body at time $t = 0$. Thus we can directly integrate the moment equation twice to obtain angular position information.

Kinematics—As written, Eqs. (7.35), (7.37), (7.38), and (7.40) are independent equations of motion. However, kinematic relationships often exist between the linear and angular motion of the rigid body. When this happens, it is usually possible to reduce the original differential equations of motion given by Eqs. (7.23) and (7.24) to a single directly integrable differential equation of motion.

The following examples illustrate the use of these integration procedures in which the linear and angular momentum principles are the result of the first time integration.

EXAMPLE 7.15

A 0.01-pound projectile strikes a 2.0-pound ballistic pendulum with a velocity of 1000 fps (Fig. 7.15a). Determine (a) the angular velocity of the ballistic pendulum the instant the projectile is fully embedded and (b) the maximum height the ballistic pendulum and projectile will rise. Neglect the weight of the bar.

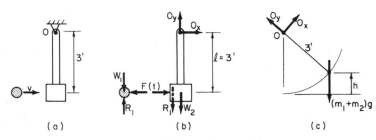

Fig. E7.15

SOLUTION

(a) Initially the pendulum is at rest just prior to the impact of the projectile while a common unknown force $F(t)$ acts on both projectile and pendulum during impact (Fig. E7.15b). The ballistic pendulum is a rigid body in pure rotation about a horizontal axis through 0, so

$(\longleftrightarrow)$ Body 1 $(+\circlearrowleft)$ Body 2 .

$$-F(t) = m_1 \frac{dv}{dt} \qquad \ell F(t) = I_0 \frac{d\omega}{dt}$$

Integration gives

$$\int_0^t F(t)\, dt = -m_1 \int_{v_0}^v dv = -m_1(v - v_0), \qquad \ell \int_0^t F(t)\, dt = I_0 \int_0^\omega d\omega = I_0 \omega$$

which combine into

$$\ell m_1 v_0 = I_0 \omega + \ell m_1 v \tag{a}$$

on elimination of $\int_0^t F(t)\, dt$. The final velocity of the projectile is the same as that of the pendulum once the projectile is firmly embedded. Thus

$$v = \ell \omega \tag{b}$$

so that Eq. (a) reduces to $\ell m_1 v_0 = (I_0 + m_1 \ell^2)\omega$. Substitution of data gives (note $I_0 = m_2 \ell^2$)

$$\frac{3(0.01)}{g}(1000) = \left[\frac{2(9)}{g} + \frac{0.01(9)}{g}\right] \omega = \frac{18.1}{g}\, \omega$$

or

$$\omega = 30/18.1 = 1.66 \text{ rad/s} \;\curvearrowright$$ *Ans.*

(b) The maximum height can be obtained from work and energy principles. From the free-body diagram (Fig. E7.15c)

$$Wk = \Delta(\text{K.E.}) = (1/2)I_0 \,(\omega_f^2 - \omega_i^2)$$

or

$$-(2.01)h = (1/2)[(18.1)/32.2]\,[0 - (1.66)^2\,]$$

or

$$h = 0.386 \text{ ft}$$ *Ans.*

since $\omega_f = 0$ at maximum height.

Note that Eq. (a) is a statement of the conservation of angular momentum of the system with respect to point 0. Verify that this is indeed so by noting that the angular momentum of a particle is the moment of its linear momentum with respect to point 0.

EXAMPLE 7.16

Two symmetrical rotors are connected by a clutch (Fig. E7.16a). Rotor 1 weighs 64.4 lb and has a radius of gyration of mass with respect to its axis of rotation of 1 ft, while rotor 2 weighs 322 lb and has a radius of gyration of 2.0 ft. Rotor 1 is brought up to a speed of 3600 rev/min while rotor 2 remains at rest. Then the clutch is rapidly engaged. Determine (a) the final angular velocity of the assembly and (b) the amount of energy dissipated in the clutch assembly due to slippage. Neglect bearing friction and the inertia of the clutch assembly.

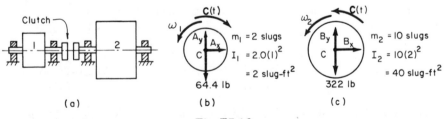

Fig. E7.16

SOLUTION

A free-body diagram of an end view of the two rotors is shown in Figs. E7.16b and E7.16c where A_x, A_y, B_x, and B_y represent the bearing reactions and $C(t)$ represents the unknown torque transmitted by the clutch assembly. Since the direction of rotation was not specified, assume this direction to be counterclockwise.

(a) To find the final angular velocity of the assembly, sum moments with respect to the axis of rotation of each body to obtain

<u>Body 1</u> $\left(\overset{+}{\curvearrowright}\right) \Sigma M_c = I_c \alpha$ <u>Body 2</u>

$$-C(t) = I_1 \frac{d\omega_1}{dt} \qquad\qquad\qquad C(t) = I_2 \frac{d\omega_2}{dt}$$

which combine to give

$$I_1 \frac{d\omega_1}{dt} + I_2 \frac{d\omega_2}{dt} = 0 \tag{a}$$

on elimination of $C(t)$. The resultant external moment on the combined system is zero. Multiplying Eq. (a) by dt and integrating gives

$$I_1 \int_{\omega_{1i}}^{\omega} d\omega_1 + I_2 \int_0^\omega d\omega_2 = 0, \qquad I_1(\omega - \omega_{1i}) + I_2(\omega - 0) = 0$$

or

$$(I_1 + I_2)\omega = I_1 \omega_{1i} \tag{b}$$

where ω is the common final angular velocity and $\omega_{1i} = (3600/60)(2\pi) = 120\pi$ rad/s is the initial angular velocity of rotor 1. Equation (b) is a statement of the conservation of angular momentum for the combined system. This result always occurs when the resultant external moment on the combined system is zero. Substituting the data into Eq. (b) and solving for ω gives

$$\omega = \frac{I_1 \omega_{1i}}{(I_1 + I_2)} = \frac{2(120\pi)}{(2 + 40)} = \frac{120\pi}{21} = 5.7\pi = 17.95 \text{ rad/s } \overset{\curvearrowright}{} \qquad Ans.$$

(b) The amount of energy dissipated in the clutch assembly due to friction while slippage occurs can be obtained from the change in the kinetic energy of the system. Thus

$$\text{Energy dissipated} = (1/2)(I_1 + I_2)\omega^2 - (1/2)I_1\omega_{1i}^2$$
$$= (42/2)(17.95)^2 - (2/2)(120\pi)^2 = 6770 - 142{,}000$$
$$= -135{,}200 \text{ ft-lb} \qquad\qquad Ans.$$

where the minus sign indicates that the energy is lost to the system. The large value of energy dissipated in the clutch assembly would cause very high temperature rises in the clutch materials. Hence oil-bath clutch assemblies are very popular for removing heat from the friction surfaces.

EXAMPLE 7.17

A 16.1-lb bowling ball is released with a linear velocity of 20 fps to the right and an angular velocity of 6 rad/s counterclockwise. The ball is 12 in. in

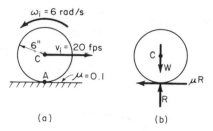

Fig. E7.17

diameter and the coefficient of friction is 0.1. Determine (a) the linear and angular velocity variation with time, (b) the time t_0 at which it begins to roll without slipping, (c) the position from the point of release at time t_0, (d) the angular position at time t_0, and (e) the translational work done on the ball during the time interval $0 < t < t_0$.

SOLUTION

The point of contact A is moving to the right relative to the floor so that the friction force is to the left (Fig. E7.17a). Since the mass center moves along a horizontal straight line, the principle of motion of the mass center gives

$$\left(\uparrow\downarrow\right) R - W = ma_y = 0$$

or $R = W$ from the free-body diagram (Fig. E7.17b). The differential equations of motion for the bowling ball can be obtained from

$$(\leftrightarrow) \ \Sigma F_x = m\ddot{x} = m\frac{dv}{dt}, \qquad (+\curvearrowleft) \ \Sigma M_C = I_C\alpha = I_C\frac{d\omega}{dt}$$

$$-0.1(16.1) = \frac{1}{2}\frac{dv}{dt}, \qquad \frac{1}{2}(0.1)(16.1) = \frac{2}{5}\left(\frac{1}{2}\right)\left(\frac{1}{2}\right)^2\frac{d\omega}{dt}$$

or

$$\frac{dv}{dt} = -3.22 \tag{a}$$

and

$$\frac{d\omega}{dt} = 16.1 \tag{b}$$

which are valid as long as the ball rolls and slips.

(a) The linear and angular velocity variation with time can be obtained by integration of Eqs. (a) and (b) to obtain

$$\int_{20}^{v} dv = -3.22\int_{0}^{t} dt, \qquad \int_{-6}^{\omega} d\omega = 16.1\int_{0}^{t} dt$$

or

$$v = 20 - 3.22t \qquad \text{(c)}$$

and

$$\omega = -6 + 16.1t \qquad \text{Ans.} \quad \text{(d)}$$

as long as the ball rolls and slips.

(b) The time t_0 at which the ball begins to roll without slipping can be obtained from the kinematic fact that $v = r\omega$ or

$$20 - 3.22t_0 = (1/2)(-6 + 16.1t_0) = -3 + 8.05t_0$$

so that

$$t_0 = (20 + 3)/(8.05 + 3.22) = 23/11.27 = 2.04 \text{ s} \qquad \text{Ans.}$$

(c) The linear position from the point of release can be obtained from Eq. (c) by integration where the initial position is taken as zero; i.e.

$$x = \int_0^x dx = \int_0^{t_0} v \, dt = \int_0^{t_0} (20 - 3.22t) \, dt = 20t_0 - 1.61t_0^2$$

$$x = 20(2.04) - 1.61(2.04)^2 = 40.8 - 6.7 = 34.1 \text{ ft to the right} \qquad \text{Ans.}$$

(d) The angular position from the point of release is obtained by integration of Eq. (d) to give

$$\theta = \int_0^\theta d\theta = \int_0^{t_0} \omega \, dt = \int_0^{t_0} (-6 + 16.1t) \, dt = -6t_0 + 8.05t_0^2$$

$$\theta = -6(2.04) + 8.05(2.04)^2 = -12.24 + 33.5 = 21.3 \text{ rad} \; \rangle \qquad \text{Ans.}$$

(e) The translational work done on the ball can be obtained from

$$Wk_{\text{trans}} = \Delta[(1/2)mv_c^2] = (1/2)m(v_{cf}^2 - v_{ci}^2)$$

where $v_{ci} = 20$ fps and v_{cf} is obtained from Eq. (c) to be

$$v_{cf} = 20 - 3.22(t_0) = 20 - 3.22(2.04) = 13.4 \text{ fps} \rightarrow$$

so that

$$Wk_{\text{trans}} = (1/2)(1/2)[(13.4)^2 - (20)^2] = -220/4 = -55 \text{ ft-lb} \qquad \text{Ans.}$$

This answer can be cross checked from the work done on the mass center by the friction force in Fig. E7.17b; i.e.

$$Wk_{\text{trans}} = \int_0^{34.1} (-1.61) \, dx = -(1.61)(34.1) = -55 \text{ ft-lb}$$

The kinematic description of the position of the mass center and the angular position of the ball were independent kinematic quantities described by their own separate equations as long as the ball was rolling and slipping. Once the "no slip" (rolling without slipping) condition was achieved, both the angular and linear equations of motion reduced to the same statement; i.e. the ball moves with constant linear and angular velocity.

EXAMPLE 7.18

The cable drum in Fig. E7.18a weighs 1932 lb and has a radius of gyration with respect to the z axis through 0 of 1.2 ft. The coefficient of friction between the brake arm and the drum is 0.5, and the weight of the brake arm may be neglected. The 1288-lb elevator carriage has a velocity of 20 fps downward at time $t = 0$ and is to be stopped in 2 s by applying a force of $P = At$. Determine (a) the constant A and (b) the distance the carriage travels in coming to rest. Neglect the effects of the cable.

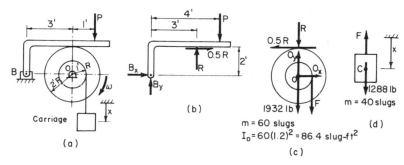

Fig. E7.18

SOLUTION

(a) Before we can develop the differential equations of motion, we must know the relationship between the forces P and R, which can be obtained by summing moments about point B of the brake arm in Fig. E7.18b to obtain

$$\overset{+}{\circlearrowleft} \ \Sigma M_B = 0 \quad (\text{since } \alpha_B = 0), \quad 4P + R - 3R = 0$$

or

$$R = 2P = 2At \tag{a}$$

where we note that the effect of the friction force is to increase the normal force R.

The linear motion of the carriage and the angular motion of the cable drum are related kinematically by

$$x = r\theta, \quad \dot{x} = r\dot{\theta} = r\omega, \quad \ddot{x} = r\ddot{\theta} = r\alpha \tag{b}$$

where $r = 1.0$ ft, x is positive downward, and θ is positive clockwise.

The angular differential equation of motion can be obtained from the free-body diagram of the cable drum in (Fig. E7.18c) to give

$$\overset{+}{\curvearrowright} \Sigma M_0 = I_0\alpha, \qquad (1)F - 2(0.5)R = I_0\alpha$$

or

$$F - 2At = I_0\alpha \tag{c}$$

on substitution of Eq. (a). Application of the principle of motion of the mass center to the cable drum will give the bearing reactions 0_x and 0_y, which are of no interest at this point.

The differential equation of motion for the carriage can be obtained from Fig. E7.18d to give

$$\overset{\downarrow}{+} \Sigma F_x = m\ddot{x}, \qquad -F + 1288 = m\ddot{x}$$

or

$$F = 1288 - m\ddot{x} \tag{d}$$

The differential equation of motion for the combined system can be obtained by substituting Eq. (d) into Eq. (c) and noting that $\alpha = \ddot{x} = dv/dt$ from Eq. (b) to give

$$1288 - m\frac{dv}{dt} - 2At = I_0\frac{dv}{dt}$$

or

$$126.4\frac{dv}{dt} = 1288 - 2At \tag{e}$$

Rearranging and integrating gives

$$\int_{20}^{v} dv = v - 20 = \int_{0}^{t} \left(10.2 - \frac{2At}{126.4}\right) dt$$

or

$$v = 20 + 10.2t - At^2/126.4 \tag{f}$$

From the specified data $v = 0$ when $t = 2$ s, so

$$A = (126.4/4)[20 + 10.2(2)] = [126.4(40.4)]/4 = 1290 \text{ lb/s} \qquad Ans.$$

(b) The total distance traveled in coming to rest can be obtained by integration of Eq. (f), giving

$$\int_{0}^{x} dx = x = \int_{0}^{t} (20 + 10.2t - 10.2t^2) dt$$

or

$$x = 20t + 5.1t^2 - 10.2t^3/3$$

where x at $t = 0$ is assumed to be zero. Substitution of $t = 2$ s gives

$$x = 20(2) + 5.1(4) - (10.2/3)(8) = 40 + 20.4 - 27.2 = 33.2 \text{ ft } \downarrow \qquad Ans.$$

We combined the two differential equations of motion before integration to eliminate the unknown force F rather than performing the first integration on both Eqs. (c) and (d) and then eliminating the unknown linear and angular impulse terms. The end result is the same. Once the constant A is determined, we can determine all other forces as functions of time, i.e., F, 0_x, 0_y, B_x, and B_y.

7.8   Velocity Dependent Forces and Couples

From Eqs. (7.25) and (7.26) the force system may be reduced to

$$F_x(v_{cx}) = m\,\frac{dv_{cx}}{dt}, \qquad F_y(v_{cy}) = m\,\frac{dv_{cy}}{dt}, \qquad M_c(\omega) = I_c\,\frac{d\omega}{dt}$$

We can illustrate the general integration procedure for these equations by writing Eq. (7.26) as

$$\int_0^t dt = I_c \int_{\omega_i}^{\omega} \frac{d\omega}{M_c(\omega)}$$

or

$$t = f_1(\omega) - f_1(\omega_i) \qquad (7.41)$$

where ω_i is the initial angular velocity and

$$f_1(\omega) = I_c \int \frac{d\omega}{M_c(\omega)}$$

Equation (7.41) gives the time t as a function of angular velocity ω. In principle we can solve for ω in Eq. (7.41) to obtain

$$\omega = \frac{d\theta}{dt} = f_2(\omega_i, t)$$

from which we can obtain the angular position θ by integration,

$$\theta = \theta_i + \int_0^t f_2(\omega_i, t)\, dt$$

where θ_i is the initial angular position.

In problems the kinematic analysis will often show that Eqs. (7.25) and (7.26) are not independent. Then a single differential equation of motion will result which may be integrable by the basic procedure outlined. The following examples will illustrate the procedure in detail.

EXAMPLE 7.19

A sphere of mass m is released from rest in a viscous fluid (Fig. E7.19a). The weight of the sphere is slightly greater than the buoyant force due to the volume of fluid displaced. Determine the motion (velocity and displacement) of the sphere if the resistance to motion is proportional to its velocity.

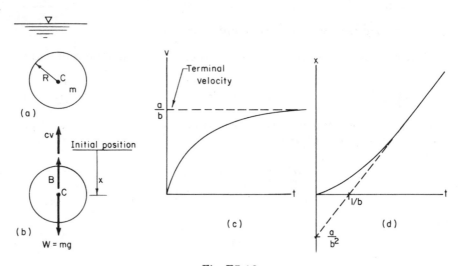

Fig. E7.19

SOLUTION

A free-body diagram of external forces is shown in Fig. E7.19b where B is the buoyant force, W is the weight, x is the position of the sphere with respect to its initial position, and cv is the resistance to motion where c is the constant of proportionality. From the free-body diagram,

$$\Sigma F = m\ddot{x} = m\frac{dv}{dt}, \qquad (W - B) - cv = m\frac{dv}{dt} \qquad \text{(a)}$$

from which we obtain

$$\int_0^t dt = \int_0^v \frac{dv}{(a - bv)} = -\frac{1}{b}\int_a^{a-bv} \frac{dz}{z} \qquad \text{(b)}$$

where $a = (W - B)/m$ and $b = c/m$ are positive constants and $z = a - bv$ is the dummy variable of integration. Integration of Eq. (b) gives

$$t = -(1/b) \, ln \, [(a - bv)/a] \tag{c}$$

or

$$v = (a/b)(1 - e^{-bt})$$

which is plotted in Fig. E7.19c where we see that the velocity asymptotically approaches a final value called the *terminal velocity*. From Eqs. (a) and (c) the terminal velocity is given by

$$v_{\text{term}} = a/b = (W - B)/c$$

The position of the sphere can be obtained from integration of Eq. (c) to give

$$\int_0^x dx = (a/b) \int_0^t (1 - e^{-bt}) \, dt$$

or

$$x = (a/b)t - (a/b^2)(1 - e^{-bt}) \tag{d}$$

A plot of Eq. (d) is shown in Fig. E7.19d where we see the motion asymptotically approaches the dashed straight line which has an intercept at $-a/b^2$.

EXAMPLE 7.20

The armature of an electric motor (Fig. E7.20a) weighs 16.1 lb and has a radius of 6 in. The stall torque for this motor is 100 ft-lb and it achieves a no-load free-running speed of 200 rad/s. The measured torque speed curve for the electric motor is shown in Fig. E7.20b. The motor is connected to a rotor with a moment of inertia of 1.9375 slug-ft^2. Determine the start-up angular velocity time relationship. Sketch the resulting relationship and neglect bearing friction in the analysis.

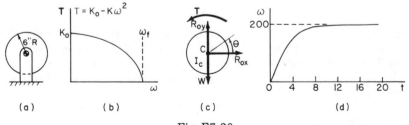

Fig. E7.20

SOLUTION

From the free-body diagram of the motor armature and attached rotor (Fig. E7.20c) we see that the body has pure rotation about its mass center.

Summation of moments about the axis of rotation gives

$$\left(\stackrel{+}{\circlearrowleft}\right) \Sigma M_c = I_c \alpha, \qquad T = K_0 - K\omega^2 = I_c \frac{d\omega}{dt} \tag{a}$$

where

$$I_c = 1.9375 + [16.1/2(32.2)](1/2)^2 = 2.0 \text{ slug-ft}^2$$
$$K_0 = 100 \text{ ft-lb}$$

and from Fig. E7.20b

$$K = K_0/\omega_f^2 = 100/(200)^2 = 0.25 \times 10^{-2} \text{ ft-lb-s}^2 \tag{b}$$

The variables of Eq. (a) can be separated to obtain

$$t = \int_0^t dt = I_c \int_0^\omega \frac{d\omega}{K_0 - K\omega^2} = \frac{I_c}{K} \int_0^\omega \frac{d\omega}{\omega_f^2 - \omega^2} \tag{c}$$

where $\omega_f^2 = K_0/K$ [see Eq. (b)] and the right-hand side of Eq. (c) is seen to be a standard integral form. Integration of Eq. (c) gives

$$t = (I_c/K)[(1/\omega_f) \tanh^{-1} (\omega/\omega_f)]_0^\omega = (I_c/K\omega_f)[\tanh^{-1} (\omega/\omega_f) - 0]$$

or

$$\omega = \omega_f \tanh [(K\omega_f/I_c) t] = 200 \tanh (0.25t) \tag{d}$$

since $K\omega_f/I_c = (0.25 \times 10^{-2}) (2 \times 10^2)/2 = 0.25 \text{ sec}^{-1}$. A plot of Eq. (d) (Fig. E7.20d) shows that the angular velocity has achieved the free-running value in approximately 12.0 s.

EXAMPLE 7.21

The 44,500-N mine skip is raised along an incline (Fig. E7.21a) by a 4450-N winch and motor combination which has a torque speed curve given by

$$T = 31,250 - 10\omega^2$$

where T is in N-m when ω is in rad/s. The radius of gyration of mass with respect to the axis of rotation is 0.35 m. The coefficient of sliding friction between the skip and its rails is 0.333. Determine (a) the final velocity of the skip, (b) the time required for the skip to come up to speed, and (c) the power required to maintain steady motion.

SOLUTION

From the free-body diagram of the skip (Fig. E7.21b),

$$\left(\stackrel{\nwarrow}{\searrow}\right) \Sigma F_y = ma_{cy} = 0, \qquad R - (3/5)W = 0$$

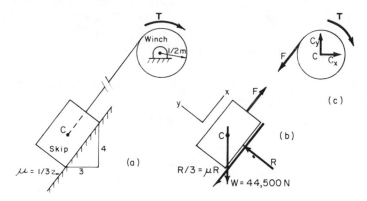

Fig. E7.21

or

$$R = (3/5)(44{,}500) = 26{,}700 \text{ N}$$

and

$$\overset{\nearrow}{\bigcirc} \Sigma F_x = ma_{cx} = m\,\frac{dv}{dt}$$

$$F - 0.333(26{,}700) - \frac{4}{5}(44{,}500) = \frac{44{,}500}{9.81}\,\frac{dv}{dt}$$

or

$$F = 8900 + 35{,}600 + 4540\,\frac{dv}{dt} = 44{,}500 + 4540\,\frac{dv}{dt} \qquad (a)$$

From the free-body diagram (Fig. E7.21c),

$$\overset{+}{\circlearrowright} \Sigma M_0 = I_0\,\frac{d\omega}{dt}$$

$$31{,}250 - 10\omega^2 - \left(\frac{1}{2}\right)(44{,}500) - \left(\frac{1}{2}\right)(4540)\,\frac{dv}{dt} = I_0\,\frac{d\omega}{dt}$$

or

$$9000 - 10\omega^2 = I_0\,\frac{d\omega}{dt} + 2270\,\frac{dv}{dt} \qquad (b)$$

where $I_0 = (4450/9.81)(0.35)^2 = 55.0$ kg-m^2. Observe that x is positive up the plane and θ is positive clockwise so that kinematically $x = r\theta$ and

$$v = r\omega = (1/2)\omega \qquad (c)$$

so that Eq. (b) reduces to

$$9000 - 10\omega^2 = (55 + 1135)\frac{d\omega}{dt} = 1190\frac{d\omega}{dt} \qquad (d)$$

(a) The final velocity of the skip can be obtained from Eq. (d) by noting that $d\omega/dt = 0$ for steady motion. Thus

$$\omega^2 = 9000/10 = 900$$

or

$$\omega = 30 \text{ rad/s} \qquad (e)$$

Therefore,

$$\mathbf{v} = (1/2)(\omega) = 15 \text{ m/s} \qquad \qquad Ans.$$

(b) To determine the time required to come up to speed, we must integrate Eq. (d). The result is

$$t = \int_0^t dt = \frac{1190}{10}\int_0^\omega \frac{d\omega}{900 - \omega^2} = \frac{119}{30}\left[\tanh^{-1}\left(\frac{\omega}{30}\right) - \tanh^{-1}(0)\right]$$

or

$$t = 3.96\left[\tanh^{-1}(\omega/30)\right]$$

The exact instant that $\omega = 30$ rad/s is not clearly defined since the angular velocity asymptotically approaches this value. However, when $\omega = 0.99(30)$, $t = 3.96(2.65) = 10.5$ s; and when $\omega = (0.998)(30)$, $t = (3.96)(3.8) = 15.1$ s. Thus for all practical purposes the skip has achieved its final speed when 10–15 s have elapsed.

(c) The power required to maintain the steady motion of the skip can be obtained from Eq. (a) when $dv/dt = 0$ and the fact that power is the time rate of doing work. Thus

$$P = \frac{dWk}{dt} = Fv = (44{,}500)(15) = 668{,}000 \text{ N-m/s (watts)} \qquad Ans.$$

when F is a constant. The power required is often expressed in terms of horsepower units where 1 hp = 550 ft-lb/s = 746 N-m/s = 746 W. Thus we could also express our answer as

$$P = 668{,}000/746 = 895 \text{ hp} \qquad Ans.$$

PROBLEMS

In kinetics problems all ropes, cords, and cables are assumed to be flexible, inextensible, and of negligible mass; while all pins, surfaces, and axles are as-

sumed to be smooth unless specified otherwise. In stating answers, the unit vectors **i** *and* **j** *are assumed to the right and upward respectively unless defined otherwise in the problem figure.*

7.1. Two particles with mass $m_1 = m$ and $m_2 = 4\,m$ have constant accelerations of $\mathbf{a}_1 = 3\mathbf{i} + 4\mathbf{j}$ and $\mathbf{a}_2 = 5.5\mathbf{i} - 16\mathbf{j}$ fps^2 respectively for 2 s (Fig. P7.1). Determine (a) the acceleration of the mass center, (b) the resultant force acting on the system of two masses, (c) the velocity of each mass at the end of 2 s, and (d) the velocity of the mass center at the end of 2 s. The initial velocity of each mass is zero.

Ans. (a) $\mathbf{a}_c = 5\mathbf{i} - 12\mathbf{j}$ fps^2,   (b) $\mathbf{R} = 25m\mathbf{i} - 60m\mathbf{j}$ lb

(c) $\mathbf{v}_1 = 6\mathbf{i} + 8\mathbf{j}$ fps, $\mathbf{v}_2 = 11\mathbf{i} - 32\mathbf{j}$ fps

(d) $\mathbf{v}_c = 10\mathbf{i} - 24\mathbf{j}$ fps

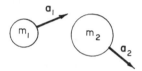

Fig. P7.1

7.2. Solve parts (c) and (d) of Problem 7.1 if the initial velocity of m_1 is $\mathbf{v}_{1i} = -6\mathbf{i} + 10\mathbf{j}$ fps and if m_2 is $\mathbf{v}_{2i} = -9\mathbf{i} + 20\mathbf{j}$ fps.

7.3. A 400-lb body rests on a horizontal plane and has an external force **F** applied (Fig. P7.3). Determine (a) the magnitude of the force **F** required to produce an acceleration of 20 fps^2 to the right if the coefficient of friction is zero and (b) the acceleration that a 324-lb force **F** would produce if the coefficient of friction $\mu = 0.25$.

Ans. (a) $F = 324$ lb,   (b) $a = 7.74$ fps$^2 \rightarrow$

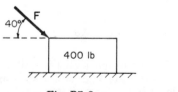

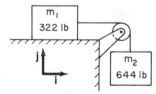

Fig. P7.3   Fig. P7.4

7.4. Two masses m_1 and m_2 start from rest. Mass m_1 moves along the horizontal frictionless plane (Fig. P7.4). Determine (a) the acceleration of mass m_1, (b) the tension in the cord, and (c) the acceleration of the mass center of masses m_1 and m_2. Neglect any effects of the pulley. How would your answers be changed if m_1 had an initial velocity of 6 fps to the left?

7.5. A commuter train has three cars and is traveling at 60 mph (Fig. P7.5). Determine the coupling force between cars when the brakes are applied if the brakes subject each car to a braking force of 6000 lb. Neglect air friction (wind resistance).

Ans. $F_{AB} = 1500$ lb T, $F_{BC} = 0$ lb

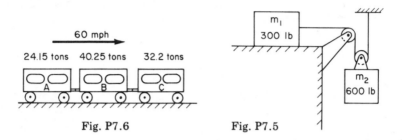

Fig. P7.6   Fig. P7.5

7.6. The two masses in Fig. P7.6 are released from rest where the coefficient of friction between mass m_1 and the horizontal surface is 0.33 and the pulleys are massless and frictionless. Determine the acceleration of each mass and the tension in the cord.

7.7. A 10,000-lb truck is traveling around a curve with a constant speed of 60 mph. The horizontal radius of curvature of the road is 400 ft. Determine the angle of road elevation θ (Fig. P7.7) so the truck has no tendency to slip; i.e., the friction force on the tires is zero.

Ans. $\theta = 31°$

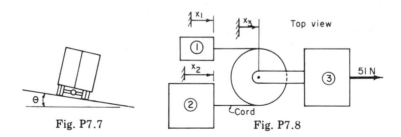

Fig. P7.7   Fig. P7.8

7.8. Three bodies are connected (Fig. P7.8) and move on a smooth horizontal plane. The masses of bodies 1, 2, and 3 are 1, 2, and 3 kg respectively. Neglecting the mass of the pulley, determine (a) the accelerations of bodies 1, 2, and 3 and (b) the acceleration of the mass center by two methods.

7.9. A 2.0-in. diameter steel (0.283 lb/in.^{3}) bar is 4.0 in. long (Fig. P7.9). Determine (a) I_x and I_z and (b) the radius of gyration k_x.

Ans. (a) $I_x = 3.83 \times 10^{-4}$ slug-ft^2, $I_z = 1.215 \times 10^{-3}$ slug-ft^2
(b) $k_x = 5.89 \times 10^{-2}$ ft

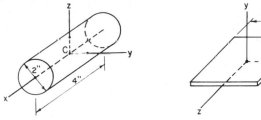

Fig. P7.9   Fig. P7.10

7.10. A 193.2-lb uniform plate is $2' \times 2' \times 1''$ thick (Fig. P7.10). Determine I_x and I_y and explain why I_{xy} is zero. Can you treat this as a thin plate?

7.11. A 20-lb slender bar 10 ft long has a 4-lb sphere 6 in. in diameter attached (Fig. P7.11). Determine (a) the location of the mass center, (b) the moment of inertia of mass with respect to the mass center about an axis perpendicular to the page, and (c) the moment of inertia of mass about an axis perpendicular to the page thru point A.

Ans.   (a) $x_c = 5.87$ ft,   (b) $I_c = 8.04$ slug-ft^2,   (c) $I_A = 33.7$ slug-ft^2

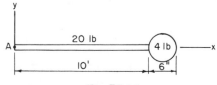

Fig. P7.11

7.12. A wheel is made from a thin ring 1 ft deep (normal to page) and four thin round spokes from 0 to the ring (Fig. P7.12). The mass of the ring is 1.0 slug, and the mass of each spoke (from 0 to ring) is 0.1 slug. Determine (a) I_x and (b) the radius of gyration k_x.

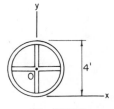

Fig. P7.12

7.13. A large bearing block is made from a steel (490 lb/ft^3) plate 1.0 ft thick

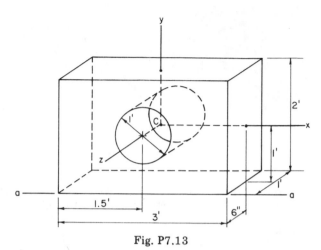

Fig. P7.13

(Fig. P7.13). Determine the moment of inertia of mass with respect to (a) the x axis and (b) the aa axis.

Ans.   (a) $I_{cx} = 36.3$ slug-ft^2,   (b) $I_a = 135.7$ slug-ft^2

7.14. Show that the moments of inertia of mass for a thin circular ring with respect to its mass center can be obtained from the moments of inertia of a solid circular cylinder.

7.15. A rigid body can be approximated as two equal 4.0-lb weights attached by light rods of negligible mass (Fig. P7.15). Determine (a) I_x, (b) I_{xy}, and (c) the product of inertia for the aa and bb axes which lie in the xy plane.

Ans.   (a) $I_x = 2.24$ slug-ft^2,   (b) $I_{xy} = 0$,   (c) $I_{aa, bb} = -4.97$ slug-ft^2

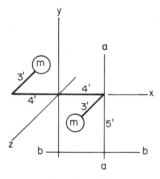

Fig. P7.15

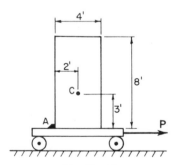

Fig. P7.16

7.16. A crated refrigerator weighs 128.8 lb and is to be moved on a 96.6-lb cart (Fig. P7.16). A stop is built into the cart at A. Determine (a) the maximum acceleration which the cart may have before the refrigerator will begin to tip and (b) the force P required to produce this acceleration.

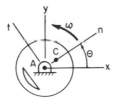

Fig. P7.17

7.17. A 483-lb flywheel is rotating at a constant angular velocity of 1800 rev/min in the horizontal plane (Fig. P7.17). The mass center C is 0.67 ft from the axis of rotation at A, and the radius of gyration of mass with respect to the mass center about an axis perpendicular to the plane of the disk is 1.2 ft. Determine (a) the n and t bearing reactions at the instant shown and (b) the x and y bearing reactions at any instant of time.

Ans. (a) $R_n = 356,000$ lb $\angle\theta$, $R_t = 0$

(b) $R_x = -356,000 \cos\theta$ i lb, $R_y = -356,000 \sin\theta$ j lb

7.18. Determine the n and t bearing reactions of the disk in Problem 7.17 if at a given instant the angular velocity is 5 rad/s clockwise (but not constant), a couple of 2840 ft-lb clockwise is applied, and the angle $\theta = 36.8°$.

7.19. A homogeneous 64.4-lb disk rotates about the mass center (Figs. P7.19a and b). In Fig. 7.19a, a constant 24.15-lb force is applied, while in Fig. P7.19b a 24.15-lb weight is attached. Determine the angular acceleration of the disk in both situations and explain why the answers are not the same.

Ans. (a) $\alpha = 12.07$ rad/s² $\curvearrowright$,   (b) $\alpha = 6.90$ rad/s² $\curvearrowright$

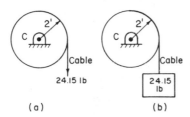

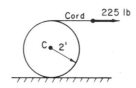

Fig. P7.19   Fig. P7.20

7.20. A 483-lb solid homogeneous cylinder rolls without slipping and has a cord of negligible mass wrapped around its midsection (Fig. P7.20). A 225-lb force is applied to the cord. Determine (a) the angular acceleration, (b) the vertical reaction of the floor on the cylinder, (c) the horizontal reaction of the floor on the surface, and (d) the minimum coefficient of friction necessary to roll without slipping.

7.21. The cylinder of Problem 7.20 is attached to a 644-lb weight by the cord passing over the frictionless support at B (Fig. P7.21). Determine (a) the acceleration of the 644-lb weight, (b) the tension in the cord, (c) the angular acceleration of the cylinder, and (d) the horizontal reaction of the floor on the cylinder. Assume the cylinder rolls without slipping.

Ans.   (a) $\mathbf{a} = 25.12$ fps² ↓,   (b) $T = 141.2$ lb T
      (c) $\alpha = 6.28$ rad/s² ↻ ,   (d) $R_x = 47.1$ lb →

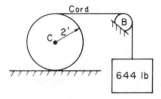

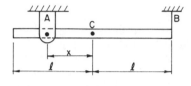

Fig. P7.21   Fig. P7.22

7.22. A slender rod of mass m and length 2ℓ is supported by a pin at A and a string at B (Fig. P7.22). Determine (a) the location of the pin so that maximum angular acceleration results when the string is cut, (b) the maximum angular acceleration, and (c) the pin reactions at A.

7.23. A 16.1-lb bowling ball is released with a forward velocity of 12 fps and angular velocity 4 rad/s counterclockwise (Fig. P7.23). Determine (a) the linear acceleration of the mass center and (b) the angular acceleration of the ball if the coefficient of friction is 0.10.

Ans.   (a) $a_c = 3.22$ fps² ←,   (b) $\alpha = 16.1$ rad/s² ↻

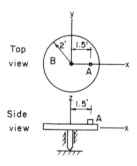

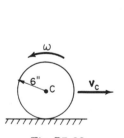

Fig. P7.23   Fig. P7.24

7.24. A small 1-lb disk A is sitting on the large 10-lb turntable B, which can rotate about the vertical z axis (Fig. P7.24). The coefficient of friction between the small disk A and turntable B is 0.419. Determine (a) the maximum constant angular velocity of the turntable before disk A begins to slide, (b) the maximum initial torque that can be applied before disk A begins to slide when the turntable is initially at rest, and (c) the maximum angular velocity of the turntable before disk A begins to slide if a constant

torque of 5.56 ft-lb is applied.   Assume that disk A is a particle and the
turntable surface is horizontal.

7.25. The homogeneous 1-m long bar AB has a mass of 10 kg and is connected
to frame C by a pin at B and a cord at A (Fig. P7.25). The frame C has a
mass of 20 kg and slides on a smooth horizontal surface. Determine (a) the
maximum force **P** so that the cord tension becomes zero and no relative
motion between bar AB and the frame occurs, (b) the horizontal and
vertical pin reactions at B on bar AB for this value of **P**, and (c) the acceleration of frame C.

Ans.   (a) **P** = 221 N $\rightarrow$,   (b) B_x = 73.5 N $\rightarrow$, B_y = 98.1 N $\uparrow$

(c) a_c = 7.35 m/s^2 $\rightarrow$

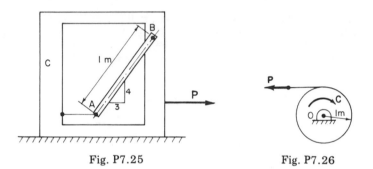

Fig. P7.25   Fig. P7.26

7.26. A homogeneous 100-kg cylinder is attached to frame C of Problem 7.25
by a cable wrapped around it (Fig. P7.26). Determine the couple **C** that
must be applied to the cylinder to produce a force **P** (=221 N) in the cable
and an acceleration of frame C of 7.35 m/s^2 to the right.

7.27. The mechanism in Fig. P7.27 consists of two slender uniform rigid bars
AB and $0A$ pinned together at A and rotating in the vertical plane. For
bar AB the mass is 12 kg, the angular velocity is 3 rad/s clockwise, and the
angular acceleration is 6 rad/s^2 counterclockwise. For bar $0A$ the angular
velocity is 4 rad/s clockwise and 5 rad/s^2 clockwise. For the position
shown, determine (a) the horizontal and vertical pin reactions at A acting
on bar AB and (b) the couple **C** that must be applied to bar AB.

Ans.   (a) A_x = 396 N $\rightarrow$, A_y = 602 N $\rightarrow$,   (b) **C** = 1300 N-m $\circlearrowright$

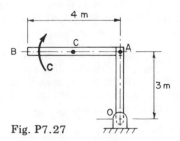

Fig. P7.27

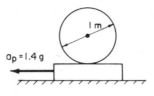

Fig. P7.28

7.28. A uniform sphere of 1-kg mass and 1.0-m diameter is supported on an accelerating platform (Fig. P7.28). Determine the angular acceleration of the sphere if the coefficient of friction between the sphere and the platform is (a) 0.50 and (b) 0.30.

7.29. A homogeneous solid cylinder of 1-m radius weighs 19.62 N and rolls without slipping on a circular surface (Fig. P7.29). When in the position shown the velocity of the center of the cylinder is 18 m/s. Determine the normal and tangential components of the reaction of the curved surface on the cylinder.

Ans. $R_n = 87.7$ N , $R_t = 3.92$ N

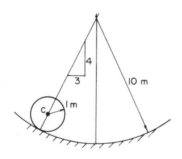

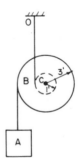

Fig. P7.29   Fig. P7.30

7.30. A 128.8-lb homogeneous cylinder B of 3-ft radius has a narrow slot of 1-ft radius for the flexible cable that supports the cylinder (Fig. P7.30). The narrow slot has negligible effect on the moment of inertia. The acceleration of the center of the cylinder is 6 fps^2 upward, and body A is moving downward. Determine (a) the weight of body A and (b) the tension in cable $0B$.

7.31. A uniform ladder of length ℓ and mass m starts from rest in the vertical position when $\theta = 0°$ and falls under the action of gravity (Fig. P7.31). The surface ends A and B of the ladder remain in contact with the smooth surfaces throughout the motion. At any instant after the ladder starts to slide, determine its angular acceleration.

Ans. $\alpha = (3g \sin \theta)/2\ell$

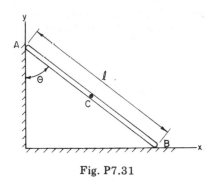

Fig. P7.31

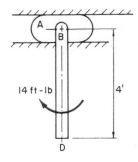

Fig. P7.32

7.32. A 96.6-lb slender homogeneous bar BD is attached to a rigid 32.2-lb block A by a pin at B (Fig. P7.32). Block A slides in a smooth track and a 14 ft-lb couple is applied to bar BD when in the position shown. For this position determine the angular acceleration of bar BD and the linear acceleration of block A. What additional information is needed to determine the vertical pin reaction at B?

7.33. A 16.1-lb uniform sphere A is in contact with a 48.3-lb body B, and both are to have only translation motion (Fig. P7.33). Body B slides on the smooth horizontal plane. Determine (a) the minimum coefficient of friction between bodies A and B, (b) the required acceleration of body B, and (c) the force **P** required for translation to occur.

Ans. (a) $\mu = 0$,   (b) $a_B = 4.65 \text{ fps}^2 \leftarrow$,   (c) $P = 9.3 \text{ lb} \leftarrow$

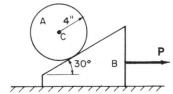

Fig. P7.33

7.34. A 128.8-lb body A has a radius of gyration of mass of 1.5 ft with respect to the horizontal axis through mass center C. (Fig. P7.34). The coefficient of friction between A and the plane is 0.1. The peg at D is frictionless. At the instant shown the angular velocity of A is 8 rad/s and the angular acceleration is 4 rad/s^2, both counterclockwise. Determine (a) the weight of body B and (b) the tension in the cable from 0 to A.

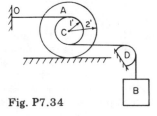

Fig. P7.34

7.35. A structure consists of a 2.0-kg rod AB, a 3.0-kg body D, and a 5.0-kg body E (Fig. P7.35). Bodies D and E can be treated as point masses and are mounted on rigid arms of negligible mass. The constant angular velocity of the assembly is 9.0k rad/s. Determine the bearing reactions at A and B on the shaft for the given position. (*Hint:* see Eq. 7.13.)

Ans. $\mathbf{A}_x = -76\mathbf{i}$ N, $\mathbf{A}_y = -335\mathbf{j}$ N, $\mathbf{A}_z = 98.1\mathbf{k}$ N

$\mathbf{B}_x = -329\mathbf{i}$ N, $\mathbf{B}_y = -150.9\mathbf{j}$ N

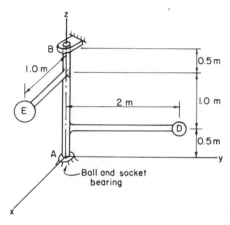

Fig. P7.35

7.36. A uniform bar AB weighs 200 lb and is connected to a spring with a modulus of 20 lb/ft (Fig. P7.36). The spring is initially stretched 10 ft when in the position shown. When the bar rotates 90° clockwise in the vertical plane, determine (a) the work done on the spring by the bar and (b) the total work done on the bar.

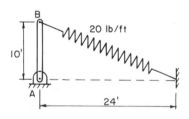

Fig. P7.36

7.37. A uniform slender rod AB weighs 64.4 lb and is connected to a 32.2-lb body D by a flexible cable which passes through two small pins E (Fig. P7.37). When in the position shown, AB has an angular velocity of 2 rad/s clockwise which increases to 5 rad/s clockwise after AB rotates 90° in the clockwise direction. Determine (a) the kinetic energy of the system when in the initial and final positions, (b) the work done on the system by rod

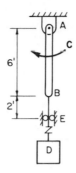

Fig. P7.37

AB and body D in moving from the initial to the final position, and (c) the work done by the constant couple $\mathbf{C}$ and the value of $\mathbf{C}$.

Ans.   (a) $T_i = 48$ ft-lb, $T_f = 588$ ft-lb,   (b) $Wk_{AB\&D} = -451$ ft-lb

(c) $Wk_C = 991$ ft-lb, $\mathbf{C} = 631$ ft-lb $\curvearrowright$

7.38. Body A weighs 64.4 lb and moves from left to right along a horizontal rod (Fig. P7.38). The spring has an unstretched length of 1.2 ft and a modulus of 120 lb/ft. Determine the initial velocity of body A if it comes to rest in the dashed position.

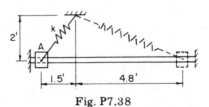

Fig. P7.38

7.39. A 100-kg body A is connected to a 1400-N/m spring and the 160-kg homogeneous cylinder B by cables (Fig. P7.39). The coefficient of friction between body A and the horizontal plane is 0.20. The spring has no elongation the instant the velocity of A is 1.0 m/s to the right. Determine the displacement of body A during which its velocity increases to 2.0 m/s to the right if a constant 800 N-m couple $\mathbf{C}$ acts on body B.

Ans. $x = 0.217$ m $\rightarrow$

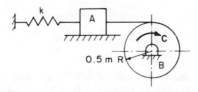

Fig. P7.39

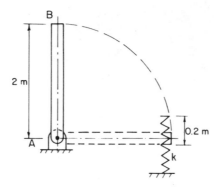

Fig. P7.40

7.40. A homogeneous rod AB has a 60-kg mass, moves in the vertical plane, and has an initial angular velocity of 1 rad/s clockwise when in the position shown in Fig. P7.40. The bar rotates clockwise until it strikes an initially unstretched spring and comes to rest in the horizontal position. Determine (a) the initial kinetic energy, (b) the spring modulus, and (c) the pin reaction at A on AB the instant the angular velocity is zero.

7.41. Body A in Fig. P7.41 weighs 257.6 lb, is rolling without slipping, and has a radius of gyration of mass of 2.12 ft with respect to the mass center. Body B weighs 96.6 lb and is attached to body A by a cord passing over a smooth pin at D. The spring has a modulus of 100 lb/ft, and is initially stretched 1 ft when body B has a downward velocity of 3 fps. Determine (a) the kinetic energy of the system in terms of the velocity of body B, (b) an expression for the velocity of body B in terms of its position y, and (c) the velocity of body B after it has moved 3 ft downward.

Ans. (a) $T = 3.0v_B^2$ ft-lb, (b) $v_B = \sqrt{9 + 21.1y - 1.85y^2}$ fps

(c) $\mathbf{v}_B = 7.46$ fps $\downarrow$

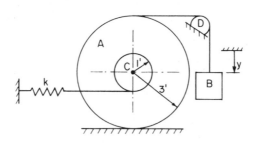

Fig. P7.41

7.42. An elevator mechanism consists of a 322-lb brake and drum assembly A with a radius of gyration of mass of 2.0 ft with respect to the axis of rotation and a 1610-lb elevator B (Fig. P7.42). When the elevator is descend-

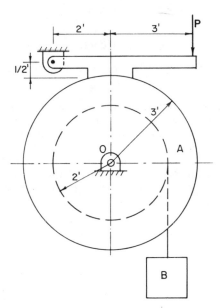

Fig. P7.42

ing with a constant velocity of 20 fps, determine (a) the power dissipated by the brake, (b) the force **P** required to maintain this rate of descent if the coefficient of friction between the brake and the drum is 0.50, and (c) the force **P** required to stop the elevator in 16 ft if the initial velocity is 20 fps and $\mu = 0.50$.

7.43. A solid homogeneous sphere of radius r rolls on a cylindrical surface of radius R (Fig. P7.43). The coefficient of friction between the sphere and the cylinder is 0.50, and the sphere starts from rest at the top of the cylinder. For this system show that (a) the kinetic energy of the sphere is given by $T = 7/10 \, mr^2\omega^2$, (b) the angular velocity of the sphere is given by

$$\omega = \sqrt{[10g(r + R)(1 - \cos \theta)]/7r^2}$$

from work and energy principles, (c) the angular acceleration of the sphere

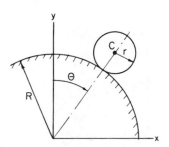

Fig. P7.43

is given by

$$\alpha = (5g \sin \theta)/7r \quad \downarrow$$

by differentiating the expression for ω, and (d) the maximum angle before the sphere begins to slip is $41.8°$.

7.44. An 80-kg cylinder A rolls without slipping and is attached to a 200-kg body B by a cable passing over a smooth pin at D (Fig. P7.44a). The system is released from rest with the nonlinear spring being initially unstretched. The spring force is given by $F_s = 450s^{1/2}$ where F_s is in Newtons when s is in meters (Fig. P7.44b). The weight B will fall onto the rigid stop. Determine (a) the kinetic energy of the system the moment before weight B strikes the stop and (b) the maximum elongation of the spring when body A comes to rest.

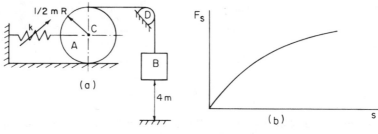

Fig. P7.44

7.45. A homogeneous rod AB of weight W and length 2ℓ is sliding on two smooth surfaces (Fig. P7.45). When in the vertical position, the bar is given a small counterclockwise angular velocity of ω_0. For this system show that (a) the velocity of the mass center can be expressed by $v_c = \ell\omega$, (b) the kinetic energy can be expressed as $T = (2M\ell^2/3)\omega^2$, (c) the angular velocity can be expressed as

$$\omega = \sqrt{\omega_0^2 + (3g/2\ell)(1 - \cos \theta)} \quad \downarrow$$

from work and energy principles, (d) the angular acceleration is given by $\alpha = (3g/4\ell) \sin \theta \downarrow$, and (e) the maximum angle at which end A leaves the vertical wall is $48.3°$ if $\omega_0 = 0$.

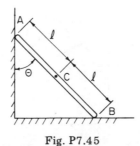

Fig. P7.45

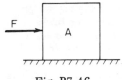

Fig. P7.46

7.46. Body A in Fig. P7.46 weighs 64.4 lb, is moving on a smooth horizontal plane, and has a force acting on it given by $F = 24t$ where F is in pounds when t is in seconds. The body starts from rest at the origin when $t = 0$ s. For the first 3 s determine (a) the linear impulse on the body, (b) the time average force, (c) the position of the body as a function of time, and (d) the position average force from the expression $Wk = \int F \, ds = (F_{ave})(\Delta s)$.

7.47. In the design of a field artillery piece one of the important parameters is the recoil spring constant. For the piece in Fig. P7.47 the barrel weighs 3220 lb and the average projectile weighs 96.6 lb, leaving the barrel with a maximum velocity of 2000 fps. Determine (a) the velocity of the barrel the moment the projectile leaves, (b) the linear impulse on the barrel, and (c) the spring modulus required to stop the barrel after traveling 8 in. Assume the friction is zero between the barrel and the horizontal plane.

Ans. (a) $\mathbf{v} = 60.0$ fps $\leftarrow$,   (b) $\hat{\mathbf{F}} = 6000$ lb-s $\leftarrow$

(c) $\mathbf{k} = 64.0 \times 10^4$ lb/ft

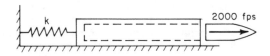

Fig. P7.47

7.48. Body A in Fig. P7.48a weighs 2 lb, is released from rest, and strikes a dynamic force gage at B. The dynamic force gage gives the contact force to be $F = 3140 \sin (1000\pi t)$ where F is in pounds when t is in seconds for $0 < t \leqslant 0.001$ s and zero for all other time (Fig. P7.48b). Determine (a) the linear impulse on A, (b) the velocity of A after impact, and (c) the energy lost during impact.

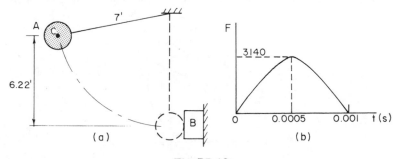

Fig. P7.48

7.49. An inextensible cord of negligible weight is wrapped around the 1.0-ft
diameter hub of 4.0-ft diameter body A (Fig. P7.49). Body A weighs
322 lb and has a radius of gyration of mass of 1.2 ft with respect to the
vertical axis of symmetry. The force in the cord varies according to
$F = 24 - 24t + 36t^2$ where F is in pounds when t is in seconds. Body A is
initially at rest on the smooth horizontal plane when t is zero. Determine
(a) the linear and angular impulse on A when $t = 2.0$ s, (b) the linear and
angular velocity as functions of time, and (c) the linear and angular displacement when $t = 2.0$ s.

Ans. (a) $\hat{\mathbf{F}} = 96$ lb-s $\rightarrow$, $\hat{\mathbf{M}}_c = 48$ ft-lb-s $\big\downarrow$

(b) $\mathbf{v} = 2.4t - 1.2t^2 - 1.2t^3$ fps$\rightarrow$, $\boldsymbol{\omega} = 0.833t - 0.417t^2 + 0.417t^3$ rad/s $\big\downarrow$

(c) $x = 6.40$ ft$\rightarrow$, $\theta = 2.22$ rad $\big\downarrow$

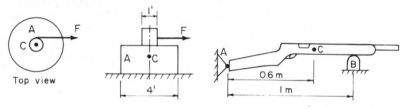

Fig. P7.49   Fig. P7.50

7.50. A 5-kg rifle is being test fired in a stand (Fig. P7.50) where the barrel
center line is 10 cm above the axis of rotation at A and the gun rests on a
support at B. The radius of gyration of mass with respect to an axis
through A is 0.894 m. The rifle fires a 0.04-kg projectile with a muzzle
velocity of 500 m/s. Assume it takes the projectile less than 0.002 s to
travel the length of the barrel. Determine (a) the linear impulse on the
projectile, (b) the angular impulse around the axis through A, and (c) the
angular velocity of the gun the instant the projectile leaves the barrel.

7.51. A 4-ft diameter homogeneous 64.4-lb cylinder is initially rolling without
slipping with a velocity of 4 fps to the right (Fig. P7.51) when it is subjected to a couple $C = 36t^2$ for a period of 5 s where C is in ft-lb when t is
in seconds. The coefficient of friction between the cylinder and the
horizontal surface is 0.746. Determine (a) the time the cylinder begins to

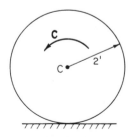

Fig. P7.51

roll and slip, (b) the linear and angular velocities when $t = 3.0$ s, and (c) the expression for the linear displacement when $t > 2$ s.

Ans.   (a) $t = 2.0$ s,   (b) $v - 36.0$ fps $\leftarrow$, $\omega - 39.0$ rad/s $\curvearrowright$
      (c) $x = 12t^2 - 36t + 24$ ft $\leftarrow$

7.52. A 2.0-ft diameter homogeneous 128.8-lb cylinder is resting on a truck bed (Fig. P7.52). The driver starts moving with a constant acceleration of 4.5 fps^2 for a period of 5 s and then moves with a uniform velocity. If the cylinder rolls without slipping, determine how far the truck moves before the cylinder is on the verge of rolling off the truck.

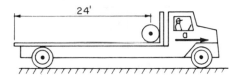

Fig. P7.52

7.53. The 30-kg sled A in Fig. P7.53 is pulled up an incline by winch B, which has a 60-kg mass and a 0.50-m radius of gyration of mass. The coefficient of friction between sled A and the incline is 0.718. A couple $C = 330t$ is applied to winch B where C is in N-m when t is in seconds. If the system is initially at rest, determine (a) the time when the sled begins to move, (b) the velocity of the sled when $t = 4$ s, (c) the linear impulse on the sled during the first 4 s, and (d) the displacement of the sled during the first 4 s.

Ans.   (a) $t_0 = 1.0$ s,   (b) $v = 33$ m/s $\angle 30°$,   (c) $\hat{F} = 990$ N-s $\angle 30°$

      (d) $x = 22.0$ m $\angle 30°$

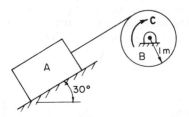

Fig. P7.53

7.54. The 128.8-lb homogeneous cylinder B in Fig. P7.54 is supported by two cords wrapped around it at A, the cords are equidistant from the mass center, the cylinder has a narrow slot cut in the center around which a cord is attached to the 64.4-lb body D. The initial velocity of body B is 6 fps upward. Determine (a) the velocity and displacement of body B as a

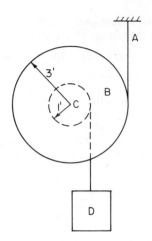

Fig. P7.54

function of time and (b) the tension in each cord at A as a function of time.

7.55. Solve Problem 7.54 if a clockwise couple of $\mathbf{C} = 124t$ is applied to body B where $\mathbf{C}$ is in ft-lb when t is in seconds. Is there a time limit where the answers are no longer valid?

Ans. (a) $v_B = -6 + 24.9t - 3t^2$ fps $\downarrow$, $y_B = -6t + 12.45t^2 - t^3$ ft $\downarrow$

(b) $T_A = 30.1 + 16t$ lb T

7.56. An electric motor A is attached to arm D (Fig. P7.56). The entire assembly is rotating about the vertical 00 axis as a rigid body with an initial angular velocity of $\omega_0 = 20$ rad/s when the motor is *not* running. When energized, the motor (rotor and disk B) achieves an angular velocity of 180 rad/s *relative* to arm D and in the same direction as ω_0. Determine the final angular velocity of arm D when (a) the bearing at E is frictionless and (b) the bearing at E has a moment of 20 ft-lb due to friction and the motor comes up to speed in 3.0 s. The following data apply to the system:

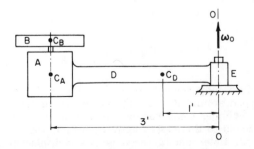

Fig. P7.56

Motor case: M_A = 1 slug, I_{CA} = 2 slug-ft^2

Rotor and disk B: M_B = 1 slug, I_{CB} = 2 slug-ft^2

Arm D: M_D – 3 slugs, I_{CD} = 5 slug-ft^2

7.57. A 120-kg merry-go-round is 4 m in diameter, has a radius of gyration of mass of 1.414 m, and rotates about the vertical axis through 0 (Fig. P7.57). The initial angular velocity of the merry-go-round is 3.0 rad/s the instant that a 30-kg child moves from 0 to A (along the straight line $0A$ relative to the merry-go-round) during a 3 s period. Determine the final angular velocity of the merry-go-round the moment the child reaches point A if (a) the bearings at 0 are frictionless and (b) a frictional moment of 60 N-m acts on the merry-go-round. What effect does the radial relative velocity of the child have on the angular momentum of the system?

Ans. (a) $\boldsymbol{\omega}$ = 2.0 j rad/s,   (b) $\boldsymbol{\omega}$ = 1.50 j rad/s

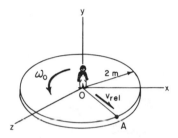

Fig. P7.57

7.58. A 193-lb skydiver in free fall will usually achieve a free-fall (terminal) velocity of approximately 120 mph. Assume that the air resistance experienced by the skydiver is given by $F = cv^2$, where F is in pounds when v is in fps. Determine (a) the time required to achieve 90% of the free-fall velocity when the diver jumps from a hovering helicopter and (b) how far the diver falls before achieving 90% of the free-fall velocity.

7.59. The elevator system in Fig. P7.59 consists of a 1932-lb counterweight A, a 1610-lb cable drum and electric motor B with a 1.414-ft, radius of

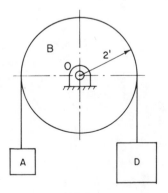

Fig. P7.59

gyration of mass with respect to the axis of rotation and a 1932-lb elevator D carrying a maximum rated load of 1288 pounds. The drive motor acts as an electrodynamic brake when the elevator is in free fall in order to prevent the buildup of excessive speed. For this motor the braking torque is given by $T_b = 257.6\omega$ where T_b is in ft-lb when ω is in rad/sec. Determine (a) the maximum free-fall velocity and (b) the time required to achieve 90% of the free-fall velocity when the elevator starts from rest.

Ans.   (a) $\mathbf{v} = 20$ fps $\downarrow$,   (b) $t = 6.61$ s

7.60. An 8050-ton (1 ton = 2000 lb) ship has a propulsion system that delivers a driving force of $F_p = 400{,}000 + 3000v$ lb (when v is in fps) when operating at 300 rpm. The resistance of the water to the motion of the ship is given by $F_R = 220v^2$ lb when v is in fps. When these two forces are equal, the ship obtains its cruising velocity v_c (Fig. P7.60). Determine (a) the cruise velocity and (b) how long it will take the ship to accelerate from a velocity of 20 fps to 40 fps.

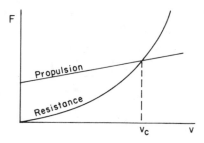

Fig. P7.60

Vibrations

8.1 Introduction

Vibrations occur in most engineering structures. This oscillatory motion must be controlled to have quiet machinery with long operating lives. When vibrations are not properly controlled, excessive sound (noise) and structural fatigue result. A dramatic structural failure caused by wind-induced vibrations was the collapse of the Tacoma Narrows suspension bridge shortly after completion. A smaller but nonetheless costly vibration occurs every day when automobiles are operated with tires out of balance. The cost in this situation includes excessive tire and shock absorber wear, general loosening and fatigue of the automobile structure, driver and passenger discomfort, operator fatigue, and possible loss of vehicle control. In addition, improperly installed machinery has caused entire building structures to be weakened.

Not all vibrations are harmful. The ability to speak is due to vibrating cords in the voice box. Ability to hear is dependent on vibration of the eardrum. The ability to faithfully record and play back sounds is dependent on the vibration characteristics of microphones and speaker systems. In fact, the entire concept of instruments to measure dynamic quantities such as force, acceleration, velocity, displacement, and pressure is dependent on the vibration characteristics of the instrument. These characteristics are the same as the single degree of freedom vibrating system described in this chapter.

The differential equation of motion for a single particle or a translating rigid body can be written as

$$m\ddot{r}_c = F(r_c, v_c, t) \tag{8.1}$$

where m is the mass of the body, r_c is the position vector locating the mass center of the body (or particle) relative to an inertial coordinate system, and F is the resultant force acting on the body. As shown in Section 7.5, the resultant force may be a function of position r_c of the body, the velocity $v_c = \dot{r}_c$ of the body, and the passage of time t. When the body moves along a straight-line path, Eq. (8.1) becomes

$$m\ddot{x} = F(x, \dot{x}, t) \tag{8.2}$$

which is a general form for rectilinear motion. In the following section we shall derive equations of motion similar to Eq. (8.2) where the function $F(x, \dot{x}, t)$ is a linear function of x and $\dot{x}$.

8.2 Equations of Motion

The problems considered here are restricted in such a way that the system can be completely described by one coordinate, i.e., a single displacement or a single angle. This type of system is called a *system of one degree of freedom.*

We first consider the problem where the force is a function of position of the body. For example, if an elastic spring is attached to a rigid body that rests on a smooth surface (Fig. 8.1a), the force exerted by the spring on the body is related to the displacement of the spring. For a *linear spring* the force is directly proportional to the displacement x when measured from its unstretched position; i.e., $F_{\text{spring}} = kx$ where k is the *spring constant.* From the free-body diagram (Fig. 8.1b) for the body when it is displaced a distance x from its equilibrium position, we can write the equations of motion as

$$\xrightarrow{+} \quad \Sigma F_x = m \frac{d^2 x}{dt^2}$$

or

$$-kx = m \frac{dx^2}{dt^2} \tag{8.3}$$

We are considering the elements of the system to be *lumped;* i.e., the distributed mass of the spring is neglected compared to the mass of the body, and the body is assumed to be rigid so that it has no springlike behavior. The negative sign on the spring force is due to the fact that this force acts in a direction opposed to the motion. Equation (8.3) can be written in the form

$$\frac{d^2 x}{dt^2} + \frac{k}{m} x = 0 \tag{8.4}$$

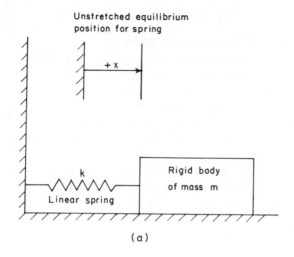

Unstretched equilibrium
position for spring

$+ x$

k

Linear spring

Rigid body
of mass m

(a)

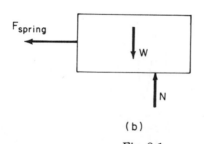

F_{spring}

W

N

(b)

Fig. 8.1

which is a *second-order, linear, ordinary* differential equation with constant coefficients. Equation (8.4) is a special form of a more general motion equation to be derived in the next section.

The solution of Eq. (8.4) is found in the following manner. Let

$$x(t) = Ae^{pt} \tag{8.5}$$

Substitution of Eq. (8.5) into (8.4) results in the characteristic equation $p^2 + k/m = 0$ which leads to $p = \pm i\sqrt{k/m} = \pm i\omega_n$ where $i^2 = -1$. The general solution to Eq. (8.4) is a linear combination of exponential functions; i.e.,

$$x(t) = a \exp(i\omega_n t) + b \exp(-i\omega_n t)$$

The relationship $e^{i\theta} = \cos\theta + i\sin\theta$ allows us to write the solution as

$$x(t) = A \cos \omega_n t + B \sin \omega_n t \tag{8.6}$$

Equation (8.6) can also be written in the form

$$x(t) = C \cos (\omega_n t - \beta) \tag{8.7}$$

where $C = \sqrt{A^2 + B^2}$ and $\beta = \tan^{-1} (B/A)$.

The type of motion described by Eq. (8.6) or (8.7) is termed *harmonic motion*, and the physical system described by the spring-mass system is often called a *harmonic oscillator*.

The quantity C in Eq. (8.7) is the *amplitude* of the motion, i.e., the maximum displacement of the mass from its equilibrium position. The *angular frequency* of the motion is ω_n and the *frequency* f in hertz (Hz) is given by $f = \omega_n /2\pi$ where ω_n is in rad/s and 1 Hz = 1 cycle/s (cps). The *period* T is the time required to complete one cycle and is given by the equations $T = 1/f$ or $T = 2\pi/\omega_n$. The angle β is called the *phase angle*. A plot of the position x versus time t is shown in Fig. 8.2.

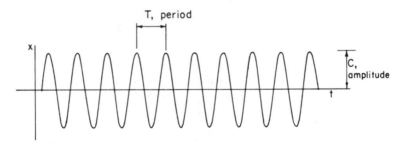

Fig. 8.2

A useful graphical interpretation of Eq. (8.6) can be given in the following manner. Consider a vector **A** of constant magnitude that is rotating with a constant angular velocity of ω_n (Figure 8.3a). At $t = 0$ let the vector be aligned with the positive x axis so that at any later time the angle between the vector and the x axis is $\omega_n t$. The projection of **A** onto the x axis is then $A \cos \omega_n t$. Since

$$\sin \omega_n t = \cos [\omega_n t - (\pi/2)]$$

$B \sin \omega_n t$ can be considered as the projection on the x axis of the rotating vector **B** of constant magnitude which "lags" the vector **A** by the angle $\pi/2$ rad (Fig. 8.3b). The projection of the resultant vector $\mathbf{C} = \mathbf{A} + \mathbf{B}$ on the x axis gives the result

$$x(t) = A \cos \omega_n t + B \sin \omega_n t$$

which is Eq. (8.6). From Fig. 8.3b we see that $x(t)$ can also be written as $x(t) = C \cos (\omega_n t - \beta)$ which corresponds to Eq. (8.7). Rotating vectors such as **A** and **B** considered above are called *phasors*.

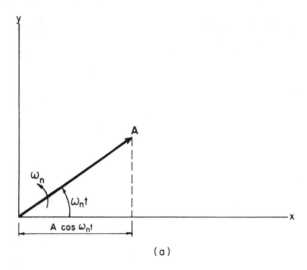

(a)

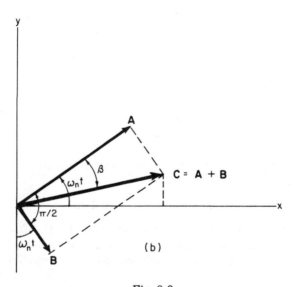

(b)

Fig. 8.3

EXAMPLE 8.1

A 50-lb motor rests on four rubber mounts. Each mount deflects 1/32 in. for an applied load of 10 lb. Determine the natural frequency of the system. Sketch a period of the deflection curve as a function of time if at a reference time $t = 0$ the deflection is zero and the velocity is a constant V_0 directed upward.

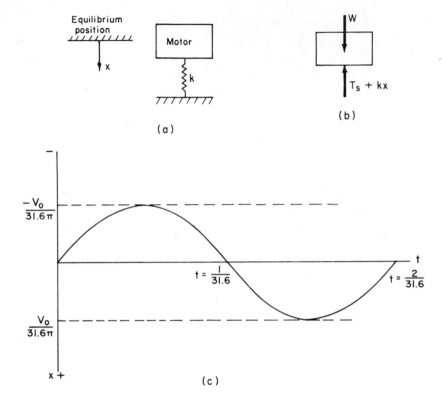

Fig. E8.1

SOLUTION

The idealized model for this system is shown in Fig. E8.1a. The effective spring constant is

$$k = 4[10/(1/32)] = 1280 \text{ lb/in.}$$

The free-body diagram for the system is shown in Fig. E8.1b. The displacement x is measured from the static equilibrium position of the system, and T_s is the compression in the spring required to hold the motor in its equilibrium position; i.e., $T_s = W$.

The equation of motion is

$$m \frac{d^2 x}{dt^2} = W - T_s - kx$$

and since $W = T_s$

$$\frac{dx^2}{dt^2} + \frac{k}{m} x = 0$$

The solution to the differential equation is $x(t) = A \cos \omega_n t + B \sin \omega_n t$ with

$\omega_n = \sqrt{k/m}$. The natural frequency of the system is

$$f = \frac{\omega_n}{2\pi} = \frac{1}{2\pi} \sqrt{\frac{k}{m}} = \frac{1}{2\pi} \sqrt{\frac{(1280)(32.2)(12)}{50}} = 15.83 \text{ Hz} \qquad Ans.$$

To find A and B we apply the initial conditions

$$x(0) = 0, \qquad \dot{x}(0) = -V_0$$

The first condition leads to $A = 0$, while the second gives

$$B = -V_0/\omega_n = -V_0/2\pi f = -V_0/31.6\pi$$

The displacement is

$$x(t) = -(V_0/31.6\pi) \sin 31.6\pi t \qquad\qquad Ans.$$

A sketch of $x(t)$ is shown in Fig. E8.1c.

The theoretical solution of the free vibration of a spring-mass system is seen to persist indefinitely (see, for example, Eq. 8.7). A simple laboratory experiment will show, however, that the motion will not continue indefinitely and the system will eventually return to its equilibrium position. This is due to the presence in the system of damping that irreversibly absorbs energy. The damping may arise from several sources such as internal damping in the spring, air friction, friction at the spring support, or from an external device explicitly added to the system to provide damping.

A common type of damping, which is developed whenever a solid moves through a fluid at a relatively low velocity, is called *viscous damping*. The magnitude of the force due to this type is directly proportional to the velocity of the solid moving through a still fluid; i.e., $F_{\text{damping}} = -cv$ where c is called the *damping coefficient* and v the magnitude of the velocity. The negative sign indicates that the damping force always opposes the motion. The damping coefficient is a function of the properties of the damping fluid and the geometry of the system forming the damping mechanism. A typical representation of a viscously damped spring-mass system is shown in Fig. 8.4a. The device used to indicate viscous damping is intended to represent a plunger moving in a viscous fluid and is called a *dashpot*. Other types of damping can be also shown to be proportional to the magnitude of the velocity at low speeds. At higher speeds the damping force may be proportional to the speed squared, which results in a nonlinear equation. The study of nonlinear problems is beyond the scope of the present study.

The free-body diagram for a linearly damped spring-mass system

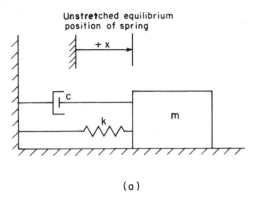

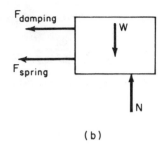

(b)

(a)

Fig. 8.4

is shown in Fig. 8.4b, and the corresponding equation of motion is

$$\oplus - c \frac{dx}{dt} - kx = m \frac{d^2 x}{dt^2}$$

or

$$\frac{d^2 x}{dt^2} + \frac{c}{m} \frac{dx}{dt} + \frac{k}{m} x = 0 \tag{8.8}$$

Equation (8.8) reduces to Eq. (8.4) if the damping is not present; i.e., $c = 0$. The mechanism whose motion is described by Eq. (8.8) is sometimes called a *damped oscillator*.

Before finding the solution to Eq. (8.8), we will derive the differential equation governing the vibration of a disk on an elastic circular shaft when twisted by a torque and then released (Fig. 8.5). The governing equation of motion for this case is the moment equation developed in Chapter 7; i.e.,

$$M_c = I_c \alpha \tag{8.9}$$

where α is the angular acceleration and I_c is the moment of inertia of the body. The free-body diagram of the shaft (Fig. 8.5b) exhibits the elastic torque due to the angle through which the shaft has been twisted, and a viscous damping torque proportional to the angular velocity is shown in Fig. 8.5c. In Chapter 10 we will show that for small angular displacements the elastic torque is linearly proportional to the angle of twist. We then have

$$M_c = -k_t \theta - c_t \frac{d\theta}{dt} \tag{8.10}$$

where the minus signs indicate that the elastic torque and viscous

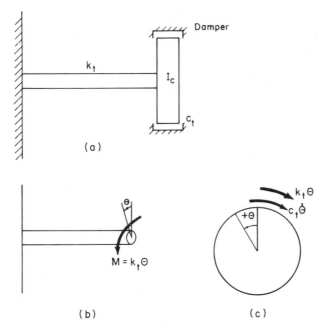

Fig. 8.5

damping act in a direction opposing the motion of the shaft. Combining Eq. (8.9) and (8.10) with $\alpha = d^2\theta/dt^2$ results in

$$I_c \frac{d^2\theta}{dt^2} + c_t \frac{d\theta}{dt} + k_t\theta = 0 \qquad (8.11)$$

which mathematically is identical to Eq. (8.8). Here we have assumed that the mass (moment of inertia) of the shaft is negligible compared to the moment of inertia I_c of the disk and that the disk is a rigid body. By appropriate identification of the constants in Eqs. (8.8) or (8.11) the solution of either equation affords the solution to the other. It can also be shown that simple electrical circuits have equations mathematically equivalent to Eqs. (8.8) or (8.11). This fact allows the modeling of mechanical vibration problems by analog electrical circuits.

In addition to the forces or torques included in the derivation of Eqs. (8.8) and (8.11), an externally applied force or torque can be considered. In many of the problems to be considered we assume these forces to be harmonic in nature. When these forces are added to the governing equations the result is

$$m \frac{d^2x}{dt^2} + c \frac{dx}{dt} + kx = F_0 \begin{Bmatrix} \cos \omega t \\ \sin \omega t \end{Bmatrix} \qquad (8.12)$$

$$I \frac{d^2\theta}{dt^2} + c \frac{d\theta}{dt} + k\theta = T_0 \begin{Bmatrix} \cos \omega t \\ \sin \omega t \end{Bmatrix} \tag{8.13}$$

where either the $\cos \omega t$ or $\sin \omega t$ can be used to describe the applied force or torque, and F_0 and T_0 are the amplitudes of the applied force and torque respectively. The motion described by Eqs. (8.12) and (8.13) is termed *forced oscillations* or *forced vibrations*.

8.3   Free-damped Vibrations

The free-damped oscillation of a system is described by Eq. (8.8) or (8.11), i.e., when the disturbing force or torque is zero. Since both equations have the same mathematical form, we will concentrate on the physical system associated with Eq. (8.8); i.e.,

$$\frac{d^2 x}{dt^2} + \frac{c}{m}\frac{dx}{dt} + \frac{k}{m} x = 0$$

We assume a solution of the form $x(t) = ae^{pt}$, and substitution into Eq. (8.8) yields the characteristic equation $p^2 + (c/m)p + (k/m) = 0$. The values of p that satisfy the characteristic equation are

$$p_{1,2} = -(c/2m) \pm \sqrt{(c/2m)^2 - (k/m)} \tag{8.14}$$

and the general solution to the differential equation is

$$x(t) = A \exp (p_1 t) + B \exp (p_2 t)$$

where the constants A and B are to be determined from the initial conditions.

Three modes of motion are possible for this system depending on the sign of the discriminant, $D = (c/2m)^2 - (k/m)$. Note that the sign of D will depend on the relative magnitudes of the ratios $c/2m$ and k/m.

Mode (a): $D = 0$. In this case $c/2m = \sqrt{k/m}$ and $p_1 = p_2$. To have two independent solutions of the differential equation, the displacement $x(t)$ must be of the form

$$x(t) = A \exp [-(c/2m)t] + Bt \exp [-(c/2m)t] \tag{8.15}$$

The motion in this case is said to be critically damped and the value of the damping coefficient is called the *critical damping* coefficient and is given by

$$c_r = 2\sqrt{km} \tag{8.16}$$

The motion of a critically damped system starting at $x = X_0$, for $t = 0$, and having an initial velocity $\dot{x}(0) = 0$ is illustrated in Fig. 8.6.

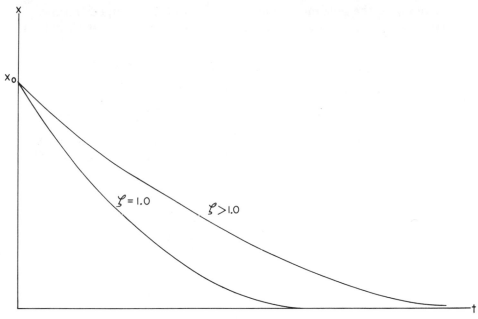

Fig. 8.6

For this mode of motion the system does not oscillate. The damping factor ζ is defined as the ratio of the damping coefficient to the critical damping coefficient, i.e., $\zeta = c/c_r$. Thus for the critically damped case $\zeta = 1.0$.

Mode (b): $D > 0$. For this *overdamped* case $c > 2\sqrt{km}$ and $\zeta > 1.0$. The motion is described by the equation

$$x(t) = \exp\left[-(c/2m)t\right] \left\{ A \exp\left[\sqrt{(c/2m)^2 - (k/m)}t\right] + B \exp\left[-\sqrt{(c/2m)^2 - (k/m)}t\right] \right\} \quad (8.17)$$

and the mass will not oscillate but simply return exponentially to its equilibrium position. A sketch of the overdamped case for

$$x(0) = X_0, \qquad \dot{x}(0) = 0$$

is shown in Fig. 8.6.

Mode (c): $D < 0$. Here the roots p_1 and p_2 are conjugate, complex numbers which can be written either in terms of m, c, and k as

$$p_{1,2} = -(c/2m) \pm i\sqrt{(k/m) - (c/2m)^2}$$

or in terms of ω_n and ζ as

$$p_{1,2} = \left(-\zeta \pm i\sqrt{1 - \zeta^2}\right)\omega_n = -\zeta\omega_n \pm i\omega_d$$

where ω_n is the natural frequency $\sqrt{k/m}$ and the damped frequency $\omega_d = \sqrt{1 - \zeta^2}\ \omega_n$ is often used. The solution for this type of motion is

$$x(t) = \exp\left(-\zeta\omega_n t\right) \left[A \exp\left(i\ \omega_d t\right) + B \exp\left(-i\ \omega_d t\right)\right]$$

or in trigonometric form

$$x(t) = \exp\left(-\zeta\omega_n t\right) \left(A \cos \omega_d t + B \sin \omega_d t\right) \qquad (8.18)$$

The motion described by Eq. (8.18) is oscillatory but is exponentially damped (Fig. 8.7) with initial conditions $x(0) = X_0$ and $\dot{x}(0) = 0$. The

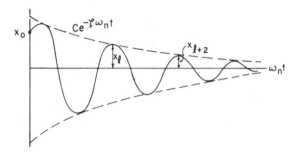

Fig. 8.7

motion for this case is said to be underdamped. Equation (8.18) can also be written in the form

$$x(t) = C \exp\left(-\zeta\omega_n t\right) \sin\left(\omega_d t - \phi\right) \qquad (8.19)$$

The damping ratio ζ is often obtained experimentally for under-damped mechanical systems from the underdamped free-oscillation curve in Fig. 8.7. The information needed to determine ζ is any two successive peak amplitudes of the displacement such as x_ℓ and $x_{\ell+2}$ that occur at times t_ℓ and $t_{\ell+2}$ respectively. The times at which these peak amplitudes occur can be found by equating the derivative of Eq. (8.19) with respect to time equal zero. This results in the equation

$$\tan\left(\omega_d t - \phi\right) = \sqrt{(1 - \zeta^2)}/\zeta$$

The values of t that satisfy the above equation are

$$\omega_d t_\ell = P + \ell\pi, \qquad \ell = 0, 1, 2, 3, \ldots$$

where

$$P = \phi + \tan^{-1}\left(\sqrt{(1 - \zeta^2)}/\zeta\right)$$

The ratio of successive peak displacements on the same side of the equilibrium position is given as

$$\frac{x_\ell}{x_{\ell+2}} = \frac{x[(P + \ell\pi)/\omega_d]}{x\{[P + (\ell + 2)\pi]/\omega_d\}} = \exp\left(\frac{2\pi\zeta\omega_n}{\omega_d}\right) \qquad (8.20)$$

The results in Eq. (8.20) depend only on the parameters of the system and not on the two particular extreme positions considered. Thus the ratio of successive maximum (or minimum) displacements is constant during the entire free motion of an underdamped system.

Taking the natural logarithm of both sides of Eq. (8.20) leads to

$$\ln(x_\ell/x_{\ell+2}) = 2\pi\zeta\omega_n/\omega_d$$

The quantity on the left side of the above equation is called the logarithmic decrement of δ and can be written as

$$\delta = 2\pi\zeta/\sqrt{1 - \zeta^2} \qquad (8.21)$$

If two successive maximum peak amplitudes are measured, the damping coefficient can be calculated from Eq. (8.21). In many physical problems it is not possible to measure successive peaks, but it is possible to measure a sequence of peak amplitudes. This sequence can be used to find the logarithmic decrement and hence the damping ratio. The details are left as an exercise for the student. For the case of a lightly damped system, $\zeta < 0.10$, Eq. (8.21) gives $\delta \approx 2\pi\zeta$.

For a lightly damped system the damped frequency ω_d and the natural frequency ω_n of the system are approximately equal. In most physical problems the damping is light ($\zeta < 0.10$); therefore damping is usually neglected in natural frequency calculations.

8.4 Forced Vibrations

Consider an undamped system, but with the addition of a harmonic external force as in Fig. 8.8. In many practical applications these oscillating forces are continuously applied to the system. For example, as the shaft of a motor rotates, an oscillating force is usually developed due to unbalance in the shaft, and other systems exposed to this oscillating force may be set into vibratory motion. We define problems of this type as *forced vibration* problems. One of the simplest systems is the spring-mass system (Fig. 8.8) where the friction force is neglected. The equation of motion is

$$\frac{d^2 x}{dt^2} + \frac{k}{m}x = \frac{F_0}{m}\sin\omega t \qquad (8.22)$$

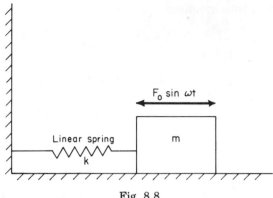

Fig. 8.8

where F_0 is the amplitude of the harmonically oscillating force and ω is the forcing frequency and in general $\omega \neq \omega_n$.

Equation (8.22) is a nonhomogeneous, second-order, linear differential equation whose complete solution is the sum of the solution of the homogeneous equation plus a particular solution.

We previously found the solution to the homogeneous equation in Section 8.2. This is called the complementary solution and is

$$x_c(t) = A \sin \omega_n t + B \cos \omega_n t \qquad (8.23)$$

A particular solution for the differential equations with a non-homogeneous term as in Eq. (8.22) is of the form

$$x_p(t) = C \sin \omega t + E \cos \omega t \qquad (8.24)$$

Substitution of Eq. (8.24) into (8.22) results in

$$-\omega^2 C \sin \omega t - \omega^2 E \cos \omega t + \omega_n^2 C \sin \omega t$$
$$+ \omega_n^2 E \cos \omega t = (F_0 / m) \sin \omega t$$

The above equation must be true for all t, and thus corresponding coefficients of $\cos \omega t$ and $\sin \omega t$ must be equal. This yields

$$(\omega_n^2 - \omega^2)C = F_0/m, \qquad (\omega_n^2 - \omega^2)E = 0$$

which gives $E = 0$ and $C = F_0/[m(\omega_n^2 - \omega^2)]$. The values of E and C substituted into Eq. (8.24) give the particular solution. The complete solution is

$$x(t) = A \sin \omega_n t + B \cos \omega_n t + \frac{F_0}{m(\omega_n^2 - \omega^2)} \sin \omega t \qquad (8.25)$$

It is common practice to call the complementary solution the *transient solution* and the particular solution the *steady-state solu-*

tion because some damping is present in all real systems and the motion due to the complementary solution damps out. After a period of time the motion is given by the particular solution

$$x(t) = \frac{F_0}{m(\omega_n^2 - \omega^2)} \sin \omega t$$

The amplitude X of the steady-state motion is $X = F_0/[m(\omega_n^2 - \omega^2)]$ which can be written in the form

$$X = (F_0/k)/[1 - (\omega/\omega_n)^2] \qquad (8.26)$$

The deflection X_0 of the mass under the action of a static force F_0 is $X_0 = F_0/k$ so Eq. (8.26) can be written as

$$X/X_0 = 1/[1 - (\omega/\omega_n)^2] \qquad (8.27)$$

The deflection X_0 is called the static deflection and is equal to the deflection of the spring due to a static load F_0. A plot of Eq. 8.27 is shown in Fig. 8.9. This curve is called a *response curve* and clearly shows how the amplitude of the motion is drastically affected by the ratio of the forcing frequency to the natural frequency. When

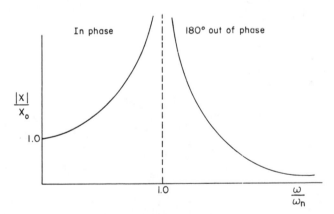

Fig. 8.9

$\omega = \omega_n$, the amplitude theoretically becomes infinite. In an actual system with damping, the amplitude is finite but can be extremely large for this so-called *resonant condition*. From Eq. (8.27) for $\omega/\omega_n < 1.0$, $X > X_0$ and has the same sign as X_0, which implies that the motion of the mass and the exciting force are in phase. For $\omega/\omega_n > 1.0$, X is negative compared to X_0 and the motion of the mass is $180°$ out of phase with the exciting force. It is apparent from Fig. 8.9 that as ω/ω_n increases, the amplitude X approaches zero.

EXAMPLE 8.2

A 75-lb instrument board is spring mounted to a panel that vibrates with a frequency of 10 Hz and an amplitude of 0.2 in. The effective spring constant is 150 lb/in. and damping is negligible. (a) Find the transient and steady-state solution, (b) sketch the relationship between the amplitude ratio (instrument board amplitude/panel amplitude) and the frequency ratio (forced frequency/natural frequency) for the steady-state solution, and (c) determine the amplitude and the magnitude of the maximum acceleration of the instrument board for the specified conditions.

SOLUTION

An idealized model of the system with the appropriate free-body diagram is shown in Figs. E8.2a and b. The harmonic motion of the panel is expressed in the form $x_1 = X_1 \cos \omega t$, and the displacement x of the instrument board is measured from its equilibrium position. If the instrument board is displaced a distance x upward, the corresponding displacement of the panel is x_1 so the effective stretch in the spring is $x - x_1$. The spring force is $k(x - x_1)$ (Fig.

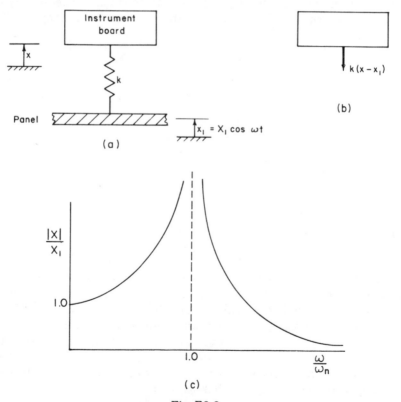

Fig. E8.2

E8.2b). The weight is not considered since we have previously shown that it will not affect the vibrational characteristics of this type of system.

The appropriate differential equation is

$$\text{(+)} \quad m \frac{d^2 x}{dt^2} = -k(x - x_1)$$

or

$$\frac{d^2 x}{dt^2} + \frac{k}{m} x = \frac{kX_1}{m} \cos \omega t$$

where m is the mass of the instrument board. The form of the equation is the same as Eq. (8.22), and we can then write

$$x_c(t) = A \sin \omega_n t + B \cos \omega_n t \qquad\qquad Ans.$$

for the transient solution, and

$$x_p(t) = \frac{kX_1}{m(\omega_n^2 - \omega^2)} \cos \omega t \qquad\qquad Ans.$$

for the steady-state solution. The steady-state amplitude X is

$$X = \omega_n^2 X_1 /(\omega_n^2 - \omega^2) = X_1 /[1 - (\omega/\omega_n)^2]$$

or

$$X/X_1 = 1/[1 - (\omega/\omega_n)^2]$$

The response curve is shown in Fig. E8.2c and is seen to be the same as that of Fig. 8.9. The natural frequency is

$$\omega_n = \sqrt{k/m} = \sqrt{(150)(32.2)(12)/75} = 27.8 \text{ rad/s}$$

The frequency ratio is $\omega/\omega_n = [(2\pi)(10)]/27.8 = 2.26$. For this frequency ratio

$$X = X_1 /[1 - (2.26)^2] = -0.0487 \text{ in.} \qquad\qquad Ans.$$

The negative sign indicates that the motion of the instrument board is $180°$ out of phase with the shaking panel.

The maximum amplitude for the steady-state acceleration is found from

$$\frac{d^2 x_p}{dt^2} = - \frac{\omega^2 kX_1}{m(\omega_n^2 - \omega^2)} \cos \omega t$$

The maximum occurs when $|\cos \omega t|$ is 1. Thus

$$\left. \frac{d^2 x_p}{dt^2} \right|_{max} = \frac{\omega^2 X_1}{1 - (\omega/\omega_n)^2} = \omega^2 |X|$$

For the above frequency ratio we then have

$$\frac{d^2 x_p}{dt^2}\bigg|_{max} = [(2\pi)(10)]^2 (0.0487) = 192.2 \text{ in./s}^2 \qquad Ans.$$

An important problem from a practical viewpoint is that of the externally excited spring-mass system with viscous damping (Fig. 8.10a). The corresponding free-body diagram is shown in Fig.

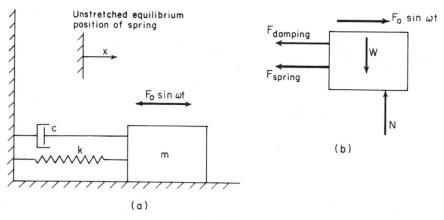

Fig. 8.10

8.10b, and the governing differential equation is

$$m\frac{d^2 x}{dt^2} + c\frac{dx}{dt} + kx = F_0 \sin \omega t \qquad (8.28)$$

The solution to this equation is composed of a complementary solution plus a particular solution. The complementary solution is given by Eqs. (8.15), (8.17), or (8.18) depending on the relative magnitude of the damping ratio. In any of the above cases the motion is exponentially damped so that the contribution of the complementary solution disappears after a short period of time. We thus need consider only the particular, or steady-state, solution.

As before we assume the particular solution to be of the form

$$x(t) = A \sin \omega t + B \cos \omega t$$

and substitute into Eq. (8.28)

$$m(-\omega^2 A \sin \omega t - \omega^2 B \cos \omega t) + c(\omega A \cos \omega t - \omega B \sin \omega t)$$
$$+ k(A \sin \omega t + B \cos \omega t) = F_0 \sin \omega t$$

The above equation must be true for all t and thus coefficients of like functions of t must be equal.

$$(-m\omega^2 + k)A - c\omega B = F_0, \qquad c\omega A + (-m\omega^2 + k)B = 0 \qquad (8.29)$$

For any real value of the frequency ω, Eqs. (8.29) have solutions for A and B which are

$$A = \frac{(k - m\omega^2)}{(k - m\omega^2)^2 + (c\omega)^2} F_0, \qquad B = \frac{-c\omega}{(k - m\omega^2)^2 + (c\omega)^2} F_0$$

Thus the steady-state solution is

$$x(t) = \frac{F_0}{(k - m\omega^2)^2 + (c\omega)^2} [(k - m\omega^2) \sin \omega t - c\omega \cos \omega t]$$

In this particular example the driving force was taken to be $F_0 \sin \omega t$; to see how the steady-state displacement is related to the driving force, we write the solution as

$$x(t) = \frac{F_0}{\sqrt{(k - m\omega^2) + (c\omega)^2}} \sin (\omega t - \alpha) \qquad (8.30)$$

where

$$\cos \alpha = \frac{k - m\omega^2}{\sqrt{(k - m\omega^2)^2 + (c\omega)^2}}$$

$$\sin \alpha = \frac{c\omega}{\sqrt{(k - m\omega^2)^2 + (c\omega)^2}}$$

and

$$\tan \alpha = c\omega/(k - m\omega^2), \qquad 0 \leqslant \alpha \leqslant \pi$$

Equation (8.30) is convenient to work with since it exhibits the phase shift between the externally applied force and the steady-state response of the system. Written in terms of the static deflection $X_0 = F_0/k$, the natural frequency ω_n, and the damping coefficient $\zeta = c/c_r$, Eq. (8.30) is

$$x(t) = \frac{X_0}{\sqrt{[1 - (\omega/\omega_n)^2]^2 + [2\zeta(\omega/\omega_n)]^2}} \sin (\omega t - \alpha) \quad (8.31)$$

The amplitude of the motion is then

$$\frac{X}{X_0} = \frac{1}{\sqrt{[1 - (\omega/\omega_n)^2]^2 + [2\zeta(\omega/\omega_n)]^2}} \qquad (8.32)$$

A plot of Eq. (8.32), showing the response curves for this system, is given in Fig. 8.11.

The type of response obtained is strongly dependent on the damping factor and the frequency ratio. The ratio $X/X_0 = M$ is

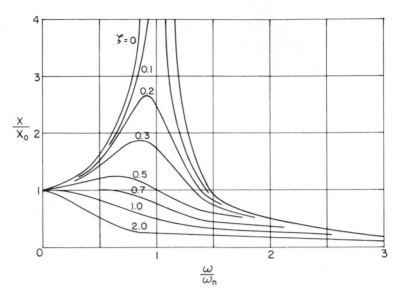

Fig. 8.11

referred to as the *amplification* or *magnification ratio*. We see from Fig. 8.11 that large amplitudes relative to the static deflection X_0 can be developed for systems with low damping when the forcing frequency is near the resonant frequency.

To find the value of the frequency ratio at which the peaks in the response curve occur, set $d(X/X_0)/d(\omega/\omega_n)$ equal to zero. This results in

$$-[1 - (\omega/\omega_n)^2](\omega/\omega_n) + 2\zeta^2(\omega/\omega_n) = 0$$

The solutions to the above equation are $\omega/\omega_n = 0$, which is of no practical interest for the forced vibration problem, and

$$(\omega/\omega_n)^2 = 1 - 2\zeta^2 \tag{8.33}$$

If $\zeta < 1/\sqrt{2}$, Eq. (8.33) has real values of ω/ω_n at which peaks will occur; while if $\zeta > 1/\sqrt{2}$, no real values of ω/ω_n exist for which the response curve has peaks. The following properties of the system being considered can be summarized as follows:

1. $X/X_0 = 1$ if $\omega/\omega_n = 0$ for any value of ζ.
2. For $0 < \zeta < 1/\sqrt{2}$, the ratio X/X_0 has a maximum which occurs at $\omega/\omega_n = \sqrt{1 - 2\zeta^2}$.
3. As ζ tends to zero, the maximum value of X/X_0 increases until for zero damping the ratio becomes infinite at $\omega/\omega_n = 1$.
4. For $\zeta > 1/\sqrt{2}$ the ratio X/X_0 decreases monotonically as ω/ω_n increases from zero.

5. For all values of ζ, X/X_0 approaches zero for $\omega/\omega_n > 1$ and ω/ω_n increasing.

The angle α given by

$$\tan \alpha = \frac{c\omega}{k - m\omega^2} = \frac{2\zeta\omega/\omega_n}{1 - (\omega/\omega_n)^2}, \qquad 0 \leqslant \alpha \leqslant \pi \qquad (8.34)$$

is called the *phase angle* or *angle of lag* of the steady-state response. The significance of α is that the displacement $x(t)$ attains its maxima α/ω units of time after the external driving force reaches its peak value. A plot of the phase angle versus the frequency ratio for different values of the damping coefficient is given in Fig. 8.12.

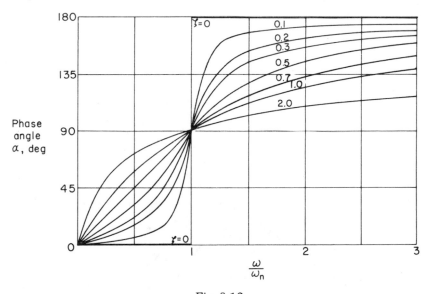

Fig. 8.12

For a given value of ζ and ω/ω_n the phase angle is a constant and can range from $0°$ to $180°$ for different values of ζ and ω/ω_n. The value of the phase angle at resonance (i.e., when $\omega/\omega_n = 1$) is $90°$ for all values of ζ. If there is no damping in the system, the phase angle is either $0°$ or $180°$.

EXAMPLE 8.3

A 60-lb motor running at 1800 rpm has an unbalanced shaft that produces a vertical shaking force with an amplitude of 18.4 lb. Select from the following choices the mounting arrangement that causes the least force to be transmitted to the foundation and determine the transmitted force: (a) motor clamped rigidly to the foundation and (b) springs of total stiffness 500 lb/in., $\zeta = 0.02$.

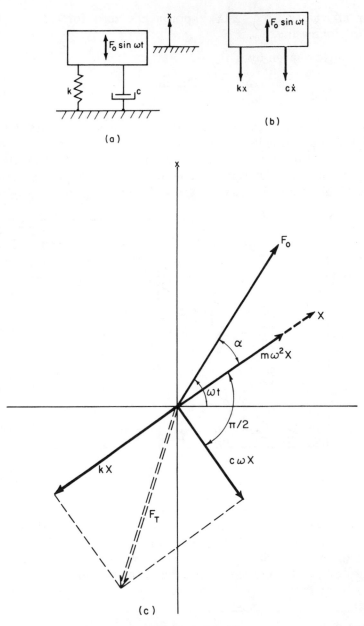

Fig. E8.3 (a), (b), (c)

SOLUTION

The idealized model is shown in Fig. E8.3a and the corresponding free-body diagram in Fig. E8.3b. The differential equation governing the motion is

$$m \frac{d^2 x}{dt^2} + c \frac{dx}{dt} + kx = F_0 \sin \omega t$$

If we consider the steady-state response as $x(t) = X \sin (\omega t - \alpha)$ and substitute this response into the differential equation, we obtain

$$-m\omega^2 X \sin (\omega t - \alpha) + c\omega X \cos (\omega t - \alpha) + kX \sin (\omega t - \alpha) = F_0 \sin \omega t$$

This equation can be written in the form

$$m\omega^2 X \sin (\omega t - \alpha) - c\omega X \sin [\omega t - \alpha + (\pi/2)] - kX \sin (\omega t - \alpha) + F_0 \sin \omega t = 0$$

and the various terms in the equation are represented on the phasor diagram (Fig. E8.3c).

With the system of vectors rotating with the angular velocity ω, the projection on the vertical x axis corresponds to a particular term in the above equation. The vector representing the steady-state amplitude X is also shown.

It is possible to determine the amplitude X by recognizing that at each instant in time the horizontal and vertical projections of the rotating force vectors must combine to give a zero net force. The phase angle α can also be found from the diagram.

For the problem considered here the force transmitted to the foundation is due to the damping force and the spring force which are seen to be $90°$ out of phase on the phasor diagram.

The resultant of these two forces F_T is shown on Fig. E8.3c, and the magnitude of F_T is

$$F_T = \sqrt{(kX)^2 + (c\omega X)^2} = kX \sqrt{1 + [2\zeta(\omega/\omega_n)]^2}$$

For (a) if the motor is rigidly attached to the foundation, the shaking force is directly transmitted to the foundation so that

$$F_T = 18.4 \text{ lb} \qquad\qquad Ans.$$

For (b)

$$\omega_n = \sqrt{k/m} = \sqrt{[(500)(12)(32.2)]/60} = 56.8 \text{ rad/s}$$

$$\omega/\omega_n = [(1800)(2\pi)]/[(60)(56.8)] = 3.32$$

The amplitude X is

$$X = \frac{18.4/500}{\sqrt{[1 - (3.32)^2]^2 + [0.04(3.32)]^2}} = 0.00367 \text{ in.}$$

and the force transmitted is

$$F_T = (500)(0.00367) \sqrt{1 + 0.0176} = 1.851 \text{ lb} \qquad\qquad Ans.$$

Arrangement (b) gives the smallest transmitted force. Although the student may be tempted to arrive at the conclusion that the rigid connection is always the poorest choice, this is not necessarily true. For example, if a stiffer spring system is used so that $\omega/\omega_n \to 1.0$, the amplitude will increase to such an extent that a much more severe condition is attained with springs than if the system is bolted directly to the foundation.

8.5   Instrumentation

The characteristics of several types of measuring instruments can be estimated from an analysis of the simple spring-mass dashpot system. For example, consider the measurement of the acceleration of a body by means of an attached *accelerometer*. Such a device is shown schematically in Fig. 8.13. The accelerometer in this case is a

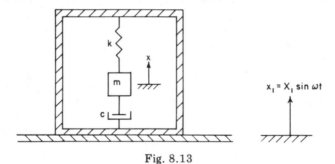

Fig. 8.13

damped spring-mass system rigidly attached to the body whose motion is to be measured. The governing equation for this system is

$$m \frac{d^2 x}{dt^2} = -c \frac{d}{dt}(x - x_1) - k(k - x_1)$$

or

$$m \frac{d^2}{dt^2}(x - x_1) + c \frac{d}{dt}(x - x_1) + k(x - x_1) = -m \frac{d^2 x_1}{dt^2} \quad (8.35)$$

We define the relative displacement $x - x_1 = z$ so that

$$m \frac{d^2 z}{dt^2} + c \frac{dz}{dt} + kz = m\omega^2 X_1 \sin \omega t \quad (8.36)$$

Equations of the form (8.36) have previously been solved and the steady-state solution is $z = Z \sin(\omega t - \alpha)$, where

$$Z = \frac{m\omega^2 X_1}{\sqrt{(k - m\omega^2)^2 + (c\omega)^2}} = \frac{\omega^2 X_1}{\omega_n^2 \sqrt{[1 - (\omega^2/\omega_n^2)]^2 + [2\zeta(\omega/\omega_n)]^2}}$$

$$= \frac{M}{\omega_n^2} \omega^2 X_1 \quad (8.37)$$

In a typical instrument either the relative displacement z or the relative velocity dz/dt is actually measured, whereas the displacement X_1 or its time derivatives are the quantities of interest. If the acceleration $\omega^2 X_1 = a_1$ is to be measured, we see from Eq. (8.37) that

$$Z = (M/\omega_n^2)(\omega^2 X_1) = (M/\omega_n^2)\, a_1 \qquad (8.38)$$

Normally, the accelerometer would be calibrated against a known acceleration. Equation (8.38) indicates that the constant of proportionality from this calibration depends on the ratio M/ω_n^2. A suitable instrument is one in which this ratio remains essentially constant over the range of interest. Since ω_n^2 is a constant for a given instrument, the variation in M is the controlling factor. The variation in M with ω/ω_n and ζ is shown in Fig. 8.14, which shows that for relatively large values of the damping ratio and at small values of ω/ω_n, M does not vary significantly as ω/ω_n changes. This suggests that an accelerometer of this type should have a high natural frequency so that $\omega/\omega_n \ll 1.0$ and a high value of damping.

The ideal amount of damping occurs when $\zeta = 0.7$ since this gives not only a nearly constant value of M over the widest frequency range (Fig. 8.14) but also a nearly linear phase shift with frequency

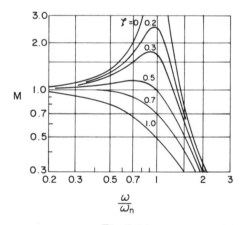

Fig. 8.14

(i.e., $\alpha \simeq 90\,\omega/\omega_n$). The linear phase shift requirement becomes important when the motion to be measured is nonsinusoidal with many frequency components; e.g., periodic motion with several significant Fourier series components. The major problem with using $\zeta = 0.7$ is that mechanical damping is generally very temperature sensitive. Hence most modern accelerometers are designed around piezoelectric crystals like quartz that exhibit low damping ($\zeta \approx 0.01$), high natural frequency (10,000–100,000 Hz), and good sensitivity and that are very small. These instruments do not exhibit significant temperature sensitivity over a wide range of temperatures, and any significant damping is achieved by electronic means outside the accelerometer.

8.6 Periodic Forcing Functions

Many spring-mass systems are excited by a forcing function that is not a harmonic function but a *periodic function*.[1] A function is periodic if $f(t + 2p) = f(t)$ for all t in the domain of definition of the function. Here $2p$ is the smallest constant for which the above equation is true and is called the period. One method used to solve these types of problems is to represent the forcing function and the solution in a Fourier series. If $f(t)$ is a periodic forcing function with period $2p$, we can represent the function in the following form

$$f(t) = a_0 + \sum_{n=1}^{\infty} a_n \cos (n\pi t/p) + b_n \sin (n\pi t/p) \qquad (8.39)$$

where

$$a_0 = \frac{1}{2p} \int_{-p}^{p} f(t) \, dt$$

$$a_n = \frac{1}{p} \int_{-p}^{p} f(t) \cos \frac{n\pi t}{p} \, dt$$

$$b_n = \frac{1}{p} \int_{-p}^{p} f(t) \sin \frac{n\pi t}{p} \, dt$$

With $f(t)$ represented in the form of Eq. (8.39), the equation of motion is

$$m \frac{d^2x}{dt^2} + c \frac{dx}{dt} + kx = a_0 + \sum_{n=1}^{\infty} a_n \cos \left(\frac{n\pi t}{p} \right) + b_n \sin \left(\frac{n\pi t}{p} \right) \qquad (8.40)$$

Since the differential equation is linear, we may solve the sequence of problems

$$m \frac{d^2x_0}{dt^2} + c \frac{dx_0}{dt} + kx_0 = a_0$$

$$m \frac{d^2x_n}{dt^2} + c \frac{dx_n}{dt} + kx_n = a_n \cos \left(\frac{n\pi t}{p} \right) + b_n \sin \left(\frac{n\pi t}{p} \right), \qquad n = 1, 2, \ldots$$

$$(8.41)$$

---

1. Study of Sections 8.6 and 8.7 requires some background in Fourier series and Laplace transforms.

and form the linear combination of the x_n to get the solution. The differential equations of Eq. (8.41) are of the form discussed in Section 8.4 and can be solved in a similar manner.

EXAMPLE 8.4

Determine the steady-state motion of the system in Fig. E8.4a.

SOLUTION

The forcing function is periodic with period $2p = 2$, and on the interval $0 < t < 2$ is given by the equation

$$F(t) = F_0 [H(t) - H(t - 1)]$$

where $H(t)$ is the unit step function defined as

$$H(t) = 1, t > 0$$
$$= 0, t < 0$$

To write $F(t)$ in the form of Eq. (8.39), we find the Fourier coefficients a_0, a_n, and b_n.

$$a_0 = \frac{1}{2} \int_{-1}^{1} F(t)\, dt = \frac{F_0}{2}$$

$$a_n = \int_{-1}^{1} F(t) \cos n\pi t\, dt = F_0 \int_{0}^{1} \cos n\pi t\, dt = 0, \qquad n = 1, 2, 3, \ldots$$

$$b_n = \int_{-1}^{1} F(t) \sin n\pi t\, dt = F_0 \int_{0}^{1} \sin n\pi t\, dt = 0 \text{ if } n \text{ is even}$$
$$= 2F_0/n\pi \text{ if } n \text{ is odd}$$

The forcing function can then be represented by

$$F(t) = \frac{F_0}{2} + 2F_0 \sum_{n=1}^{\infty} \frac{1}{(2n - 1)\pi} \sin (2n - 1)\pi t$$

The differential equation governing the motion of the system is

$$m \frac{d^2x}{dt^2} + c \frac{dx}{dt} + kx = \frac{F_0}{2} + 2F_0 \sum_{n=1}^{\infty} \frac{1}{(2n - 1)\pi} \sin (2n - 1)\pi t$$

The steady-state solution corresponds to the particular solution of the differential equation. We assume the following form for the particular solution,

$$x_p(t) = f_0 + \sum_{n=1}^{\infty} f_{2n-1} \sin (2n - 1)\pi t + \sum_{n=1}^{\infty} d_{2n-1} \cos (2n - 1)\pi t$$

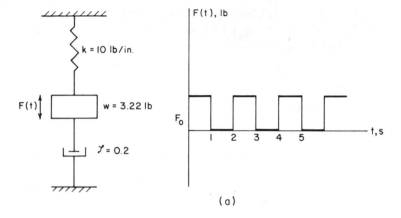

(a)

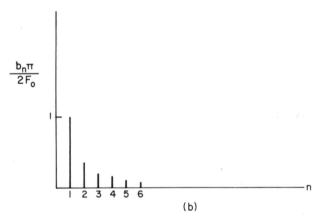

(b)

Fig. E8.4 (a), (b)

and substitute into the equation of motion.

$$kf_0 + \sum_{n=1}^{\infty} [-mf_{2n-1}(2n-1)^2\pi^2 - cd_{2n-1}(2n-1)\pi + kf_{2n-1}]$$

$$\cdot \sin(2n-1)\pi t + \sum_{n=1}^{\infty} [-md_{2n-1}(2n-1)^2\pi^2 + cf_{2n-1}(2n-1)\pi + kd_{2n-1}]$$

$$\cdot \cos(2n-1)\pi t = \frac{F_0}{2} + 2F_0 \sum_{n=1}^{\infty} \frac{1}{(2n-1)\pi} \sin(2n-1)\pi t$$

The last equation is valid for all t, thus coefficients of like terms must be equal. This leads to

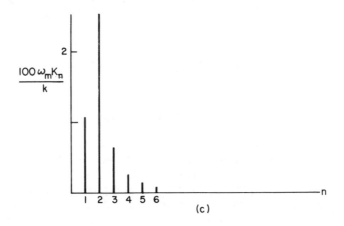

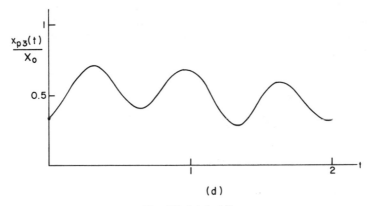

Fig. E8.4 (c), (d)

$$f_0 = F_0/2k = X_0/2$$
$$-md_{2n-1}(2n-1)^2\pi^2 + cf_{2n-1}(2n-1)\pi + kd_{2n-1} = 0$$
$$-mf_{2n-1}(2n-1)^2\pi^2 - cd_{2n-1}(2n-1)\pi + kf_{2n-1} = 2F_0/[(2n-1)\pi]$$

A solution to the last two questions gives

$$f_{2n-1} = \frac{2X_0\omega_m^2[\omega_m^2 - (2n-1)^2\pi^2]}{(2n-1)\pi\{[\omega_m^2 - (2n-1)^2\pi^2]^2 + 4\zeta^2\omega_m^2(2n-1)^2\pi^2\}}$$

$$d_{2n-1} = \frac{-4\zeta\omega_m^3 X_0}{[\omega_m^2 - (2n-1)^2\pi^2]^2 + 4\zeta^2\omega_m^2(2n-1)^2\pi^2}$$

Knowing f_{2n-1}, f_0, d_{2n-1}, the steady-state solution can be written as

$$x_p(t) = (X_0/2) + (\omega_m/k)\sum_{n=1}^{\infty} K_n b_n \sin[(2n-1)\pi t - \phi_n]$$

where

$$K_n = \frac{1}{\sqrt{[\omega_m^2 - (2n - 1)^2 \pi^2]^2 + 4\omega_m^2 \zeta^2 (2n - 1)^2 \pi^2}}$$

and

$$\tan \phi_n = [2\omega_m \zeta (2n - 1)\pi]/[\omega_m^2 - (2n - 1)^2 \pi^2]$$

The equation for $x_p(t)$ is written to demonstrate how the coefficients b_n of the forcing function and the "transfer function" K_n affect the solution.

A plot of $b_n \pi/2F_0$ versus n is shown in Fig. E8.4b, and a plot of $100 \, \omega_m K_n/k$ versus n is shown in Fig. E8.4c for the ω_m and ζ computed from the data in Fig. E8.4a. Note that the product of $K_2 b_2$ is the largest amplitude of the terms in the series. The combination of natural frequency ω_m, the damping ratio ζ, and the value of n determine which of the K_n is the largest. Using the product of K_n with b_n, we are then able to determine how many terms from the series for $x_p(t)$ need to be retained to get a good approximation to the solution.

The first three terms of the steady-state solution, denoted by $x_{p3}(t)$, are

$$x_{p3}(t)/X_0 = (1/2) + (2\omega_m/\pi)[K_1 \sin (\pi t - \phi_1) + (K_2/3) \sin (3\pi t - \phi_2)]$$

For the data in Fig. E8.4a

$$\omega_m = 10 \text{ rad/s} \qquad K_2 = 0.025$$
$$\zeta = 0.2 \qquad \phi_1 = 7.95° = 0.139 \text{ rad}$$
$$K_1 = 0.011 \qquad \phi_2 = 73.95° = 1.29 \text{ rad}$$

and

$$x_{p3}(t)/X_0 = 0.5 + 6.36 \, [0.011 \sin (\pi t - .139) + 0.025 \sin (3\pi t - 1.29)]$$

A plot of the response curve $x_{p3}(t)/X_0$ is given in Fig. E8.4d.

Implicit in the solution of this problem was the assumption that the Fourier series for the forcing function and the solution converge almost everywhere. In most physical problems the conditions under which these series converge are usually satisfied. For a discussion of the conditions of convergence of the Fourier series the interested student is referred to the many texts treating this subject.

8.7 Nonperiodic Forcing Functions

Many problems involving a spring-mass system are not loaded by an external periodic force. These types of problems can be solved by finding the particular solution corresponding to the forcing function and adding the complementary solution. Another solution technique is the method of *Laplace transforms*.

The Laplace transform is defined as

$$L[f(t)] = \int_0^\infty f(t)e^{-st}\,ds = \tilde{f}(s) \qquad (8.42)$$

where the function $f(t)$ must be of such a nature that

$$\int_0^\infty e^{-at}\,|f(t)|\,dt$$

is convergent. In what follows we will assume that all functions we work with have Laplace transforms. The inverse transform of $\tilde{f}(s)$ is defined by a complex integral and reproduces $f(t)$. Thus

$$L^{-1}[\tilde{f}(s)] = f(t) = \frac{1}{2\pi i} \int_{a-i\infty}^{a+i\infty} \tilde{f}(s)e^{st}\,ds \qquad (8.43)$$

Many tables of both the transforms and their inverses exist, and as such the integrals given by Eqs. (8.42) and (8.43) can be easily found in many instances. Table (8.1) gives the Laplace transform of some common functions.

To demonstrate the use of the Laplace transforms, let us consider

Table 8.1. $L[f(t)] = \int_0^\infty f(t)e^{-st}\,dt = \tilde{f}(s)$

	$f(t)$	$\tilde{f}(s)$	
1	e^{-at}	$1/(s+a)$	
2	$\cos bt$	$s/(s^2+b^2)$	
3	$\sin bt$	$b/(s^2+b^2)$	
4	t^n	$n!/s^{n+1}$	n = positive integer
5	$H(t)$	$1/s$	$H(t)$ = the unit step function
6	$e^{-at}\cos bt$	$(s+a)/[(s+a)^2+b^2]$	
7	$e^{-at}\sin bt$	$b/[(s+a)^2+b^2]$	
8	$e^{-at}t^n$	$n!/(s+a)^{n+1}$	n = a positive integer
9	$\cosh bt$	$s/(s^2-b^2)$	
10	$\sinh bt$	$b/(s^2-b^2)$	
11	$f'(t)$	$s\tilde{f}(s) - f(0^+)$	
12	$f''(t)$	$s^2\tilde{f}(s) - sf(0^+) - f'(0^+)$	
13	$\int_0^t f(t-\lambda)g(\lambda)\,d\lambda$	$\tilde{f}(s)\tilde{g}(s)$	convolution integral

the problem of the damped harmonic oscillator with an externally applied force given by $F_0 \sin \omega t$. The differential equation is given by Eq. (8.28).

$$m \frac{d^2 x}{dt^2} + c \frac{dx}{dt} + kx = F_0 \sin \omega t$$

We multiply both sides of (8.28) by e^{-st} and integrate from 0 to ∞ with respect to t.

$$m \int_0^\infty \frac{d^2 x}{dt^2} e^{-st} \, dt + c \int_0^\infty \frac{dx}{dt} e^{-st} + k \int_0^\infty xe^{-st} \, dt$$

$$= F_0 \int_0^\infty \sin \omega t \, e^{-st} \, dt \qquad (8.44)$$

We define $\tilde{x}(s)$ to be

$$\tilde{x}(s) = \int_0^\infty x(t)e^{-st} \, dt$$

Consider the second integral in Eq. (8.44) and integrate by parts

$$\int_0^\infty \frac{dx}{dt} e^{-st} \, dt = e^{-st}x(t) \Big|_0^\infty + s \int_0^\infty x(t)e^{-st} \, dt = -x(0) + s \, \tilde{x}(s) \qquad (8.45)$$

where we assume $x(t)$ is bounded as t goes to infinity and $x(0)$ is the initial displacement of the system.

Integrate the first integral in Eq. (8.44) by parts to obtain

$$\int_0^\infty \frac{d^2 x}{dt^2} e^{-st} \, dt = -\dot{x}(0) + s \int_0^\infty \frac{dx}{dt} e^{-st} \, dt = -\dot{x}(0) - sx(0) + s^2 \, \tilde{x}(s) \qquad (8.46)$$

where $\dot{x}(0)$ is the initial velocity of the system. The transform of $\sin \omega t$ is (from Table 8.1)

$$\int_0^\infty \sin \omega t \, e^{-st} \, dt = \frac{\omega}{s^2 + \omega^2}$$

Eq. (8.44) then transforms into the algebraic equation

$$m[-\dot{x}(0) - sx(0) + s^2 \, \tilde{x}(s)] + c[-x(0) + s\tilde{x}(s)] + k\tilde{x}(s) = F_0 \, \omega/(s^2 + \omega^2) \qquad (8.47)$$

Let the initial conditions be zero; i.e., $\dot{x}(0) = x(0) = 0$. Solving (8.47) for $\tilde{x}(s)$ results in

$$\tilde{x}(s) = F_0\,\omega/[(s^2 + \omega^2)(ms^2 + cs + k)] \tag{8.48}$$

The Laplace transform of the original differential equations has reduced the problem to an algebraic equation for the transformed variable. In a very general sense this reduction to an algebraic expression is the reason for using transforms. Also, the initial conditions are automatically incorporated into the transformed solution.

To complete the problem, the inverse transform must now be computed. Equation (8.43) can be used, but another approach is to write Eq. (8.48) as a sum of functions whose inverses are easily recognized. By the use of partial fractions Eq. (8.48) can be written as

$$\tilde{x}(s) = \frac{F_0}{k^2}\,\omega M^2 \left\{ \frac{-cs}{s^2 + \omega^2} + \frac{k - \omega^2 m}{s^2 + \omega^2} \right.$$
$$+ \frac{c(s + c/2m)}{[s + (c/2m)]^2 + (k/m) - (c^2/4m^2)} \tag{8.49}$$
$$\left. + \frac{(c^2/2m) - k + m\omega^2}{[s + (c/2m)]^2 + (k/m) - (c^2/4m^2)} \right\}$$

where

$$M = \frac{1}{\sqrt{[1 - (\omega/\omega_n)^2]^2 + [2\zeta(\omega/\omega_n)]^2}}$$

The inverse transform of each term of Eq. (8.49) can now be found from Table 8.1 and results in

$$x(t) = \frac{F_0\,\omega M^2 m}{k^2} \left\{ -2\zeta\omega_n \cos \omega t + \frac{(\omega_n^2 - \omega^2)}{\omega} \sin \omega t \right.$$
$$+ 2\zeta\omega_n \exp(-\zeta\omega_n t) \cos \omega_d t + \frac{\exp(-\zeta\omega_n t)}{\omega_d} \tag{8.50}$$
$$\left. \cdot [\omega_n^2 (2\zeta^2 - 1) + \omega^2] \sin \omega_d t \right\}$$

The last two terms in Eq. (8.50) give the transient solution with the constants of integration evaluated for the initial conditions

$$x(0) = \dot{x}(0) = 0$$

The first two terms in Eq. (8.50) give the steady-state solution due to the forced motion. If the transient solution is neglected, Eq. (8.50) can be written in the same form as Eq. (8.30).

The Laplace transform technique leads to the same solution of our harmonically forced oscillation problem as before. The advantage of this method is the incorporation of the initial conditions into the problem as well as transforming the problem into an algebraic equation.

Many forced vibration problems do not have a periodic forcing function associated with them. There are cases where the forcing function acts only for a finite period of time; i.e., increasing from zero to a maximum and then decreasing to zero again for all time. For example, if we consider the damped spring-mass system discussed in Section 8.4 with an arbitrary forcing function $F(t)$, Eq. (8.28) becomes

$$m \frac{d^2 x}{dt^2} + c \frac{dx}{dt} + kx = F(t) \tag{8.51}$$

The Laplace transform of Eq. (8.51) yields

$$\tilde{x}(s) = \frac{\tilde{F}(s)}{ms^2 + cs + k} + \frac{x(0)(ms + c)}{ms^2 + cs + k} + \frac{m\dot{x}(0)}{ms^2 + cs + k} \tag{8.52}$$

where

$$\tilde{F}(s) = \int_0^\infty F(t)e^{-st} \, dt$$

If we take $x(0) = \dot{x}(0) = 0$ as initial conditions, the inverse transform is accomplished by employing the convolution formula,

$$L[f(t)]\, L[g(t)] = L\left[\int_0^t f(\tau)g(t - \tau)\, d\tau\right] = L\left[\int_0^t f(t - \tau)g(\tau)\, d\tau\right]$$

Hence $x(t)$ is

$$x(t) = \int_0^t F(t - \tau)L^{-1}\left(\frac{1}{ms^2 + cs + k}\right) d\tau$$

The inverse transform of the bracketed expression is

$$L^{-1}\left(\frac{1}{ms^2 + cs + k}\right) = \frac{1}{m\omega_d}\exp\left(-\zeta\omega_n t\right)\sin\omega_d t$$

and $x(t)$ becomes

$$x(t) = \frac{1}{m\omega_d}\int_0^t F(t - \tau)\exp(-\zeta\omega_n\tau)\sin\omega_d\tau \, d\tau \tag{8.53}$$

Hence if $F(t)$ is known, Eq. (8.53) can be integrated either numerically or analytically to give the appropriate response.

In this last type of problem the transient solution is not present due to our choice of initial conditions. If the initial conditions are not taken to be zero, the solution would exhibit the transient solution that would damp out in a short period of time.

Suppose that $F(t)$ is given by the function

$$F(t) = F_0 H(t) - F_0 H(t - t_0)$$

where $H(t)$ is the unit step function

$$H(t) = 1 \text{ if } t > 0$$
$$= 0 \text{ if } t < 0$$

A graph of the above forcing function is shown in Fig. 8.15. Substi-

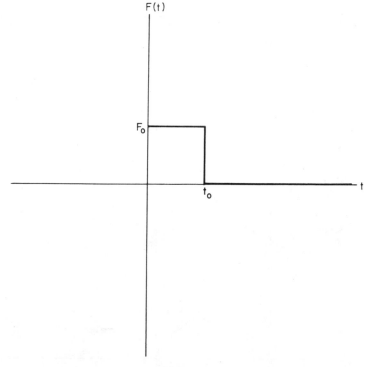

Fig. 8.15

tuting for $F(t)$ into Eq. (8.53) yields

$$x(t) = \frac{F_0}{m\omega_d} \int_0^t [H(t - \tau) - H(t - \tau - t_0)] \exp(-\zeta\omega_n\tau) \sin \omega_d\tau \, d\tau$$

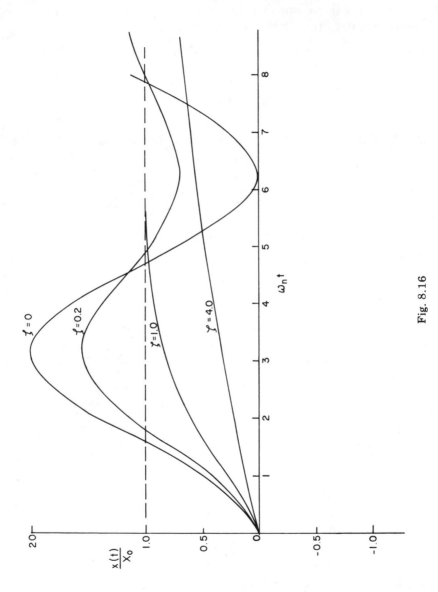

Fig. 8.16

Integrating the above equation and noting the definition of $H(t)$ yields

$$x(t) = (X_0/\sqrt{1-\zeta^2})[\sqrt{1-\zeta^2} - \exp(-\zeta\omega_n t)$$
$$\cdot (\zeta \sin \omega_d t + \sqrt{1-\zeta^2} \cos \omega_d t)], \qquad 0 < t < t_0 \qquad (8.54)$$

$$x(t) = [X_0 \exp(-\zeta\omega_n t)/\sqrt{1-\zeta^2}] \{\exp(\zeta\omega_n t_0)$$
$$\cdot [\zeta \sin \omega_d(t-t_0) + \sqrt{1-\zeta^2} \cos \omega_d(t-t_0)] \qquad (8.55)$$
$$- (\zeta \sin \omega_d t + \sqrt{1-\zeta^2} \cos \omega_d t)\}, \qquad t > t_0$$

where X_0 is the static deflection F_0/k. In the above we have implicitly implied that $\zeta < 1$. If $\zeta > 1$, Eq. (8.54) results in the solution

$$x(t) = X_0 - [X_0 \exp(-\zeta\omega_n t)/\sqrt{\zeta^2-1}]$$
$$\cdot (\zeta \sinh \omega_d t + \sqrt{\zeta^2-1} \cosh \omega_d t), \qquad 0 < t < t_0, \zeta > 1 \qquad (8.56)$$

where in Eq. (8.56) $\omega_d = \omega_n\sqrt{\zeta^2-1}$. The solution for $t > t_0$ and $\zeta > 1$ can be obtained from Eq. (8.55) by replacing the cos and sin terms by the cosh and sinh and replacing $\sqrt{1-\zeta^2}$ by $\sqrt{\zeta^2-1}$. For both $\zeta < 1$ and $\zeta > 1$ the motion is seen to be exponentially damped. The case $\zeta = 1$ is obtained from either Eq. (8.54) or (8.55) by letting ζ tend to one in the limit. This results in

$$x(t) = X_0 - X_0[(\omega_n t + 1)\exp(-\omega_n t)], \qquad 0 < t < t_0, \zeta = 1 \qquad (8.57)$$

$$x(t) = -X_0 \omega_n t_0 \exp[-\omega_n(t-t_0)], \qquad t > t_0, \zeta = 1$$

The special case of no damping (i.e., $\zeta = 0$) is easily found from Eq. (8.54) and (8.55).

$$x(t) = X_0(1 - \cos \omega_n t), \qquad 0 < t < t_0, \zeta = 0 \qquad (8.58)$$

$$x(t) = X_0[\cos \omega_n(t-t_0) - \cos \omega_n t], \qquad t > t_0, \zeta = 0$$

With $\zeta = 0$ the motion theoretically continues forever. Illustrations of the response curve for a loading function of the form $F(t) = F_0 H(t)$ is shown in Fig. 8.16. The displacement for $\zeta < 1$ is seen to be a damped oscillation about the static deflection and quite rapidly approaches the static deflection. For $\zeta > 1$ the deflection asymptotically approaches the static deflection without exceeding X_0. The larger the damping coefficient the more slowly the steady-state value is reached.

EXAMPLE 8.5

The suspension system of an automobile may be modeled by a spring-mass dashpot system (Fig. E8.5). If an external force $F(t) = F_0 H(t)$ is given to the

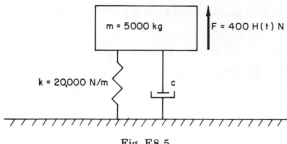

Fig. E8.5

system at rest, find the damping necessary so that the system reaches and remains within 5% of its equilibrium configuration at the end of 5 s.

SOLUTION

The differential equation describing the motion of the system is

$$m\frac{d^2x}{dt^2} + c\frac{dx}{dt} + kx = F_0 H(t)$$

and the solution is given by Eq. (8.54),

$$x(t) = [X_0 \exp(-\zeta\omega_n t)/\sqrt{1-\zeta^2}\,]$$
$$\cdot\, [\sqrt{1-\zeta^2} - \exp(-\zeta\omega_n t)(\zeta \sin \omega_n \sqrt{1-\zeta^2}\,t + \sqrt{1-\zeta^2} \cos \omega_n\sqrt{1-\zeta^2}\,t)]$$

We wish to find c such that

$$X(5) = 0.05X_0$$

or

$$(X_0 e^{-10\zeta}/\sqrt{1-\zeta^2})[\sqrt{1-\zeta^2} - e^{-10\zeta}(\zeta \sin 10\sqrt{1-\zeta^2} + \sqrt{1-\zeta^2} \cos 10\sqrt{1-\zeta^2})]$$
$$= 0.05 X_0$$

Thus

$$0.05e^{10\zeta} = 1 - e^{-10\zeta}[(\zeta/\sqrt{1-\zeta^2}) \sin 10\sqrt{1-\zeta^2} + \cos 10\sqrt{1-\zeta^2}]$$

To solve this equation exactly, at least a desk calculator is needed. A quick
simple method is to graph both sides of the equation, and the points of intersection of the two curves are the solutions.

A reasonable approximation is to neglect the terms on the right side in-
volving ζ since $e^{-10\zeta}$ can be expected to be small. Hence we have $0.05e^{10\zeta} = 1$.
Solving for ζ gives $\zeta = 0.3$, which gives

$$c = (3/10)c_r = 6000 \text{ kg/s} \qquad\qquad\qquad Ans.$$

PROBLEMS

8.1. An airplane radio unit supported on four rubber mounts weighs 40 lb and
has a natural frequency of 400 cpm. The amount of damping present in
the system is negligible. If the amplitude of vibration of the radio support
is 0.1 in. at a frequency of 1500 cpm, what is the amplitude of vibration
of the radio?

Ans.   0.00764 in.

8.2. The spring-mass system in Fig. P8.2 is bolted to a shaker table vibrating in
such a way that its displacement in feet is given by $x = 0.01 \sin 50t$. Find
the amplitude of the displacement of the mass. The mass is 0.01 slugs and
each spring constant is 15 lb/ft.

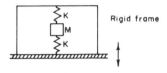

Fig. P8.2

8.3. A string carrying a mass m is restrained to move in the vertical plane
(Fig. P8.3). The string is under a constant tension for small displacements.
For small oscillations write the equation of motion and find the natural
frequency of the vertical vibration of the string.

Ans. $m\ddot{x} + T\{(x/b) + [x/(L - b)]\} = 0,$ $\omega_n = \sqrt{TL/[mb(L - b)]}$ rad/s

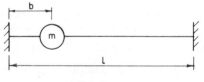

Fig. P8.3

8.4. Determine an expression for the natural frequency of the undamped
system of Fig. P8.4 for small oscillations. The mass of the rod supporting
the weight W is negligible.

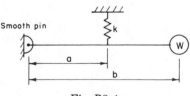

Fig. P8.4

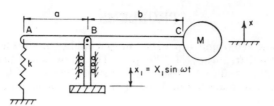

Fig. P8.5

8.5.   A weightless rod is connected to a smooth pin at B, is attached to a spring at A, and has a mass M acting at C (Fig. P8.5). If the fulcrum is given a vertical displacement $x_1 = X_1 \sin \omega t$, derive the equation of vertical motion of the mass M.

   Ans. $M\ddot{x} - (a^2/b^2)kx = [(a/b) + (a/b)^2]\,kX_1 \sin \omega t$

8.6.   For the damped spring-mass system of Fig. P8.6 determine (a) the natural frequency, (b) the damped frequency, (c) the critical damping, and (d) the logarithmic decrement.

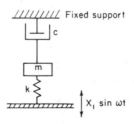

$k = 40$ lb/in.

$W = 4$ lb

$c = 0.2$ lb-sec/in.

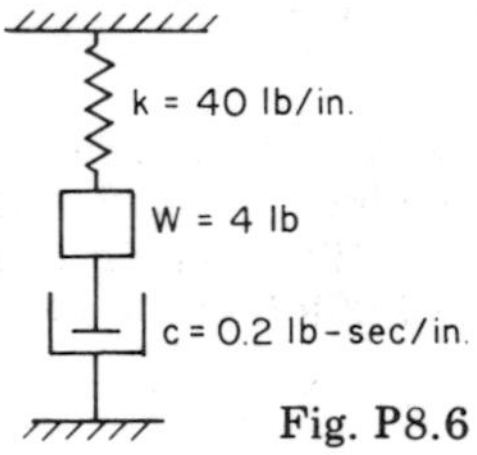

Fig. P8.6

8.7.   The amplitude ratio between two successive cycles of a viscously damped spring-mass system is 1.25. Determine the damping coefficient.

   Ans. $\zeta = 0.0355$

8.8.   The base of the spring-mass dashpot system in Fig. P8.8 oscillates harmonically at a frequency of 50 Hz with an amplitude of 0.1 in. Determine the amplitude of the force transmitted to the fixed support. The natural frequency of the system is 25 Hz, and the logarithmic decrement was found to be 0.3. Express your answer in terms of the damping coefficient c.

Fixed support

c

m

k

$X_1 \sin \omega t$

Fig. P8.8

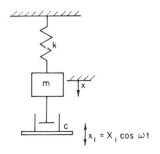

Fig. P8.9

8.9. In Figure P8.9 the weights of the spring and dashpot may be neglected. (a) Develop the differential equation of motion of the mass of the system, (b) determine the expression for the steady-state maximum amplitude of vibration of the mass, and (c) determine the expression for the phase angle between the displacement of the mass and the vibratory motion x_1.

Ans. (a) $m\ddot{x} + c\dot{x} + kx = c\dot{x}_1$,   (b) $X = c\omega X_1/\sqrt{(k - m\omega^2)^2 + c^2\omega^2}$

      (c) $\tan\beta = (m\omega^2 - k)/c\omega$

8.10. The natural frequency of the spring-mass dashpot system shown in Fig. P8.9 is 10 rad/s. If the amplitude of the applied displacement is 0.5 in. at a frequency of 10 rad/s, what force is transmitted to the support if the mass weighs 386.4 lb.

8.11. For a forced-vibration system governed by Equation (8.32), show that for $\zeta > 0.707$ no peak will occur in the response curves, whereas for $\zeta < 0.707$ the peak will occur at $\omega/\omega_n = \sqrt{1 - 2\zeta^2}$.

8.12. An accelerometer consists of a spring-mass dashpot (Fig. P8.12) and is attached to a body that is vibrating harmonically. (a) Derive the differential equation of motion for this system with the relative displacement between the mass and the body taken as the dependent variable. (b) If the accelerometer is to be used in the frequency range 0–2,000 Hz, which of the following two possible designs would you recommend?

    (1) $\omega_n = 5000$ Hz, $\zeta = 0.5$ or (2) $\omega_n = 8000$ Hz, $\zeta = 0.3$

Explain and support your answer with the necessary computations.

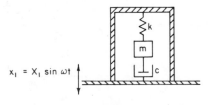

Fig. P8.12

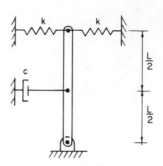

Fig. P8.13

8.13. A homogeneous slender rod of mass m has one end attached to a spring and the other end to a smooth pin (Fig. P8.13). Derive the differential equation of motion under the assumption of small displacements and find (a) the undamped natural frequency and (b) the damping ratio.

Ans. $\ddot{\theta} + (3/4)(c/m)\dot{\theta} + [6(k/m) - (3/2)(g/L)]\theta = 0$

(a) $\omega_n = \dfrac{\sqrt{6(k/m) - (3/2)(g/L)}}{2\pi}$

(b) $\zeta = \dfrac{(3/8)(c/m)}{\sqrt{6(k/m) - (3/2)(g/L)}}$

8.14. A slender homogeneous rod has a mass m and length L and is supported and connected to a spring (Fig. P8.14). If the system is in equilibrium when the rod is vertical, determine for small angular displacements (a) the differential equation of motion and (b) the amplitude of motion for steady-state oscillation.

8.15. For the system in Fig. P8.15 find the relative motion between the base and the mass.

Ans. $z = m\omega^2 X/\sqrt{(k - m\omega^2)^2 + (c\omega)^2}$

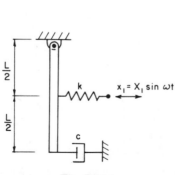

Fig. P8.14

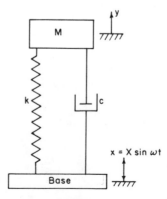

Fig. P8.15

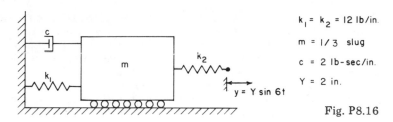

$k_1 = k_2 = 12$ lb/in.

$m = 1/3$ slug

$c = 2$ lb-sec/in.

$Y = 2$ in.

Fig. P8.16

8.16. The spring-mass dashpot system is excited by a forcing displacement $y = Y \sin 6t$ applied at the end of the spring (Fig. P8.16). For the steady-state condition find (a) the amplitude of motion of the mass, (b) the amplitude of total force transmitted to the walls, and (c) the phase angle between the spring force k_1 and the total force transmitted to the wall.

8.17. The mass in Fig. P8.17 weighs 2 lb, and the natural frequency of the system (without damping) is 1 cps. When the two dashpots are added and the mass displaced vertically from its equilibrium position, it is found that a damped oscillation results and the ratio of two successive amplitudes is 2. If $c_1 = 0.01$ lb-s/ft, what is the value of c_2?

Ans. $c_2 = 0.0760$ lb-s/ft

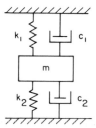

Fig. P8.17

8.18. An accelerometer of the type illustrated in Fig. 8.14 of Section 8.5 is to be used to measure the amplitude of the acceleration of a panel to which it is attached. The accelerometer has a damping factor ζ of 0.7 and a natural frequency ω_n of 1000 Hz. It is desired to relate the acceleration to the relative amplitude Z by the equation, panel acceleration $= CZ$, where C is a constant. What is the appropriate value for C for values of ω less than 100 Hz?

8.19. A device has been built for subjecting various specimens to longitudinal vibrations (Fig. P8.19). A mechanical device supplies a harmonic dis-

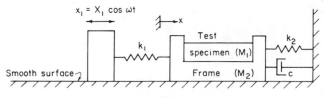

Fig. P8.19

placement to the spring k_1 as illustrated. (a) Derive the differential equation of motion that describes the motion of a specimen rigidly attached to the frame. (b) Experimentally it was determined that the natural frequency of the system was 1 cps when a specimen weighing 160 lb was mounted on the 300-lb frame. The spring constant k_2 is 100 lb/ft and the damping coefficient c is 10 lb-s/ft. Determine k_1 and the ratio of the amplitude of vibration of the specimen to that of the exciting displacement X_1 for $\omega = 2$ rad/s.

Ans.   (a) $(M_1 + M_2)\ddot{x} + c\dot{x} + (k_1 + k_2)x = k_1 x_1 \cos \omega t$,   (b) 0.921

8.20. A periodic excitation is applied to the mass of a spring-mass dashpot system (Fig. P8.20). Determine the steady-state motion of the mass m if $k = 30$ lb/in., $m = 20$ lb-s^2/in., $c = 10$ lb-s/in., $F_0 = 300$ lb.

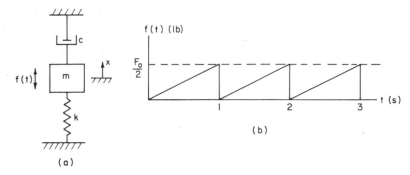

Fig. P8.20

8.21. Find the steady-state motion of the relative displacement between the mass and the body for the accelerometer in Fig. P8.12 if the forcing function applied to the body is that in Fig. P8.21.

Ans. $z_p = \dfrac{F_0}{2k} \left\{ 1 - \dfrac{2\omega_n}{\pi^2} \sum_{\ell=1}^{\infty} \right.$

$$\left. \cdot \left(\dfrac{[(2\ell - 1)^2 \pi^2 - \omega_n^2] \cos (2\ell - 1)\pi t + 2\zeta w_n (2\ell - 1)\pi \sin (2\ell - 1)\pi t}{(2\ell - 1)^2 \{[(2\ell - 1)^2 \pi^2 - \omega_n^2]^2 + 4\zeta^2 \omega_n^2 (2n - 1)^2 \pi^2 \}} \right) \right\}$$

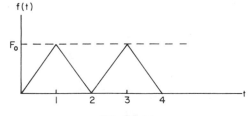

Fig. P8.21

8.22. For the spring-mass dashpot system of Fig. P8.9 determine the motion of the system if the applied displacement is

$$x_1 = Xt, \quad 0 < t < 1$$
$$= X, \quad t > 1$$

The configuration is taken to be initially at rest in an equilibrium state.

8.23. Given the spring-mass system in Fig. P8.5, determine the motion of the system if the vertical displacement on the fulcrum is $x_1 = Xe^{-\alpha t} \sin \beta t$, α and β real and positive. The configuration is initially at rest.

Ans. $x(t) = \dfrac{[1 + (a/b)]X}{k\{[(a/b)\omega_n + \alpha]^2 + \beta^2\}\{[(a/b)\omega_n - \alpha]^2 + \beta^2\}}$

$$\cdot \left[\frac{a}{b} \omega_n \left(\beta^2 - \alpha^2 + \frac{a^2}{b^2} \omega_n^2 \right) e^{-\alpha t} \sin \beta t + 4\alpha \frac{a}{b} \omega_n e^{-\alpha t} \cos \beta t \right.$$

$$\left. + 2\beta \left(\beta^2 + \alpha^2 + \frac{a^2}{b^2} \omega_n^2 \right) \sinh \beta t - 4\alpha\beta \frac{a}{b} \omega_n \cosh \beta t \right]$$

part V

ENGINEERING MATERIALS
AND APPLICATIONS

chapter 9

Properties of Materials

9.1 Introduction

In Chapter 3 the concept of stress at a point was developed by using the physical concept of equilibrium of forces on a small volume of material surrounding the point. In Chapter 6 the concept of strain at a point was developed by using the physical concept of continuity of displacements (no holes or voids were formed during deformation) together with considerations of the deformations associated with a small volume of material surrounding the point. In many engineering design situations the quantities of interest are the stresses, while the quantities that can be visualized and measured are the strains; thus a need arises for a means of relating the two.

Our knowledge of the physics of solids has not yet been developed to the point where force-deformation relationships for a given material can be predicted from considerations of its atomic or molecular structure. As a result, such relationships must be developed from experimental observations.

The major characteristics of the response of a material to a system of forces can be determined by observing the behavior of standardized samples (test specimens) of the material when subjected to tensile, compressive, bending, or twisting loads. It is readily observed that different materials respond in different ways and to different degrees to these applied forces. For example, a gentle pull on a slender rod of steel will cause no appreciable change in the length of the rod, whereas the same force applied to an identical rod of rubber will cause a much greater change. When accurate measurements of load and deformation are available for a particular material from one or more of these simple tests, an attempt can be made to draw general conclusions regarding the characteristics of the material. Mathematical models are extremely useful at this stage since they provide a mathematical statement of the observed physical behavior which can be used for analysis. The common structural materials such as steel, glass, and aluminum exhibit a physical behavior that can be expressed by simple mathematical relationships. Other materials such as rubber and plastics exhibit a more complicated behavior and are difficult to characterize with simple observations and mathematical models. Several of the experiments used to characterize material behavior are discussed in the following sections.

9.2 Tension Test

The most basic test for determining load-deformation relationships for a given material is the *tension test*. For this test a cylindrical bar with threaded ends is often used and an axial force is applied (Fig. 9.1a). The data recorded are the axial force, the change

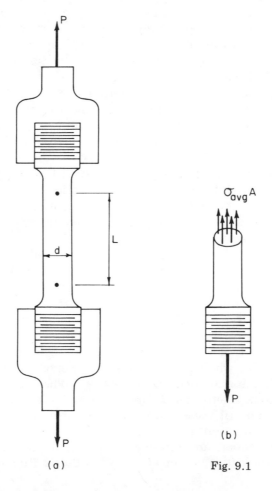

(a)

(b)

Fig. 9.1

in length between two points in the uniform section known as the *gage length*, and the change in diameter in the uniform section. Experimental observations indicate that the specimen increases in length and decreases in diameter as the load increases. The change in length depends upon the material, the applied force, the gage length, and the cross-sectional area of the specimen. The dependence of the load-deformation relationship on the geometry of the specimen is not desirable and can be eliminated by considering stress and strain rather than load and deformation.

For the simple case of uniaxial tension existing in the tensile test specimen, an average normal stress σ_{avg} in the gage section can be obtained by dividing the load P by the original cross-sectional area A (Fig. 9.1b). Thus

$$\sigma_{avg} = P/A \qquad (9.1)$$

In a similar manner an average axial strain ϵ_a and an average transverse strain ϵ_t can be obtained by dividing the change in length and the change in diameter by the original length and the original diameter respectively. Thus

$$\epsilon_a = \Delta L/L \qquad (9.2)$$

$$\epsilon_t = \Delta d/d \qquad (9.3)$$

A plot of σ_{avg} versus ϵ_a is known as a *stress-strain diagram*. A typical curve (Fig. 9.2) could be obtained from an aluminum or steel alloy and is useful for defining some of the terms frequently used to characterize a material.

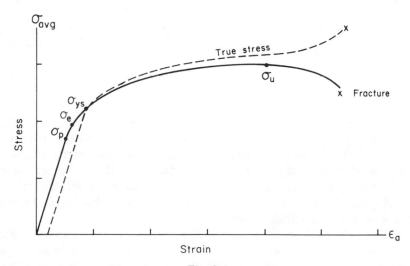

Fig. 9.2

For many materials the initial part of the curve is a straight line; thus the relationship between stress and strain in this region can be expressed by the linear equation

$$\sigma_{avg} = E\epsilon_a \tag{9.4}$$

The constant of proportionality E is the slope of the straight-line portion of the curve and is a property of the material known as the *modulus of elasticity*, or *Young's modulus*. The greatest stress for which the stress remains proportional to strain is defined as the *proportional limit* σ_p.

If we continue from the proportional limit along the typical stress-strain curve (Fig. 9.2), the next point of interest reached is the stress known as the *elastic limit* σ_e. This is the stress beyond which the loading and unloading curves for the material will not coincide. For most metals the proportional limit and the elastic limit are relatively close (Fig. 9.3a). For other materials such as rubber, however, the proportional limit is far below the elastic limit or does not exist (Fig. 9.3b). Both materials in Fig. 9.3 are referred to as elastic

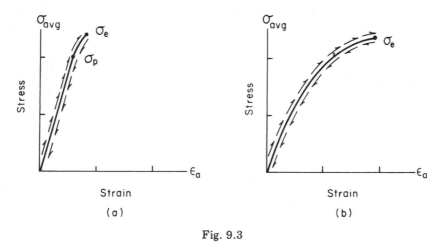

Fig. 9.3

materials since any deformation is completely recoverable upon removal of the load. The material in Fig. 9.3a would be referred to as a *linearly elastic material* below the proportional limit. The material in Fig. 9.3b would be considered a *nonlinear elastic material;* therefore, Eq. (9.4) could not be used to characterize its behavior over any portion of the stress-strain diagram.

If a linear elastic material is loaded to a stress beyond the elastic limit, some permanent deformation will be observed upon removal of the load (Fig. 9.4a). In general, the unloading curve from a stress be-

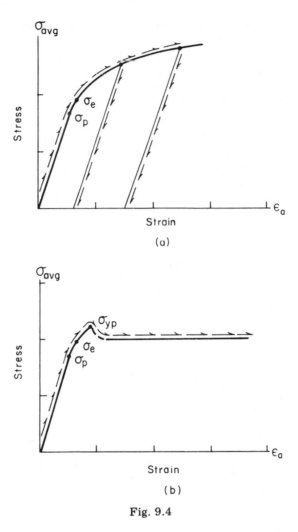

Fig. 9.4

yond the elastic limit will be approximately a straight line parallel to the initial portion of the stress-strain diagram (Fig. 9.4a). The behavior of a material which results in a permanent deformation is termed *yielding*. Materials such as mild steel, bronze, and brass have a well-defined stress level where considerable elongation occurs with no increase in stress. This stress level is referred to as the *yield point* σ_{yp}. In the case of mild steel the stress level decreases slightly at the onset of yielding and the elongation per unit length may reach values as great as 2%. This type of material behavior is illustrated in Fig. 9.4b. Other materials such as alloy steels and aluminum do not exhibit a well-defined yield point, and some other measure is needed to define the transition from elastic to plastic behavior. For these materials a quantity known as the *yield strength* σ_{ys} is often used and is defined as the stress required to produce a certain specified permanent or plastic strain when unloaded. The value of the yield strength is obtained by passing a straight line (parallel to the straight-line portion of the stress-strain diagram) through the point on the strain axis representing the designated value of the permanent strain. The stress at the point where this line intersects the stress-strain curve is the yield strength. The value of the permanent strain must be specified and is usually in the range 0.0005–0.002. This value is frequently given as a percentage "offset," e.g., 0.05% offset. The construction to obtain the yield strength for a 0.05% offset is shown in Fig. 9.5.

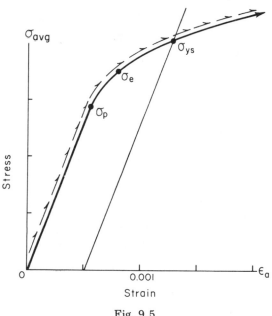

Fig. 9.5

In most instances the values of the proportional limit, elastic limit, yield point, and yield strength are essentially the same, and any can be used as a measure of the transition from linear elastic to plastic behavior.

Returning now to the typical stress-strain diagram in Fig. 9.2, we see that the next point of interest beyond the yield strength is the maximum stress that can be carried by the specimen of material. This stress is defined as the *ultimate strength* σ_u of the material. If the loading is continued after the ultimate strength has been reached,

Table 9.1. Typical properties of some common materials

		Specific weight (pcf)	Elastic modulus		Poisson's ratio	Yield strength*		Ultimate strength	
			Tension (psi)†	Shear (psi)		Tension (psi)	Shear (psi)	Tension (psi)	Shear (psi)
Steel, structural	hot rolled	490	29×10^6	11×10^6	0.30	36,000	21,000	66,000	45,000
Steel, 0.2% carbon	cold rolled	490	30×10^6	11.6×10^6	0.30	60,000	36,000	90,000	60,000
Steel, 0.4% Carbon	hot rolled	490	30×10^6	11.6×10^6	0.30	50,000	25,000	80,000	55,000
Steel, 0.8% Carbon	hot rolled	490	30×10^6	11.6×10^6	0.30	70,000	42,000	120,000	105,000
Stainless steel— 18% Ni, 8% Cr	annealed	490	28×10^6	12.5×10^6	0.30	35,000	...	85,000	60,000
Stainless steel— 18% Ni, 8% Cr	cold rolled	490	28×10^6	12.5×10^6	0.30	170,000	...	200,000	...
Cast iron	gray	450	15×10^6	6×10^6	0.25	...	...	25,000	30,000
Cast iron	malleable	460	25×10^6	12.5×10^6	0.25	35,000	25,000	55,000	48,000
Aluminum	cast 195-T6	175	10.3×10^6	3.8×10^6	0.33	24,000	...	36,000	20,000
Aluminum	alloy 2024-T4	175	10.6×10^6	4×10^6	0.33	42,000	24,000	64,000	40,000
Titanium	annealed	295	15×10^6	6×10^6	0.34	135,000	...	155,000	100,000

Note: Properties may vary widely depending on exact composition, heat treatment, etc.
*0.2% offset.
†1 psi = 6.895×10^4 dynes/cm^2 = 6.895×10^3 N/m^2.

actual fracture of the specimen will occur. For brittle materials such as cast iron or glass the fracture will usually occur abruptly, whereas for ductile materials such as low-carbon steels there will be a considerable reduction in the cross section of the test specimen ("necking") prior to the actual fracture. The "apparent" drop in the stress beyond the ultimate strength is due to this reduction in area. Recall that the calculation of stress for this diagram was based on the original cross-sectional area. If the actual area is used, which is sometimes done, the curve will continue to rise until fracture occurs (as shown by the dotted curve labeled true stress). For less ductile materials there is little change in cross-sectional area before fracture occurs.

Typical values of the characteristic stresses noted and moduli of elasticity for several common materials are listed in Table 9.1. If material properties such as those listed in Table 9.1 are independent of direction, the material is *isotropic*; otherwise it is *anisotropic*. Similarly, if the material properties are independent of location in the material, the material is said to be *homogeneous*; otherwise it is *nonhomogeneous*.

Other important observations can be made from the simple tension test. The specimen both increases in length and decreases in diameter as the load is applied. A comparison of the axial strain ϵ_a, computed from Eq. (9.2), with the transverse strain ϵ_t, computed from Eq. (9.3), for isotropic materials within the elastic range of behavior indicates that the strains are linearly related by the equation

$$\epsilon_t = -\nu\epsilon_a \qquad (9.5)$$

The constant of proportionality ν is known as *Poisson's ratio* and is a property of the material. The negative sign indicates that the axial and transverse strains are of opposite sign. For most common metals Poisson's ratio has a value of approximately 0.33.

From Eqs. (9.4) and (9.5) the axial normal stress σ_{avg} can be related to both the axial normal strain ϵ_a and the transverse normal strain ϵ_t. Thus for stresses below the proportional limit

$$\epsilon_a = \sigma_{avg}/E$$
$$\epsilon_t = -\nu\epsilon_a = -\nu\sigma_{avg}/E \qquad (9.6)$$

We will find these relationships useful in section 9.5 when more general stress-strain relations are discussed.

9.3 Compression Test

The *compression test* differs from the tension test only in the direction of the applied force. As a result of the application of an axial compressive force, the length of the specimen decreases and the diameter of the specimen increases. Thus if we follow the sign conventions previously established for normal stresses and normal strains, signs will be reversed in the compression test from those in the tension test. For most of the common engineering metals, stress-strain diagrams from compression tests give the same modulus of elasticity and Poisson's ratio as tensile test diagrams. The proportional limit, yield point, and yield strength in compression also have approximately the same values as in tension. For values of stress beyond yield, however, the stress-strain diagrams in tension and compression can be significantly different, especially for very ductile materials. Compression tests are widely used for brittle materials such as stone, concrete, and cast iron which have relatively little strength in tension.

9.4 Torsion Test

Thus far we have been concerned primarily with relationships between normal stresses and normal strains. Relationships between shearing stresses and shearing strains are usually obtained from tests of thin circular tubes subjected to twisting moments (Fig. 9.6a). If the tube and loading fixture are properly designed, only shear stresses will develop on a cross section of the tube in the gage section (Fig. 9.6b). If the wall thickness dimension of the tube is small with respect to the diameter, it can be shown that the shear stresses do not vary significantly through the thickness. Thus an average shear stress τ_{avg} may be used as a good approximation of the true state of stress in the tube. The magnitude of the average shear stress can be computed from the dimensions of the tube and the applied load in the following manner. Consider the small element of area of the cross section in Fig. 9.6c. The torque transmitted by this small portion can be expressed as $dT = \tau_{avg} R \, dA$. Since the shear stress τ_{avg} and the mean radius R are the same for all elements of area, the total torque transmitted is obviously $T = \tau_{avg} R \int_A dA = \tau_{avg} R A$. For the thin tube the area of the cross section can be expressed as $A = 2\pi R t$. Thus the average shear stress can be expressed in terms of the applied load and the dimensions of the tube as

$$\tau_{avg} = T/2\pi R^2 t \tag{9.7}$$

The strains associated with the twisting of the tube can be determined from careful measurements of the deformations. The length and diameter of the tube do not change as the torque is applied. This suggests that the axial and transverse normal strains are negligible. It is readily observed, however, that cross sections of the

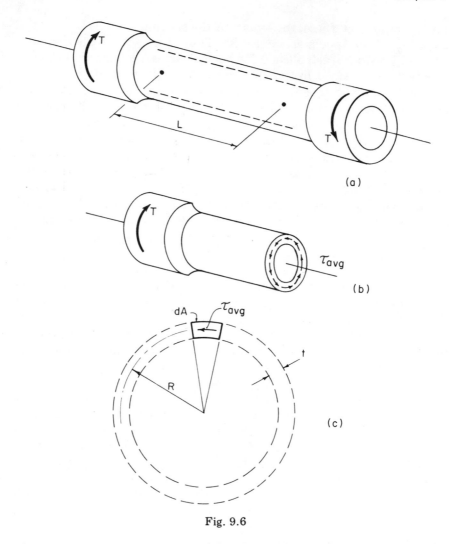

Fig. 9.6

tube rotate with respect to one another and that the rotation per
unit length is constant (Fig. 9.7a). This deformation produces a
shear strain as indicated in Fig. 9.7b where a typical element removed
from the wall of the tube is shown. An average shear strain γ_{avg} can
be computed from the angle of twist θ of one end of the gage section with respect to the other (which can be measured easily) and the
length L and radius R of the tube. Thus $R\theta = \gamma_{\text{avg}}L$ or

$$\gamma_{\text{avg}} = R\theta/L \qquad (9.8)$$

A typical plot of average shear stress τ_{avg} versus average shear strain
γ_{avg} computed from Eqs. (9.7) and (9.8) is shown in Fig. 9.8. The

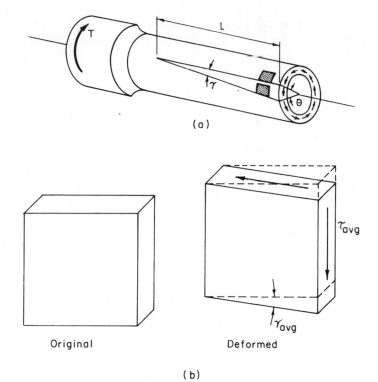

(a)

Original   Deformed

(b)

Fig. 9.7

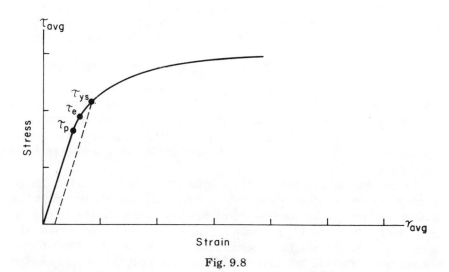

Fig. 9.8

curve is very similar to the one obtained from the tension test; the initial portion of the curve is a straight line. Thus the relationship between stress and strain for this portion of the curve can be expressed by the linear equation

$$\tau_{\text{avg}} = G\gamma_{\text{avg}} \tag{9.9}$$

where the constant of proportionality G is a material property known as the *shear modulus*. A proportional limit, elastic limit, and yield strength can be determined in the same way these quantities are determined in the tension test. Since the values for these three quantities are essentially the same, a *shear yield strength* τ_{ys} (which is approximately one-half the tensile yield strength σ_{ys} for most ductile metals) is used to define the transition from linear elastic to plastic behavior. Torsion tests are seldom continued much beyond the yield strength of the material since the thin tubes used for this type of test become unstable and useful data for τ_{avg} or γ_{avg} cannot be obtained. For cases where tests have been successfully conducted in the large strain range, the shape of the curve for torsion is similar to the curve obtained from the tension test except for the phenomenon of "necking" that does not occur.

9.5   Generalized Hooke's Law

From the results of tension and torsion tests we have been able to establish the behavior of a material under the action of a single normal stress (σ_{avg} in the tension test) and a single shear stress (τ_{avg} in the torsion test). In both cases the relationship between stress and strain [Eqs. (9.6) and (9.9)] was linear for values of stress below a level known as the proportional limit for the material. This linear relationship is known as *Hooke's law*.

In Chapter 3 it was observed that six independent components of stress (σ_x, σ_y, σ_z, τ_{xy}, τ_{yz}, and τ_{zx}) can exist at an arbitrary point of a loaded body. In Chapter 6 it was shown that the deformations associated with the loads could result in six components of strain (ϵ_x, ϵ_y, ϵ_z, γ_{xy}, γ_{yz}, and γ_{zx}) at the same point. The equations that relate stresses to strains in this general case are known as *generalized Hooke's law*.

From tension and torsion tests conducted for conventional engineering materials (steel, aluminum, glass, etc.) we observe that the strains associated with stresses below the yield strength are extremely small (less than 1% in general). Since the strains are small, the geometry or shape of a body is not changed significantly when forces are applied. Thus if two different loads are applied to a body either simultaneously or independently, the same stress and strain distribution will result. This occurs because the deformations associated with the first of the two loads did not produce a change in geometry sufficient to affect the stresses or strains associated with the second load. The procedure by which a gross effect is represented as the sum of a number of individual effects is known as the *principle of superposition.*

Application of the principle of superposition to a small element of isotropic material acted upon by a general state of stress can provide the general relationships between stress and strain needed for analysis. For example, consider the deformation of the element associated with the normal stress σ_x in Fig. 9.9. This state of stress is

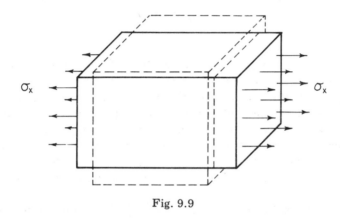

Fig. 9.9

obviously associated with a typical element of material in the gage section of the tension specimen. Thus from Eqs. (9.6) we may write

$$\epsilon_x = \sigma_x/E, \qquad \epsilon_y = \epsilon_z = -\nu\sigma_x/E$$

We also observe in the tensile test that the rectangular sides of the element remain rectangles. Since the $90°$ angles between the sides of the element do not change, $\gamma_{xy} = \gamma_{yz} = \gamma_{zx} = 0$. Thus it can be seen that a uniaxial stress produces three components of normal strain but no shear strains. Similar relationships are associated with the normal stresses σ_y and σ_z.

Consider now the deformation of the element associated with the shear stresses $\tau_{xy} = \tau_{yx}$ in Fig. 9.10. This state of stress is obviously associated with a typical element of material in the gage section of the torsion specimen. Thus from Eq. (9.9) we may write $\tau_{xy} = G\gamma_{xy}$. Since measurements of deformation in the torsion specimen indicate

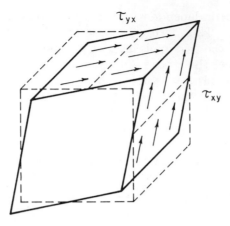

Fig. 9.10

that length, diameter (and therefore circumference), and wall thickness remain constant, $\epsilon_x = \epsilon_y = \epsilon_z = 0$. We can also visualize that cross sections of the tube would not distort as they rotate with respect to one another. Thus the rectangular faces of the element on which the shear stresses act would remain rectangles and $\gamma_{yz} = \gamma_{zx} = 0$. From this we see that a shear stress component produces a companion shear strain component but no additional shear or normal strain components. Similar relationships are associated with the shear stresses τ_{yz} and τ_{zx}.

From these stress-strain relationships for a typical normal stress and a typical shear stress we can by superposition express generalized Hooke's law as

$$\epsilon_x = (\sigma_x/E) - (\nu/E)(\sigma_y + \sigma_z)$$

$$\epsilon_y = (\sigma_y/E) - (\nu/E)(\sigma_z + \sigma_x)$$

$$\epsilon_z = (\sigma_z/E) - (\nu/E)(\sigma_x + \sigma_y) \qquad (9.10)$$

$$\gamma_{xy} = \tau_{xy}/G, \qquad \gamma_{yz} = \tau_{yz}/G, \qquad \gamma_{zx} = \tau_{zx}/G$$

Eqs. (9.10) are the general stress-strain relations for a linear, elastic, isotropic material. Quite often we find it necessary to compute the stress components at a point in a body from knowledge of the strain components. For example, the strains may be determined experimentally. Solving Eqs. (9.10) simultaneously for the stresses in terms of the strains yields

$$\sigma_x = \frac{E}{(1 + \nu)(1 - 2\nu)} [(1 - \nu)\epsilon_x + \nu(\epsilon_y + \epsilon_z)]$$

$$\sigma_y = \frac{E}{(1 + \nu)(1 - 2\nu)} [(1 - \nu)\epsilon_y + \nu(\epsilon_z + \epsilon_x)]$$ (9.11)

$$\sigma_z = \frac{E}{(1 + \nu)(1 - 2\nu)} [(1 - \nu)\epsilon_z + \nu(\epsilon_x + \epsilon_y)]$$

$$\tau_{xy} = G\gamma_{xy}, \qquad \tau_{yz} = G\gamma_{yz}, \qquad \tau_{zx} = G\gamma_{zx}$$

Again these are general expressions applicable to any state of stress in a linear, elastic, isotropic material.

The plane state of stress is frequently encountered in practical engineering problems. This state of stress occurs on the free surfaces of all types of machine components and structural elements where electrical resistance strain gages are used to determine strains and stresses. For the plane state of stress $\sigma_z = \tau_{zx} = \tau_{zy} = 0$ where the z axis is taken normal to the surface at the point of interest. For this state of stress a reduced form of Eqs. (9.10) can be obtained which simplifies strain gage computations. Thus

$$\epsilon_x = (\sigma_x/E) - (\nu\sigma_y/E)$$

$$\epsilon_y = (\sigma_y/E) - (\nu\sigma_x/E)$$

$$\epsilon_z = -(\nu/E)(\sigma_x + \sigma_y)$$ (9.12)

$$\gamma_{xy} = \tau_{xy}/G, \qquad \gamma_{yz} = 0, \qquad \gamma_{zx} = 0$$

An alternate form of expression that relates strains to stresses is obtained by solving Eqs. (9.12) simultaneously for σ_x and σ_y. Thus

$$\sigma_x = [E/(1 - \nu^2)](\epsilon_x + \nu\epsilon_y)$$

$$\sigma_y = [E/(1 - \nu^2)](\epsilon_y + \nu\epsilon_x)$$ (9.13)

$$\sigma_z = 0, \qquad \tau_{xy} = G\gamma_{xy}, \qquad \tau_{yz} = 0, \qquad \tau_{zx} = 0$$

Equations (9.10) and (9.11) suggest the existence of three elastic constants (E, G, and ν) for a material. In Chapter 3 we showed that the maximum shearing stress τ_{max} at a point is related to the principal stresses σ_1 and σ_3 (Eq. 3.16). In a similar manner in Chapter 6 it was shown that the maximum shear strain γ_{max} was related to the principal strains ϵ_1 and ϵ_3 (Eq. 6.13). With the above information it is possible to show that the three constants are not independent.

From Eq. (6.13) we note that

$$\gamma_{max} = \epsilon_1 - \epsilon_3$$ (a)

From Eqs. (9.10), which are applicable for any set of Cartesian coordinate axes including the principal axes, we may write

$$\gamma_{max} = \tau_{max}/G \qquad \text{(b)}$$

and

$$\epsilon_1 - \epsilon_3 = [(\sigma_1/E) - (\nu/E)(\sigma_2 + \sigma_3)] - [(\sigma_3/E) - (\nu/E)(\sigma_1 + \sigma_2)]$$
$$= [(1 + \nu)/E](\sigma_1 - \sigma_3)$$

However from Eq. (3.16) $\sigma_1 - \sigma_3 = 2\tau_{max}$; therefore

$$\epsilon_1 - \epsilon_3 = [(1 + \nu)/E](2\tau_{max}) \qquad \text{(c)}$$

Substituting Eqs. (b) and (c) into (a) yields

$$\tau_{max}/G = [(1 + \nu)/E](2\tau_{max})$$

Thus

$$G = E/[2(1 + \nu)] \qquad (9.14)$$

Equation (9.14) indicates that the three elastic constants are not independent; therefore if two are known the third can be computed.

EXAMPLE 9.1

Three strain gages are mounted on the surface of an aluminum ($E = 10 \times 10^6$ psi and $\nu = 0.33$) machine component (Fig. E9.1). When loads are applied to the component the following strain readings are recorded

$$\epsilon_a = 1500 \ \mu\text{in./in.}, \qquad \epsilon_b = -800 \ \mu\text{in./in.}, \qquad \epsilon_c = -500 \ \mu\text{in./in.}$$

Determine the normal stress components σ_x and σ_y and the shear stress component τ_{xy} at the strain gage location.

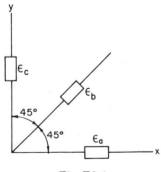

Fig. E9.1

SOLUTION

The Cartesian components of strain ϵ_x, ϵ_y, and γ_{xy} can be determined from the strain gage data by using Eqs. (6.15). For the special case of the rectangular rosette

$$\epsilon_x = \epsilon_a = 1500 \ \mu\text{in./in.}, \qquad \epsilon_y = \epsilon_c = -500 \ \mu\text{in./in.}$$

$$\gamma_{xy} = 2\epsilon_b - \epsilon_a - \epsilon_c = -2600 \ \mu\text{in./in.}$$

On the surface of the machine component where the strain gages were placed no normal or shear stresses act; therefore a plane state of stress exists. For the plane state of stress a reduced form of generalized Hooke's law is applicable as expressed by Eqs. (9.13). Thus

$$\sigma_x = \frac{E}{1 - \nu^2} (\epsilon_x + \nu\epsilon_y) = \frac{(10)^7}{1 - (0.33)^2} [0.0015 + (0.33)(-0.0005)] = 15{,}000 \text{ psi}$$
$$\text{Ans.}$$

$$\sigma_y = \frac{E}{1 - \nu^2} (\epsilon_y + \nu\epsilon_x) = \frac{(10)^7}{1 - (0.33)^2} [-0.0005 + (0.33)(0.0015)] = 0 \text{ psi} \quad \text{Ans.}$$

$$\tau_{xy} = G\gamma_{xy}$$

Substituting the value for G in terms of E and ν as given by Eq. (9.14),

$$\tau_{xy} = \frac{E}{2(1 + \nu)} \gamma_{xy} = \frac{(10)^7}{2(1 + 0.33)} (-0.0026) = -9780 \text{ psi} \qquad \text{Ans.}$$

EXAMPLE 9.2

A thick-walled cylindrical pressure vessel will be used to store gas under a pressure of 14,000 psi. During initial pressurization of the vessel electrical resistance strain gages were used to measure axial and hoop components of strain on the inside and outside surfaces. On the inside surface the gage readings were $\epsilon_{ai} = 1000 \ \mu\text{in./in.}$ and $\epsilon_{hi} = 1500 \ \mu\text{in./in.}$ On the outside surface the gage readings were $\epsilon_{ao} = 1000 \ \mu\text{in./in.}$ and $\epsilon_{ho} = 895 \ \mu\text{in./in.}$ Determine the axial and hoop components of stress associated with these strains if $E = 30 \times 10^6$ psi and $\nu = 0.30$.

SOLUTION

On the outside surface of the vessel where the strain measurements were made, no normal or shear stresses are present; therefore a plane state of stress exists. For a plane state of stress a reduced form of generalized Hooke's law is applicable as expressed by Eqs. (9.13). Since the in-plane axial and hoop strains are in perpendicular directions, the axial and hoop components of stress are determined as

$$\sigma_{ao} = \frac{E}{1 - \nu^2} \, (\epsilon_{ao} + \nu\epsilon_{ho})$$

$$= \frac{30(10)^6}{1 - (0.30)^2} \, [0.001000 + 0.30(0.000895)] = 41,800 \text{ psi} \qquad Ans.$$

$$\sigma_{ho} = \frac{E}{1 - \nu^2} \, (\epsilon_{ho} + \nu\epsilon_{ao})$$

$$= \frac{30(10)^6}{1 - (0.30)^2} \, [0.000895 + 0.30(0.001000)] = 39,400 \text{ psi} \qquad Ans.$$

On the inside surface of the vessel a component of stress normal to the surface exists; therefore stresses are related to strains by generalized Hooke's law as expressed by Eqs. (9.10). Thus

$$\epsilon_{ai} = (\sigma_{ai}/E) - (\nu/E)(\sigma_{hi} + \sigma_{ri}), \qquad \epsilon_{hi} = (\sigma_{hi}/E) - (\nu/E)(\sigma_{ai} + \sigma_{ri}) \qquad \text{(a)}$$

On the inside surface of the vessel, the stress component σ_r which acts normal to the surface is known and can be expressed as

$$\sigma_{ri} = -p = -14,000 \text{ psi} \qquad \text{(b)}$$

Solving Eqs. (a) for σ_{ai} and σ_{hi} after substituting Eq. (b) into Eq. (a) yields

$$\sigma_{ai} = \frac{E}{1 - \nu^2} \, (\epsilon_{ai} + \nu\epsilon_{hi}) - \frac{\nu}{1 - \nu} \, p$$

$$= \frac{(30(10)^6}{1 - (0.30)^2} \, [0.001000 + 0.30(0.001500)] - \frac{0.30}{1 - 0.30} \, (14,000)$$

$$= 41,800 \text{ psi} \qquad\qquad Ans.$$

$$\sigma_{hi} = \frac{E}{1 - \nu^2} \, (\epsilon_{hi} + \nu\epsilon_{ai}) - \frac{\nu}{1 - \nu} \, p$$

$$= \frac{30(10)^6}{1 - (0.30)^2} \, [0.001500 + 0.30(0.001000)] - \frac{0.30}{1 - 0.30} \, (14,000)$$

$$= 53,400 \text{ psi} \qquad\qquad Ans.$$

9.6 Simple Shear Test for Fluids

A second type of test that can be used in principle to determine the behavior of a material such as a fluid under simple shear is illustrated in Fig. 9.11a. The material to be tested is placed between two large parallel plates, one of which is fixed and the other movable. The material in contact with either plate is assumed to have the same displacement as the plate. For solids this can be accomplished by bonding, whereas this "no-slip" condition at a rigid boundary is a

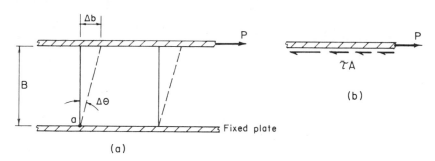

Fig. 9.11

fundamental characteristic of real fluids. A load P applied to the upper plate will cause the material to be displaced by an amount Δb (Fig. 9.11a). The shearing strain γ has been defined (Eq. 6.3) as the change in angle at point a and is given by the equation (for small $\Delta\theta$).

$$\gamma = \Delta\theta \approx \tan \Delta\theta = \Delta b/B \tag{9.15}$$

As shown in Fig. 9.11b, the shearing stress τ is determined from the expression

$$\tau = P/A \tag{9.16}$$

where A is the area of the material in contact with the upper plate. Thus we see from this type of test that the applied shearing stress and the corresponding shearing strain can be determined and a shear stress-strain diagram plotted.

If a fluid such as water is placed between the parallel plates and a load P applied, the upper plate will attain a constant velocity V in the direction of the force P. This observed behavior is commonly used as a basis for the definition of a *fluid*, i.e., a material set in motion under the action of any applied shearing stress. This definition was used in Chapter 4 in connection with the study of fluids at rest. A short time interval Δt after the motion starts, the displacement Δb is given by the equation $\Delta b = V \Delta t$. In this case the displacement of the upper plate is a function not only of the applied force P (which controls V) but also of time. Thus it is not reasonable to attempt to relate the applied shearing stress, which is still given by Eq. (9.6), to the displacement Δb. The rate $\dot{\gamma}$ at which the shearing strain is changing is given by the equation

$$\dot{\gamma} = (\Delta b/B)(1/\Delta t) = V/B \tag{9.17}$$

and since V and B are both constant, the rate of strain is independent of time. This result suggests that the rate of strain rather than the strain may correlate with shearing stress for a fluid.

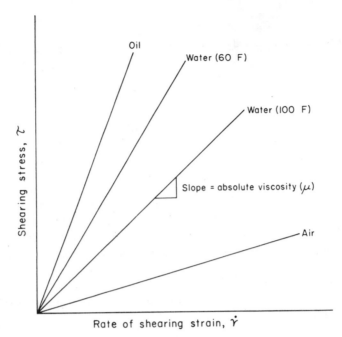

Fig. 9.12

A plot of the applied shearing stress τ versus the rate of shearing strain $\dot{\gamma}$ reveals that for common fluids such as water, air, oil, etc., the relationship is linear (Fig. 9.12). The slope of the $\tau\dot{\gamma}$ curve is called the *absolute viscosity*, which is a property of the fluid strongly dependent on the type of fluid and its temperature. The viscosity of oil is much higher than that of water (Fig. 9.12). Also, the viscosity of a given liquid decreases as the temperature increases. Fluids that

Table 9.2. Typical properties of some common fluids

Fluid	Temperature (F)	Specific weight (pcf)	Absolute viscosity†	
			$lb\text{-}s/ft^2$	$dyne\text{-}s/cm^2$
Air*	59	0.076	0.37×10^{-6}	0.018×10^{-2}
Water	32	62.42	37.46×10^{-6}	1.790×10^{-2}
	60	62.37	23.59×10^{-6}	1.128×10^{-2}
	100	62.00	14.24×10^{-6}	0.681×10^{-2}
	200	60.12	6.37×10^{-6}	0.305×10^{-2}
Glycerine	68	78.6	$18,000 \times 10^{-6}$	860×10^{-2}

*"Standard air" at 59 F and atmospheric pressure (14.7 psi).
†The unit combination lb-s/ft^2 is equivalent to slugs/ft-s. The unit combination dyne-s/cm^2 is commonly referred to as a *poise* (478.8 poise = 1 lb-s/ft^2). Also, 1 lb-s/ft^2 = 4.788 × 10 N-s/m^2.

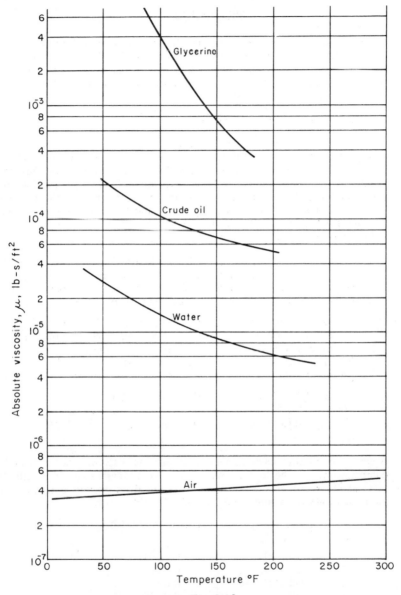

Fig. 9.13

obey this type of linear, stress rate of strain relationship are known as *Newtonian* fluids. Some typical properties of common Newtonian fluids are given in Table 9.2. The variation of viscosity with temperature is shown in Fig. 9.13.

PROBLEMS

9.1. A 0.5050-in. diameter bar is subjected to a tensile load of 8000 lb. During
loading the distance between two marks on the bar increases from 2.000
in. to 2.008 in. The diameter of the bar decreases to 0.5043 in. while
loaded. Determine the modulus of elasticity E and Poisson's ratio v.
Assume linear elastic behavior.

Ans. $E = 10 \times 10^6$ psi, $v = 0.35$

9.2. Two electrical resistance strain gages were placed on an alloy steel tensile
test specimen having an initial diameter of 0.505 in. One gage was oriented
in the axial direction and the other in the circumferential direction.
From the output of the gages it was determined that strain remained proportional to load up to a load of 18,000 lbs, at which time the axial gage
indicated a tensile strain of 3100 μin./in. and the circumferential gage a
compressive strain of 900 μin./in. Determine (a) the proportional limit,
(b) the modulus of elasticity, and (c) Poisson's ratio.

9.3. The stress-strain diagram in Fig. P9.3 was obtained from a tensile test of a
metal specimen. The original diameter of the specimen was 0.505 in., the
gage length was 2.000 in., and the final diameter at the location of the
fracture was 0.425 in. Determine the following if possible:

(a) modulus of elasticity	(f) yield strength (0.2% offset)
(b) Poisson's ratio	(g) yield strength (0.05% offset)
(c) proportional limit	(h) ultimate strength
(d) elastic limit	(i) fracture stress
(e) yield point	(j) true fracture stress

Ans. (a) E = 26,800,000 psi   (h) σ_{ult} = 73,500 psi
      (c) σ_p = 40,000 psi   (i) $\sigma_{fracture}$ = 65,000 psi
      (f) σ_{ys} = 47,000 psi   (j) $\sigma_{fracture\ true}$ = 91,600 psi
      (g) σ_{ys} = 44,000 psi

9.4. A concrete cylinder having a 6-in. diameter and a 12-in. length is tested in
compression. As the load is applied, the following changes in length (decreases) are recorded. Plot the stress-strain diagram and determine the
modulus of elasticity and the 0.05% offset yield strength.

Load (lbs)	Change in length (in.)
0	0
16,900	0.0024
35,200	0.0048
52,200	0.0072
69,000	0.0096
84,500	0.0120
97,500	0.0144
107,000	0.0168
113,000	0.0192
117,000	0.0216
120,000	0.0240

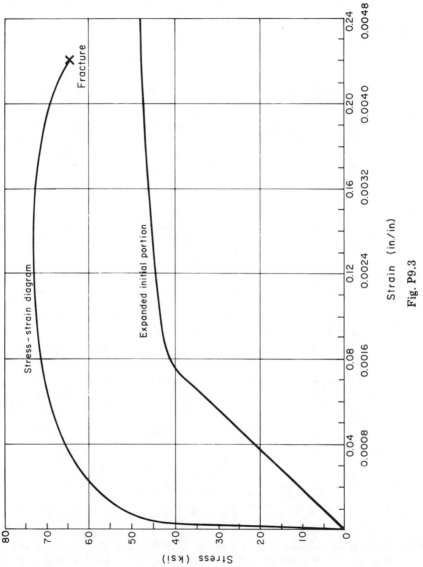

Stress – strain diagram

Fracture

Expanded initial portion

Stress (ksi)

Strain (in/in)

Fig. P9.3

9.5. A thin-walled cylindrical tube of material is being used for a torsion test. The mean radius of the tube is 2 in. and the wall thickness is 0.25 in. When a torque of 200,000 in.-lb is applied, two cross sections of the tube 10 in. apart rotate 2.3° with respect to one another. Determine (a) the average shear stress in the tube, (b) the average shear strain associated with this stress, and (c) the shear modulus of the material.

> *Ans.* (a) τ_{avg} = 31,900 psi,   (b) γ_{avg} = 0.008 rad
>
> (c) $G = 3.96 \times 10^6$ psi

9.6. What strains would occur at a point in a steel ($E = 30 \times 10^6$ psi, $\nu = 0.30$) machine component where the following stresses occur: $\sigma_x = 20,000$ psi, $\sigma_y = -10,000$ psi, $\sigma_z = \tau_{zx} = \tau_{xy} = \tau_{yz} = 0$.

9.7. At a point on the surface of an aluminum ($E = 10 \times 10^6$ psi, $\nu = 1/3$) machine component a state of plane stress ($\sigma_z = \tau_{zx} = \tau_{zy} = 0$) exists. Determine the stresses at the point associated with the following strains: $\epsilon_x = 1500\ \mu$in./in., $\epsilon_y = 600\ \mu$in./in., $\gamma_{xy} = 400\ \mu$rad.

> *Ans.* σ_x = 19,125 psi, σ_y = 12,375 psi, τ_{xy} = 1500 psi

9.8. At a point in the interior of an aluminum ($E = 10 \times 10^6$ psi, $\nu = 1/3$) machine component a state of plane strain ($\epsilon_z = \gamma_{zx} = \gamma_{zy} = 0$) exists. Determine the stresses at the point associated with the following strains: $\epsilon_x = 1500\ \mu$in./in., $\epsilon_y = 600\ \mu$in./in., $\gamma_{xy} = 400\ \mu$rad.

9.9. A rectangular rosette strain gage is mounted on a power transmission shaft with its central gage oriented along the axis of the shaft (Fig. P9.9). When the shaft is under full load, the gage readings are: $\epsilon_a = 600\ \mu$in./in., $\epsilon_b = 0$, $\epsilon_c = -600\ \mu$in./in. Determine the principal stresses and the directions in which they act. Use $E = 30 \times 10^6$ psi, $\nu = 0.30$.

> *Ans.* σ_1 = 13,860 psi along a, $\sigma_2 = 0$, $\sigma_3 = -13,860$ psi along c

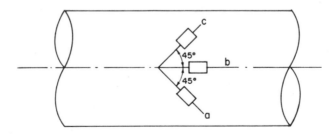

Fig. P9.9

9.10. Develop an expression for the strain ϵ_z in terms of the strains ϵ_x and ϵ_y and the elastic constants of the material E and ν for the case of plane stress ($\sigma_z = 0$).

9.11. On the outside surface of a cylindrical vessel under internal pressure the axial and hoop stresses are known to be 20,000 psi and 40,000 psi respectively. Electrical resistance strain gages are used to measure axial and hoop components of strain on the surface of 267 μin./in. and 1130 μin./in. respectively. Determine the modulus of elasticity and Poisson's ratio of the material, assuming linear elastic behavior.

Ans. $E = 30.1 \times 10^6$ psi, $\nu = 0.301$

9.12. A cube of aluminum ($E = 10 \times 10^6$ psi, $\nu = 1/3$) is subjected to a uniform external pressure of 30,000 psi. If the length of an edge is originally 6 in., determine the change in volume, assuming the material behavior remains linear and elastic.

Solid Mechanics

10.1 Introduction

In this chapter we will consider the problem of relating the external loads on a deformable material to the stresses and deformations produced in the material. The analysis of various structures and structural elements can become exceedingly complex due to the complicated geometry and loadings that may be encountered. Thus as in other areas of mechanics, we attempt to simplify problems by making suitable approximations and considering these more complicated problems as the sum of several simpler problems. The three basic structural elements to be considered in detail in this book are illustrated in Fig. 10.1 and include (1) tension (or compression) of bars, (2) torsion of circular shafts, and (3) bending of relatively long (compared to lateral dimensions) structural members called *beams*. In a complex structure a given member may be subjected to all three types of loading at the same time (Fig. 10.1d).

In the design and analysis of various structures and machines the stresses induced by the applied loads must be evaluated so that it can be determined whether the structure will be strong enough and whether the deformations will be excessive. Thus we will focus on developing suitable relationships for predicting the stresses and deformations in the three simple structural elements described above.

We will first consider the stresses developed in a thin-walled pressure vessel, however, since this practical problem represents one of the simplest examples in which we can readily demonstrate the relationship between an applied load and the resulting stress.

10.2 Thin-walled Pressure Vessels

The types of vessels to be considered are those having the shape of a body of revolution (Fig. 10.2a) and whose thickness is small

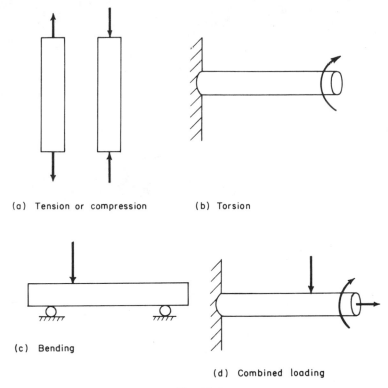

(a) Tension or compression   (b) Torsion

(c) Bending

(d) Combined loading

Fig. 10.1

compared to the principal radii of curvature of the vessel. Examples of these types of containers are water tanks, railroad tank cars, gas storage tanks, etc.

The only stresses of significance for this class of vessels are those which act in a direction tangent to the middle surface of the wall. Experiments indicate it is reasonable to assume these stresses to be uniformly distributed through the thickness of the wall of the vessel. Since no shearing stresses act on the planes on which these normal stresses act, we conclude that the stresses along a meridian or parallel are principal stresses.

A free-body diagram of a small element bounded by two meridians and two parallels is shown in Fig. 10.2b. The tensile stress along the meridian we denote as σ_m, and the tangential tensile stress perpendicular to the meridian we denote σ_t (commonly called the *hoop stress*). The radius of curvature of the meridian at A is r_m, and the radius of curvature perpendicular to the meridian is r_t. If the pressure inside the tank is p and the wall thickness t, force equilibrium in

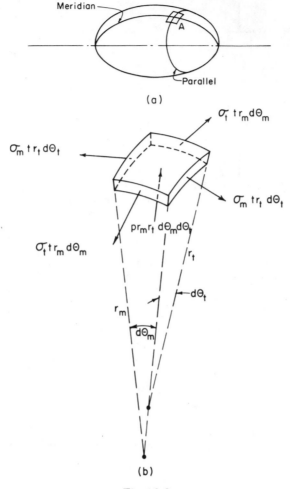

Fig. 10.2

a direction normal to the tank surface at A requires that

$$pr_m r_t \, d\theta_m \, d\theta_t = 2\sigma_t t r_m \, d\theta_m \sin\frac{d\theta_t}{2} + 2\sigma_m t r_t \, d\theta_t \sin\frac{d\theta_m}{2} \qquad (10.1)$$

For small angles

$$\sin\frac{d\theta}{2} \approx \frac{d\theta}{2}$$

so we can simplify Eq. 10.1 to

$$p/t = (\sigma_t/r_t) + (\sigma_m/r_m) \qquad (10.2)$$

Equation (10.2) represents the general relationship between the internal pressure and the stresses for thin-walled pressure vessels of revolution. No use was made of a stress-strain relation, so this problem can be considered as *statically determinate*.

EXAMPLE 10.1

Find the stresses in a thin-walled sphere under uniform internal pressure.

SOLUTION

For this problem we have spherical symmetry due to the type of load and geometry. Thus

$$\sigma_t = \sigma_m = \sigma, \qquad r_t = r_m = r \quad \text{(radius of sphere)}$$

and therefore from Eq. (10.2)

$$\sigma = pr/2t \qquad\qquad\qquad Ans.$$

EXAMPLE 10.2

Find the stresses in a thin-walled cylinder with flat ends under uniform internal pressure.

Fig. E10.2

SOLUTION

From the geometry we have

$$r_t = r \quad \text{(radius of cylinder)}, \qquad r_m = \infty$$

Thus the term $\sigma_m/r_m \to 0$ in Eq. (10.2), and we have

$$\sigma_t = pr/t \qquad\qquad\qquad Ans.$$

The meridianal component of stress does not contribute to equilibrium in a direction normal to the surface of the vessel; however, it can be evaluated by considering equilibrium in a direction parallel to the generators of the cylinder (Fig. E10.2). Summing forces in a direction parallel to the axis of the cylinder, we have $p\pi r^2 = \sigma_m 2\pi rt$. Thus

$$\sigma_m = pr/2t \qquad\qquad\qquad Ans.$$

10.3 Extension of a Bar by an Axial Load

Consider a solid bar having a constant but arbitrary cross section loaded in a direction along the axis of the bar (Fig. 10.3). We orient the bar with respect to a rectangular Cartesian coordinate system (Fig. 10.3). The equations of the bases are then given by $x = 0$, $x = L$

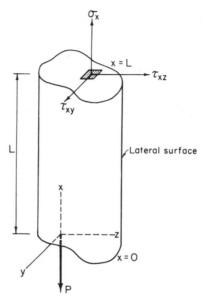

Fig. 10.3

and the lateral surface by some function $f(y, z) = 0$. Although the axially loaded bar represents a relatively simple problem, we will study it from a mathematically rigorous viewpoint to illustrate the various factors that must be considered in the analysis of deformations of solids. We let the tensile force be applied to the base of the bar in such a way that the bar undergoes only extension in the x direction; i.e., no bending is produced by the load. If we pass a plane through the bar parallel to the base $x = 0$ (Fig. 10.3), at each point on this plane we may have stresses as shown. In Fig. 10.3 we placed the origin of the coordinate system at the point of application of the load.

For the bar to be in equilibrium, the resultant force must be zero. On the upper surface the resultant in the x direction is given by Eq. (3.26) as

$$R_x = \int_A \sigma_x \, dA$$

therefore since we require that $\Sigma F_x = 0$, it follows that

$$P = \int_A \sigma_x \, dA \qquad (10.3)$$

Similarly, for the y and z directions we have from Eqs. (3.26)

$$\int_A \tau_{xy} \, dA = 0 \qquad (10.4)$$

$$\int_A \tau_{xz} \, dA = 0 \qquad (10.5)$$

Also, we must be sure that the stress distribution is such that the resultant moment on the cut section is zero since there is no moment on the base at $x = 0$. Thus $\Sigma M_x = 0$, which from Eq. (3.26) gives

$$\int_A y\tau_{xz} \, dA - \int_A z\tau_{xy} \, dA = 0 \qquad (10.6)$$

Similarly, by summing moments about the y and z axes respectively, we obtain the equations

$$\int_A z\sigma_x \, dA = 0 \qquad (10.7)$$

$$\int_A y\sigma_x \, dA = 0 \qquad (10.8)$$

Equations (10.3) through (10.8) represent the six equilibrium equations that must be satisfied on any parallel surface passing through the bar and perpendicular to the generators of the bar. Many different choices of stress distribution can be chosen to satisfy these six equations. To obtain additional information regarding the stresses, we consider the type of deformations that may be induced.

Suppose that on the lateral surface of the bar we mark three different reference lengths L_1, L_2, and L_3 prior to loading (Fig. 10.4). We also draw longitudinal lines parallel to the longitudinal axis of the bar and perpendicular to the previous gage length markings. When the load is applied so that only extension occurs, we observe that (1) the three original lengths L_1, L_2, and L_3 change by an amount ΔL_1, ΔL_2, and ΔL_3 and (2) the angles between the lines parallel to the longitudinal axis and the lines perpendicular to this axis remain unchanged, i.e. remain at $90°$. If we examine the ratios

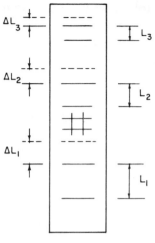

Fig. 10.4

$\Delta L_1/L_1$, $\Delta L_2/L_2$, $\Delta L_3/L_3$, we note they are all equal; i.e.,

$$\Delta L_1/L_1 = \Delta L_2/L_2 = \Delta L_3/L_3 = e \quad \text{(constant)}$$

However, e is simply the normal strain in the x direction ϵ_x, and we can reasonably assume that this strain is uniform throughout the bar.

On the lateral surface of the bar we note that as e changes, the lateral dimensions also change in the y and z directions. Thus, if we measure an initial characteristic lateral dimension D_y of the bar in the y direction before loading, then after loading examine the ratio of the change in this dimension to the original dimension, we find

$$\epsilon_y = \Delta D_y/D_y = -\nu e = -\nu\epsilon_x \qquad (10.9)$$

where ν is Poisson's ratio, which is constant for any one material. In a similar manner we find the strain in the z direction given by

$$\epsilon_z = \Delta D_z/D_z = -\nu e = -\nu\epsilon_x \qquad (10.10)$$

We have thus observed a set of normal strains (actually average strains) for the body as it deforms.

The second observation indicates that the shearing strains γ_{xz} and γ_{xy} are zero. In a manner similar to the above, if two perpendicular lines on a plane parallel to the bases are observed, one sees that γ_{yz} is also zero.

We now have the set of six average strains on the surface of the bar. Since the strains are continuous functions of position, we do not expect much change as we move a small distance into the interior of the body. If the geometry of the bar is such that the lateral dimensions are small compared to the length, it seems physically reasonable to assume that the strains in the interior of the body are the same as those observed on the surface. We thus take as our distribution of strain in the body the following:

$$\epsilon_x = e, \quad \epsilon_y = -\nu e, \quad \epsilon_z = -\nu e, \qquad \gamma_{xy} = \gamma_{yz} = \gamma_{zx} = 0 \qquad (10.11)$$

If we now assume that the material we are considering is an isotropic, homogeneous, linearly elastic material, then the constitutive Eqs. (9.11) can be used to find the stresses corresponding to this set of strains. It immediately follows that

$$\sigma_x = E\epsilon_x = Ee \qquad (10.12)$$

and

$$\sigma_y = \sigma_z = \tau_{xy} = \tau_{yz} = \tau_{zx} = 0 \qquad (10.13)$$

We now take these expressions for the stresses and substitute into the previously determined equilibrium equations. From Eq. (10.3)

$$P = \int_A \sigma_x \, dA = EeA = \sigma_x A \qquad (10.14)$$

and Eqs. (10.4), (10.5), and (10.6) are identically satisfied. However, for Eqs. (10.7) and (10.8) to be satisfied,

$$\int_A z\sigma_x \, dA = Ee \int_A z \, dA = 0$$

and

$$\int_A y\sigma_x \, dA = Ee \int_A y \, dA = 0$$

To satisfy the last two equations, we must choose the origin of the coordinate system at the centroid of the cross section; thus the point of application of the load must also be at the centroid of the cross section.

The total elongation ΔL of the bar, which has an initial length L, is

$$\Delta L = \epsilon_x L = eL \qquad (10.15)$$

and the change in the lateral dimension in the y direction ΔD_y is

$$\Delta D_y = -\nu e D_y \qquad (10.16)$$

and in the z direction

$$\Delta D_z = -veD_z \tag{10.17}$$

In summary, for a bar of constant cross section constructed of a linearly elastic, isotropic, homogeneous material and loaded with an axial load passing through the centroid of the cross section we conclude that

$$\epsilon_x = \sigma_x / E = P/EA \tag{10.18}$$

with the elongation given by

$$\Delta L = PL/EA \tag{10.19}$$

The results given above are also applicable to relatively short bars with compressive loads. However, for long slender rods under compressive loading the analysis is not applicable since "buckling" may occur. This problem will be considered in Section 10.10.

To obtain the solution to the problem of the axially loaded bar, the equations of equilibrium were satisfied, the geometry of the deformations were taken into account, and the type of material was characterized through the constitutive equations. In general these three factors:

1. *Equilibrium.*
2. *Geometry.*
3. *Materials.*

must be considered in the analysis of most solid mechanics or strength of materials problems.

EXAMPLE 10.3

The bar in Fig. E10.3a is of constant cross section and is held rigidly between the walls. An axial load P is applied to the bar at a distance $L/3$ from the left end. Neglecting localized effects in the vicinity of the applied load, determine the reactions of the walls. The bar is loaded in such a manner that the resultant P passes through the centroid of the cross section.

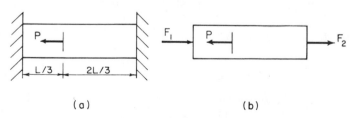

(a)   (b)

Fig. E10.3

SOLUTION

We first draw a free-body diagram showing the applied load P together with the reactions of the wall on the bar (Fig. E10.3b). There is only one equation of static equilibrium, namely, $F_1 - P + F_2 = 0$. This equation contains two unknowns F_1, F_2; thus the problem is *statically indeterminate*. An additional equation is needed, which is based on the deformation of the bar. If the portion of the bar to the left of the applied load P is shortened, the portion to the right of P must be lengthened by the same amount. This fact is expressed by the deformation equation

$$F_1(L/3)/EA = F_2(2L/3)/EA$$

and therefore $F_1 = 2F_2$. We now substitute into the equilibrium equation to obtain

$$F_1 = (2/3)P, \qquad F_2 = (1/3)P \qquad\qquad\qquad Ans.$$

directed as shown in Fig. E10.3b.

10.4 Torsion of a Circular Shaft

The problem of determining the stresses and deformations produced in a cylinder of arbitrary cross section by twisting it about its longitudinal axis is beyond the scope of this book. However, the special case of torsion of a cylinder of circular cross section is relatively simple and provides results applicable to the design and analysis of a wide variety of mechanical and electromechanical equipment.

We will again employ three factors to analyze this problem:

1. *The satisfaction of the equilibrium conditions.*
2. *Observations and assumptions regarding deformations.*
3. *Material behavior.*

For the case of torsion of a circular shaft it is convenient to first study deformations.

Consider a solid circular shaft (Fig. 10.5a) loaded with a couple at its end, where the couple is commonly referred to as the *torque* applied to the shaft. For a shaft loaded in this manner experimental observations indicate the following:

1. The length of the shaft between arbitrarily spaced cross sections does not change for small angles of twist of the shaft.
2. The shaft remains circular and the outside diameter does not change.
3. Cross sections of the shaft rotate relative to one another.
4. A diameter remains a straight line after twisting.

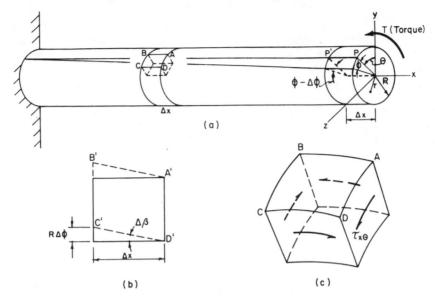

Fig. 10.5

On the basis of these observations we may assume that plane cross sections originally perpendicular to the axis of the shaft remain plane and perpendicular to the axis and suffer no deformation or distortion within the plane. The only deformation thus suffered by the shaft as a result of the twisting moment is for cross sections to rotate with respect to one another. Thus in Fig. 10.5a, the point P will rotate through an angle ϕ, called the *angle of twist*, whereas the nearby point P' will rotate through the angle $\phi - \Delta\phi$. A small element on the surface of the shaft such as $ABCD$ (Fig. 10.5a) will deform into the shape $A'B'C'D'$ (Fig. 10.5b), giving rise to an average shearing strain $\gamma(R)_{\text{avg}}$ such that

$$\gamma(R)_{\text{avg}} = \Delta\beta \approx \tan \Delta\beta = \frac{R\Delta\phi}{\Delta x}$$

and in the limit

$$\gamma(R) = \lim_{\Delta x \to 0} \frac{R\Delta\phi}{\Delta x} = R\frac{d\phi}{dx} \qquad (10.20)$$

For similar elements in the interior of the shaft

$$\gamma(r) = r\frac{d\phi}{dx} \qquad (10.21)$$

Since no other distortion takes place, the only stress component is a shearing stress required to deform the element into the shape $A'B'C'D'$ (Fig. 10.5c). Since this stress is acting on a plane that is normal to the x axis and is perpendicular to the radius r, we can designate it as $\tau_{x\theta}$. For linearly elastic, isotropic, and homogeneous materials the shearing stress is related to the corresponding shearing strain through the equation

$$\tau_{x\theta} = G\gamma = Gr\frac{d\phi}{dx} \tag{10.22}$$

We see from Eq. (10.22) that the shearing stress varies linearly with distance from the axis of the shaft. Since the torque developed at any cross section must be the same as the applied torque T, the rate of change of the angle of twist $d\phi/dx$ must be a constant along the shaft.

We are now in a position to use equilibrium considerations to relate the applied torque T to the shearing stress $\tau_{x\theta}$ and the angle of twist ϕ. Consider the section of shaft in Fig. 10.6. To hold the shaft

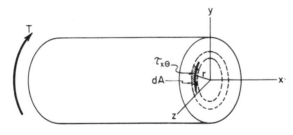

Fig. 10.6

in equilibrium, a torque equal to the applied torque must develop at each cross section due to the distribution of $\tau_{x\theta}$. Thus

$$\Sigma M_x = 0, \quad T = \int_A r\tau_{x\theta}\, dA = \int_A r^2 \tau_{x\theta}\, d\theta\, dr$$

where the integration is taken over the cross-sectional area A of the shaft. We now make use of the constitutive relationship, Eq. (10.22), to obtain

$$T = \int_A Gr^3 \frac{d\phi}{dx}\, d\theta\, dr = G\frac{d\phi}{dx}\int_A r^2\, dA \tag{10.23}$$

The integral in Eq. (10.23) is the polar moment of inertia commonly denoted as J and is equal to $\pi R^4/2$ for a circular cross section. Equation (10.23) can now be integrated to give

$$\frac{d\phi}{dx} = \frac{T}{JG}$$

$$\phi = \int_0^L \frac{T}{JG}\, dx = \frac{TL}{JG} \tag{10.24}$$

The magnitude of the shearing stress can be obtained from Eqs. (10.22) and (10.23) and is

$$\tau_{x\theta} = Tr/J \tag{10.25}$$

Thus through the use of Eqs. (10.24) and (10.25) we can determine the deformation (angle of twist) and the corresponding stress distribution for a straight, circular shaft that is twisted with a torque T. It should be emphasized that the above relationships are valid only if the stresses in the shaft remain below the proportional limit of the material.

EXAMPLE 10.4

A 2-in. diameter steel shaft is being used to transmit a torque of 525 ft-lb. Determine the maximum shearing stress in the shaft and the angle of twist in a 60-in. length.

SOLUTION

We see from Eq. (10.25) that the maximum stress will occur at the outside surface. Thus

$$(\tau_{x\theta})_{\max} = \frac{(525)(12)(1)}{(\pi/2)(1)^4} = 4010 \text{ psi} \qquad\qquad \textit{Ans.}$$

The angle of twist is obtained from Eq. (10.24) and is

$$\phi = \frac{(525)(12)(60)}{(\pi/2)(1)^4(12)(10)^6} = 0.0200 \text{ rad} \qquad\qquad \textit{Ans.}$$

where we have used $G = 12 \times 10^6$ psi for the modulus of rigidity of the steel.

EXAMPLE 10.5

A circular shaft fixed at both ends is acted on by a torque T (Fig. E10.5a). Determine the torques developed at the fixed ends.

SOLUTION

We first draw a free-body diagram of the shaft (Fig. E10.5b). From equilibrium consideration we have $T = T_R + T_L$. This is the only nontrivial equi-

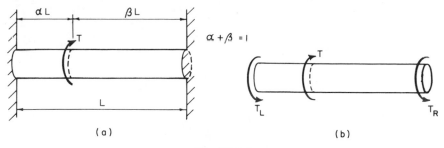

Fig. E10.5

librium equation and contains two unknowns T_R and T_L. The problem is thus statically indeterminate. To complete the solution of the problem, we must write a deformation equation. The angle of twist at the cross section where the torque T is applied must be the same for both segments of the shaft. Thus $\phi_L = \phi_R$ which leads to $T_L \alpha L / JG = T_R \beta L / JG$. Here J and G are the same for each segment, and we have $T_L = (\beta / \alpha) T_R$. Substitution into the equilibrium equation results in

$$T_L = \beta T, \qquad T_R = \alpha T \qquad\qquad Ans.$$

10.5   Torsion of Thin-walled Tubes

The problem of a thin-walled tube acted upon by a torque T is often encountered in the design of aircraft and other structures. It is possible to derive simple formulas that yield approximate values for the shear stress and angle of twist of these tubes.

We consider a tube of variable thickness t, so constructed that the cross section is uniform along the tube (Fig. 10.7). Upon applica-

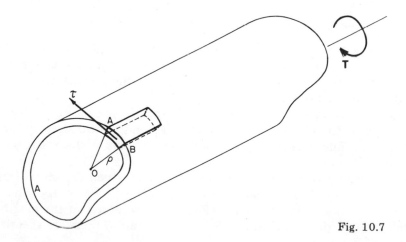

Fig. 10.7

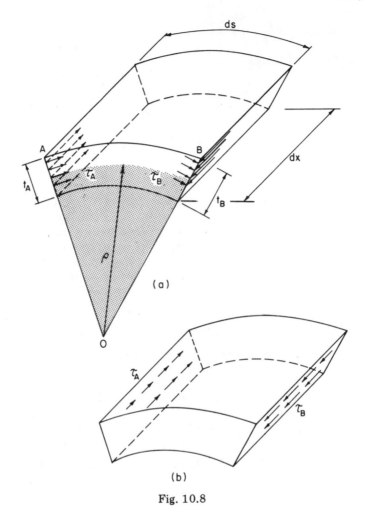

Fig. 10.8

tion of the torque T, the tube develops a shearing stress τ on the cross section in a direction parallel to the tangent to the lateral surface of the shell. The moment developed by the shear stresses must equal the applied torque T.

We now isolate an infinitesimal element of the tube (Fig. 10.8a). Corner A of the element has a thickness t_A and a shear stress τ_A acting on it, while corner B has a thickness t_B and a shear stress τ_B acting on it.

We have assumed that since the thickness is small the shear stress τ does not vary through the thickness at each of the corners. Shearing stresses must appear in pairs (Eq. 3.7); therefore on the longitudinal faces of the infinitesimal segment at A and B there are shearing stresses τ_A and τ_B (Fig. 10.8a). Since all cross sections are identical, the shearing stresses along the two longitudinal faces are constant and are equal to τ_A and τ_B (Fig. 10.8b). If we sum forces in the longitudinal direction, we have $\tau_A t_A \, dx - \tau_B t_B \, dx = 0$ or

$$\tau_A t_A = \tau_B t_B = q \qquad (10.26)$$

which indicates that the product of the shear stress and thickness at any point in the cross section is a constant. The constant q is called the *shear flow* and has dimensions of force per unit length.

The moment of the shear force on any cross section must be equal to T, the applied torque; i.e., $T = \int_s \rho(\tau t \, ds) = q\int_s \rho \, ds$ where ρ is the radius of curvature of the median line between the outer and inner surface of the thin tube. From Fig. 10.8a we see that $\rho \, ds$ is equal to two times the area of the shaded triangle. Thus

$$T = 2qA \qquad (10.27)$$

where A is the area enclosed within the median line of the tube. Hence the shearing stress is given by

$$\tau = T/2At \qquad (10.28)$$

Equation (10.28) gives the average shearing stress across the thickness t; and if t is small, this value of τ agrees reasonably well with experimental results.

To compute the angle of twist for a thin-walled tube, we subdivide the cross section into n equal segments. The angle of twist for the ith segment can then be calculated from Eq. (10.21) if we assume that all cross sections behave identically. Hence $\phi_i = \gamma_i L/\rho_i$. From Hooke's law and Eq. (10.28) we have

$$\phi_i = \tau_i L/G\rho_i = (TL/2GA)(1/\rho_i t_i)$$

If we then assume that the average angle of twist of the thin-walled section is the average of the angles of twist of each segment,

$$\phi = \frac{1}{n}\sum_{i=1}^{n}\phi_i = \frac{TL}{2G}\sum_{i=1}^{n}\left(\frac{1}{\rho_i t_i}\right)\frac{1}{n}$$

But $\rho_i \approx 2\,A_i/\Delta S_i$, and $n = A/A_i$; therefore

$$\phi = \frac{TL}{4GA}\sum_{i=1}^{n}\frac{\Delta s_i}{A_i t_i}\frac{A_i}{A} = \frac{TL}{4GA^2}\sum_{i=1}^{n}\frac{\Delta S_i}{t_i} \qquad (10.29)$$

In the limit as the number of subdivisions increase, Eq. (10.29) can be written as

$$\phi = \frac{TL}{4GA^2} \int_0^S \frac{ds}{t} \qquad (10.30)$$

If the thickness is constant, Eq. (10.30) can be integrated immediately and gives

$$\phi = (TL/4GA^2)(S/t) \qquad (10.31)$$

where S is the mean circumference of the tube. In other types of applications the tube may be composed of a number of plates, each of a constant thickness, and the integral will reduce to a sum of the lengths of these plates divided by their thickness.

EXAMPLE 10.6

A thin-walled tube (Fig. E10.6) is to be used to transmit a torque T. Find the maximum value T can have if the maximum allowable shear stress is 10 N/mm^2, and the maximum allowable angle of twist is 0.01 rad. The length of the tube is 10 m, and $G = 7 \times 10^4$ N/mm^2.

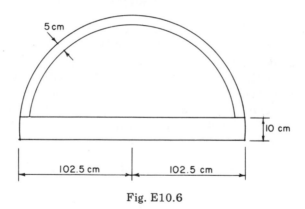

Fig. E10.6

SOLUTION

The area enclosed within the median line is

$$A = 5\pi \times 10^5 \text{ mm}^2$$

In the circular area the shear stress is

$$\tau = T/[2(5\pi \times 10^5)(50)] = (2T/\pi) \times 10^{-8}$$

If we allow a shear stress of 10 N/mm^2, we then have

$$T = 5\pi \times 10^8 \text{ N-mm}$$

For the torque computed above, the shear stress in the straight section is

$$\tau = 5\pi \times 10^8 / [2(5\pi \times 10^5)(100)] = 5 \text{ N/mm}^2$$

Thus as far as restriction on shear stress is concerned the torque can be taken as

$$T = 5\pi \times 10^8 \text{ N-mm}$$

To see if this torque gives an angle of twist larger than the allowed value, we compute ϕ as follows:

$$\phi = \frac{TL}{4GA^2} \left\{ \int_{\text{arc}} \frac{ds}{t} + \int_{\text{bar}} \frac{ds}{t} \right\} = \frac{TL}{4GA^2} \left\{ \frac{1}{50} \pi \cdot (1000) + \frac{1}{100} (2000) \right\}$$

$$\phi = \frac{(5\pi \times 10^8)(10^4)}{(4)(7 \times 10^4)[(\pi^2/4) \times 10^{12}]} (20\pi + 20) = 0.001883 \text{ rad}$$

The allowable torque is controlled by the shear stress and is

$$T = 5\pi \times 10^8 \text{ N-mm} \qquad\qquad Ans.$$

10.6 Pure Bending of Beams with Symmetrical Cross Sections

A common structural element is a bar supported in some manner and loaded transversely so that *bending* is induced (Fig. 10.1c). Structural members of this type are usually called *beams*. In this section we will consider the stresses and deformations induced in a beam that is loaded by a couple or bending moment. To keep the problem as simple as possible, the following assumptions are made: (1) the beam is initially straight; (2) the cross section of the beam does not vary along its length; (3) the beam is made of an isotropic, homogeneous, linearly elastic material; and (4) the cross section of the beam has a longitudinal plane of symmetry (LPS). Furthermore, if the beam is subjected only to an applied bending moment which lies in the LPS (see Fig. 10.9 in which the LPS is the xy plane), the

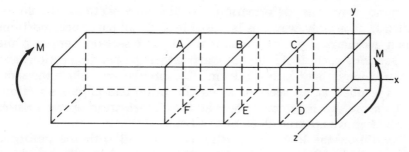

Fig. 10.9

beam is said to be in pure bending; i.e., the possibility of the beam twisting is eliminated. The LPS in this case is called the *plane of bending*. For simplicity the cross section of the beam is shown as rectangular; however, the analysis is applicable to any cross section having an LPS.

To investigate this problem, we will proceed in a way similar to our analysis of the tension and torsion problems, i.e., by considering the equations of equilibrium, geometry of deformations, and the relationship between stress and strain through the constitutive equations.

We first consider the longitudinal deformations of the beam, using reasonable assumptions and experimental evidence. Consider three planes that are equally spaced and perpendicular to the LPS of the undeformed beam (Fig. 10.9). The traces of these planes on the LPS are labeled AF, BE, and CD. After the application of the moment M, the beam will be deformed into the shape shown in Fig. 10.10, which is exaggerated since the bending deflections are nor-

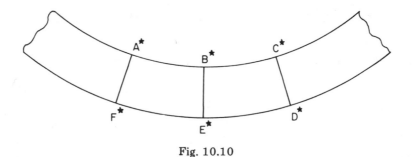

Fig. 10.10

mally very small. The section of the undeformed beam between AF and BE in Fig. 10.9 is identical in length and cross section with the section between BE and CD, and the net internal bending moment is constant along the beam. Since both these portions are loaded in the same way, the deformations of the two sections should be identical. The only way this is possible and still have these portions remain geometrically contiguous is if (1) plane sections remain plane (i.e., sections such as AF do not warp under the bending action) and (2) the beam is bent into the form of a circular arc. Since the same load is applied at all sections along the beam and all cross sections are the same, we may conclude that the deformation characteristics described above are true all along the beam.

To obtain the longitudinal strain associated with the deformation specified above, consider a small portion of length Δx of the

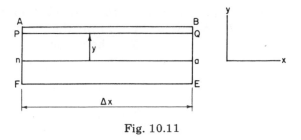

Fig. 10.11

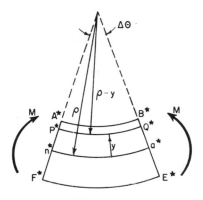

Fig. 10.12

initially undeformed beam (Fig. 10.11). Under the loading by the moment, this portion will become a curved piece (Fig. 10.12). Since there is no resultant force acting on the cross section, some fibers must shorten and some lengthen so that both compressive and tensile stresses develop. From Fig. 10.12 we see that the fibers on the top of the beam tend to shorten while those on the bottom elongate. Since the deformations in the beam must be continuous, one fiber exists whose length must remain unchanged. In fact, in the three-dimensional beam a plane of these fibers exists which remains unchanged in length under the deformation. This plane is called the *neutral surface*, and its trace on the LPS is called the *neutral axis*. (The intersection of the neutral surface with the plane of the cross section is also commonly referred to as the neutral axis.) For the type of assumptions made, it can be shown that the neutral surface is perpendicular to the LPS. The exact location of this surface will be discussed later. Let us consider the fiber *na* to be the neutral axis, and let y be the distance from *na* to some other parallel plane *PQ* in the undeformed element. Assume that under the deformation these planes remain the same distance apart. If we let ρ be the radius of curvature of the neutral axis $n*a*$ in the deformed body, then the radius of curvature of $P*Q*$ is $\rho - y$. From the geometry of Fig. 10.12 we have $\Delta x = na = n*a* = \rho\,\Delta\theta$. The change in length of the PQ fiber is then

$$P*Q* - PQ = (\rho - y)\Delta\theta - \Delta x = (\rho - y)\Delta\theta - \rho\Delta\theta = -y\Delta\theta$$

Thus the longitudinal strain ϵ_x is given by

$$\epsilon_x = \lim_{\Delta x \to 0} \frac{P*Q*}{PQ} = \lim_{\Delta x \to 0} \frac{-y\,\Delta\theta}{\rho\,\Delta\theta} = -\frac{y}{\rho} \qquad (10.32)$$

We have specified that the beam deforms to the shape of a circular arc so that ρ is a constant and thus the strain ϵ_x is directly proportional to the distance y from the neutral axis. For the situation depicted in Fig. 10.12, $\rho > 0$; therefore the negative sign indicates that the fibers above the neutral axis are in compression ($y > 0$) and those below the neutral axis are in tension ($y < 0$).

To specify the stress distribution in the beam, we must also consider the other strains that may develop. If we perform an experiment in which a grid of orthogonal lines is scribed on various exposed surfaces of the beam (Fig. 10.13), we find that the lines remain

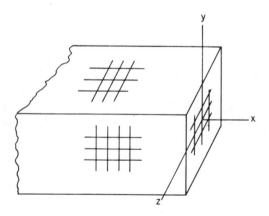

Fig. 10.13

orthogonal after the beam is bent. This observation indicates that the shearing strains γ_{xz}, γ_{yz}, and γ_{xy} may be assumed to be zero. Furthermore, we note from this experiment that the longitudinal strains in the y and z direction are simply induced through Poisson's ratio; i.e.,

$$\epsilon_y = -\nu\epsilon_x, \qquad \epsilon_z = -\nu\epsilon_x$$

These results when substituted into the general constitutive Eqs. (9.11) for linearly elastic solids reveal that

$$\sigma_x = E\epsilon_x \qquad (10.33)$$

and $\sigma_y = \sigma_z = \tau_{xz} = \tau_{yz} = \tau_{xy} = 0$.

We are now in a position to consider the equilibrium equations for the beam. For any given section, such as in Fig. 10.14, the resul-

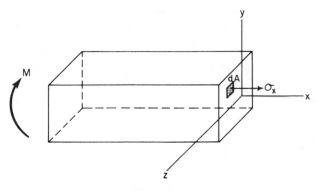

Fig. 10.14

tant force due to the internal stress distribution must be zero since the beam is loaded only by a couple. Also, the moment about the z axis must be equal to the applied couple, whereas the moment with respect to the y and x axes must be zero. To ensure that no resultant force is developed we have only to satisfy

$$\int_A \sigma_x \, dA = 0$$

where A is the cross-sectional area since there are no stresses in the y or z directions. From Eqs. (10.32) and (10.33) we obtain

$$-\frac{E}{\rho} \int_A y \, dA = 0 \qquad (10.34)$$

Since y is measured from the neutral axis, Eq. (10.34) implies that the neutral surface passes through the horizontal centroidal axis of the cross section.

Equilibrium of moments about the y axis requires that

$$-\frac{E}{\rho} \int_A yz \, dA = 0 \qquad (10.35)$$

This equation is satisfied identically since the cross section has an LPS. The other moment equilibrium equation (with respect to the z axis) gives

$$M = - \int_A y\sigma_x \, dA = \frac{E}{\rho} \int_A y^2 \, dA \qquad (10.36)$$

We note that the integral in Eq. (10.36) is the second moment of the cross-sectional area, or the moment of inertia I of the area with respect to the horizontal centroidal axis. Thus we can write

$$1/\rho = M/EI \qquad (10.37)$$

This equation is known as the *Euler-Bernoulli equation*. With the moment-curvature relation given by Eq. (10.37), the stress σ_x becomes

$$\sigma_x = -My/I \qquad (10.38)$$

which is called the *elastic flexure formula*. Note that for $M > 0$ (as shown in Fig. 10.9) σ_x is a compressive stress for $y > 0$ and a tensile stress for $y < 0$.

EXAMPLE 10.7

Consider the simply supported rectangular beam in Fig. E10.7a. Determine an expression for the stress distribution between the two loads.

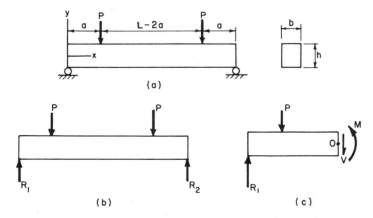

(a)

(b)   (c)

Fig. E10.7

SOLUTION

A free-body diagram of the entire beam is given in Fig. E10.7b, and it is readily shown that $R_1 = R_2 = P$. We next take a cut of the beam at an arbitrary section between the loads and draw a free-body diagram (Fig. E10.7c). At the cut section we have in general a shearing force V and a moment M, both due to the internal stress distribution at the cut section. Summation of forces in the y direction yields $R_1 - P - V = 0$, but $R_1 = P$; therefore $V = 0$.

Summation of moments about 0 gives $R_1 x - P(x - a) - M = 0$, or $M = Pa$. Since the point of the cut was arbitrary, the moment is constant in the section between the loads. Thus the beam is in pure bending in this region. For a beam of rectangular cross section the moment of inertia is $bh^3/12$, and the stress distribution through the thickness of the beam is

$$\sigma_x = -12yPa/bh^3 \qquad\qquad Ans.$$

at any cross section located between the loads. We see from this solution that the maximum fiber stress will be on the surface of the beam, whereas it will go to zero at the neutral axis.

10.7 Pure Bending of Beams with Nonsymmetrical Cross Sections

To analyze beams of this nature we proceed as in Section 10.6. The assumptions for analyzing the deformation are exactly as before except we no longer have a longitudinal plane of symmetry. The resultant external force is a moment about a line in the cross section of the beam. This moment can be decomposed into two moments about the perpendicular x and y coordinate axes. Here the axes are arbitrarily chosen (Fig. 10.15). Under the assumption of pure bend-

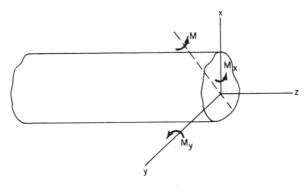

Fig. 10.15

ing (i.e., no twisting of the beam) and the further assumption that a plane normal to the longitudinal axis of the beam before bending remains plane after bending, the strain due to the moments M_x and M_y is $\epsilon_z = -y/\rho_y$ and $\epsilon_z = -x/\rho_x$ respectively. Here $1/\rho_x$ and $1/\rho_y$ are the radii of curvature of the neutral axis when bending is due either to M_y or M_x. The beam is assumed to be made of a homogeneous linearly elastic material and the induced stresses do not exceed the proportional limit of the material. Furthermore the displacements are assumed small. The strain in the z direction is then

$$\epsilon_z = - \left[(x/\rho_x) + (y/\rho_y) \right]$$

and the only nonzero stress is

$$\sigma_z = -E[(x/\rho_x) + (y/\rho_y)] \qquad (10.39)$$

The force equilibrium equations are $\Sigma F_x = \Sigma F_y = 0$, which are satisfied identically, and

$$\Sigma F_z = \int_A \sigma_z \, dA = -E \left(\frac{1}{\rho_x} \int_A x \, dA + \frac{1}{\rho_y} \int_A y \, dA \right) = 0$$

The above equation implies that $\int_A x \, dA = \int_A y \, dA = 0$, which says that the x and y axes must be chosen as centroidal axes of the cross section.

The moment equilibrium conditions are $\Sigma M_z = 0$, which is satisfied identically, and

$$\Sigma M_x = M_x = \int_A y \sigma_z \, dA = -E \left(\frac{1}{\rho_x} \int_A xy \, dA + \frac{1}{\rho_y} \int_A y^2 \, dA \right)$$

$$\Sigma M_y = M_y = \int_A x \sigma_z \, dA = -E \left(\frac{1}{\rho_x} \int_A x^2 \, dA + \frac{1}{\rho_y} \int_A xy \, dA \right)$$

or

$$M_x = -E[(I_{xy}/\rho_x) + (I_x/\rho_y)] \qquad (10.40)$$

$$M_y = -E[(I_y/\rho_x) + (I_{xy}/\rho_y)] \qquad (10.41)$$

where I_x, I_y are the moments of inertia of the cross section and I_{xy} is the product of inertia of the cross section. If we solve Eqs. (10.40) and (10.41) for ρ_x, ρ_y and substitute into Eq. (10.39), we obtain the stress-moment relation

$$\sigma_z = \frac{1}{I_x I_y - I_{xy}^2} [(M_y I_x - M_x I_{xy})x + (M_x I_y - M_y I_{xy})y] \qquad (10.42)$$

Equation (10.42) thus represents the flexure formula for a beam of nonsymmetric cross section. This equation can be simplified by noting that a set of centroidal axes exists for which the product of inertia is zero (principal axes of inertia). Referred to this set of axes, Eq. (10.42) becomes

$$\sigma_z = (M_\alpha \beta/I_\alpha) + (M_\beta \alpha/I_\beta) \qquad (10.43)$$

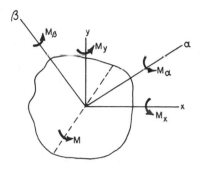

Fig. 10.16

where M_α, M_β are the moments referred to the principal axes of inertia (Fig. 10.16). A further simplification of Eq. (10.43) is possible if the applied moment is about one of the principal axes, say the α axis. Then

$$\sigma_z = M_\alpha \beta / I_\alpha \qquad (10.44)$$

which is identical in form to the previously derived flexure formula (Eq. 10.38). We thus conclude that the flexure formula as given by Eq. (10.38) is valid for nonsymmetric cross sections if (1) the in-plane coordinate axes pass through the centroid of the cross section, (2) the in-plane coordinate axes are chosen as the principal axes of inertia, and (3) the plane of the external moment is perpendicular to one of the principal axes of the cross section.

The location of the neutral axis can be determined from Eq. (10.42) or Eq. (10.43). On the neutral axis $\sigma_z = 0$; thus the trace of the neutral surface on a cross section (referred to the principal axes of inertia) is $\alpha = -(I_\beta/I_\alpha)(M_\alpha/M_\beta)\beta$.

10.8 Flexure and Cross Shear in Beams with an LPS

We have restricted attention to beams in which the bending moment is constant along the length, i.e., the case of pure bending. As a simple example of a situation for which the moment is not constant, consider the beam that is loaded at one end with a concentrated load P and rigidly attached to a base at the opposite end (Fig. 10.17a). A beam supported in this manner is called a *cantilever beam*. We now cut the beam at some arbitrary section and draw a free-body diagram (Fig. 10.17b). Summing forces in the vertical direction we have from our equilibrium equation $V = -P$, where V is the shearing force developed at the section. This force is the *cross shear*. Similarly, the sum of the moments about the point 0 must be zero to ensure equi-

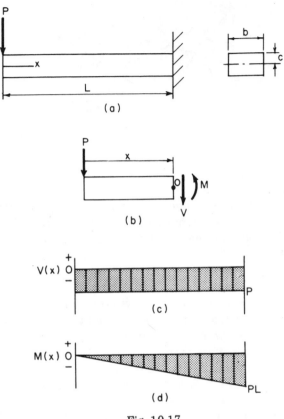

Fig. 10.17

librium and thus $M = -Px$. We can now plot the variation of V and M along the beam (Figs. 10.17c and d). These plots are the shear and moment diagrams discussed in Section 2.6. Thus we have a moment that changes at each cross section and a shear force at each cross section. The presence of the cross shear requires the existence of a shearing stress, and as a consequence planes originally perpendicular to the LPS no longer remain plane or perpendicular. This is a simple example of the type of problem we wish to consider, i.e., beams with loads on the lateral surface rather than beams loaded simply with couples.

We previously derived relations that exist between the shear force and distributed load and the moment and shear force, i.e.,

$$\frac{dV}{dx} = -w(x) \qquad (10.45)$$

and

$$\frac{dM}{dx} = V(x) \tag{10.46}$$

For the case of the cantilever beam we have $w(x) = 0$, $V(x) = -P$ (a constant), and $M(x) = -Px$.

The first assumption we make is that the normal stress on a cross section of the beam is given by the formula derived for pure bending, i.e.,

$$\sigma_x = -M(x)y/I \tag{10.47}$$

We realize that Eq. (10.47) is not quite correct since for pure bending it was derived on the basis of planes remaining plane, which is not true when shearing forces are developed on the cross section. Experimental evidence indicates that the errors introduced are insignificant for long slender beams.

Since a shearing force $V(x)$ exists at each cross section, shearing stresses must also exist on cross sections of the beam perpendicular to the axis of the beam. We wish now to develop an equation for the intensity of these shearing stresses in terms of the shearing force at the section and the properties of the cross section. The expression to be developed applies only to a cross section of rectangular shape. The results are commonly used, however, to give approximate values of the shearing stress for other cross sections having an LPS.

Consider an element of length Δx cut from a beam (Fig. 10.18a). Since the bending moment varies along the beam, on the left side we have a bending moment M and on the right side $M + \Delta M$. The normal stresses caused by these moments are respectively

$$\sigma_x = -My/I, \qquad \sigma_x + \Delta\sigma_x = -[(M + \Delta M)y]/I$$

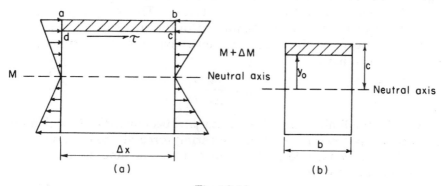

Fig. 10.18

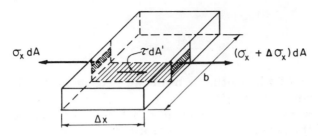

Fig. 10.19

where y is measured from the neutral axis (Fig. 10.18b). If we consider equilibrium of the portion $abcd$ cut from the top of the beam, the forces exerted on this element are shown in Fig. 10.19 where τ is used to designate the average shearing stress on the horizontal shaded area dA'. We have shown tensile normal stresses but they may be compressive, depending on the moment. We see that the element will not be in equilibrium unless a force on the bottom is included. Thus, summing forces in the horizontal direction over the two vertical and one horizontal areas, we have

$$- \int_A \sigma_x \, dA + \int_A (\sigma_x + \Delta\sigma_x) \, dA + \int_{A'} \tau \, dA' = 0 \qquad (10.48)$$

Note that $\Delta\sigma_x = -\Delta M \, y/I$ where ΔM is a function of x, but not of the cross-sectional dimensions.

Thus we have from Eq. (10.48)

$$\frac{\Delta M}{I} \int_A y \, dA = \int_{A'} \tau \, dA' = \tau_{\text{avg}} b \, \Delta x \qquad (10.49)$$

where τ_{avg} is the average shearing stress over the surface $b \, \Delta x$. In equating the two integrals involving the shearing stress, we have made use of the mean value theorem for integrals. The integral on the left side of Eq. (10.49) is the first moment of the area above the cut surface with respect to the neutral or centroidal axis. We define this to be Q. We then have

$$\tau_{\text{avg}} = \frac{\Delta M}{\Delta x} \frac{Q}{Ib}$$

If we take the limit as $\Delta x \to 0$, then $\Delta M/\Delta x$ becomes dM/dx which by Eq. (10.46) is equal to $V(x)$, the shearing force at the cross section. Dropping the average designation on τ but understanding that it is in fact an average, we have

$$\tau = VQ/Ib \qquad (10.50)$$

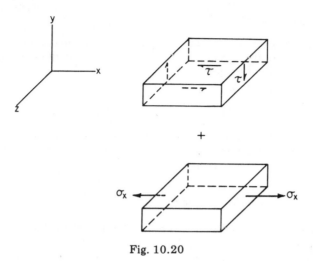

Fig. 10.20

This is a longitudinal shearing stress; but since from moment equilibrium we know that shearing stresses must appear in pairs, a shearing stress of equal magnitude must also be present on a transverse face. Thus the complete stress picture at a point for a beam not under constant moment is a shearing stress $\tau_{xy} = \tau$ plus a normal stress σ_x (Fig. 10.20). The results of this analysis compare favorably with experimental results provided the beam is narrow; i.e., $b \ll L$ where b is the beam width and L is the length of the beam.

EXAMPLE 10.8

For the rectangular beam in Fig. 10.17 determine the maximum normal stress and the maximum transverse shear stress on a plane perpendicular to the LPS.

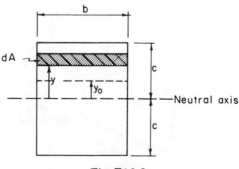

Fig. E10.8

SOLUTION

Since the maximum moment occurs at $x = L$, the maximum normal stress will occur at this location. For a rectangular cross section the moment of inertia is $I = (1/12)(b)(2c)^3 = (2/3)bc^3$. Thus $\sigma_x = 3PLy/2bc^3$. The maximum normal stress occurs at $y = \pm c$, thus

$$\sigma_x = 3PL/2bc^2 \quad \text{(tension or compression)} \qquad Ans.$$

Since the shear force V is constant along the beam, the choice of cross section for the maximum transverse shear stress determination is immaterial. We first find Q from the integral definition

$$Q = \int_A y \, dA$$

where the area A is taken as the area between the level y_0 at which the shearing stress is required and the outer surface of the beam (Fig. E10.8). Thus we see that if $y_0 = c$, $Q = 0$, whereas if $y_0 = 0$ (i.e., if τ is evaluated at the neutral axis), $Q = bc^2/2$, its maximum value. Thus the maximum value of transverse shearing stress occurs on the neutral surface and is

$$\tau = 3P/4bc \qquad Ans.$$

10.9   Beam Deflections Due to Bending

In addition to the stress distribution in a beam, the displacement or deflection of the beam from its initially undeformed state is of interest. In describing the deflection of a beam, we will consider the behavior of the neutral surface and neglect any deflection due to shearing stresses. For our reference we will let the x axis coincide with the straight neutral axis of an unstressed beam (Fig. 10.21) having an LPS. As the beam is loaded, the neutral axis forms a curve having a radius of curvature ρ, which in general will vary along the span of the beam in accordance with the Euler-Bernoulli law (Eq. 10.37), $1/\rho = M(x)/EI$. This curve is called the *elastic curve*. Here we

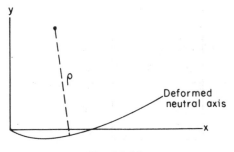

Fig. 10.21

are assuming that this equation, developed for the case of pure bending, will be valid for other loading conditions as well. We also know from calculus that

$$\frac{1}{\rho} = \frac{d^2 y/dx^2}{[1 + (dy/dx)^2]^{3/2}}$$

and with small deflections for which $dy/dx \ll 1$,

$$\frac{1}{\rho} \approx \frac{d^2 y}{dx^2}$$

Thus we can relate deflections to the moment distribution along the beam through the differential equation

$$\frac{d^2 y}{dx^2} = \frac{M(x)}{EI} \tag{10.51}$$

Equation (10.51) is commonly used to obtain deflections in beams, and it is found to give good results if the following conditions are satisfied: (1) the material of the beam is linearly elastic, isotropic and homogeneous, (2) the deflections are small, and (3) shearing deformations are small. This last condition is usually satisfied except for relatively short deep beams.

Numerous methods have been devised for predicting deflections. In the following example we will illustrate one method based on the integration of Eq. (10.51).

EXAMPLE 10.9

Determine the maximum deflection of the cantilever beam in Fig. E10.9.

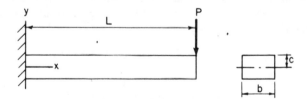

Fig. E10.9

SOLUTION

With the coordinate system at the fixed end of the beam the moment equation is $M(x) = -P(L - x)$. Thus from Eq. (10.51) we have

$$\frac{d^2 y}{dx^2} = -\frac{P}{EI}(L - x)$$

The first integration yields

$$EI\frac{dy}{dx} = -PLx + \frac{Px^2}{2} + C_1$$

At the fixed end $x = 0$ we specify that the slope is zero; i.e., $dy/dx = 0$ so that $C_1 = 0$. The second integration gives

$$EIy = -(PLx^2/2) + (Px^3/6) + C_2$$

Since at $x = 0$, $y = 0$ it follows that $C_2 = 0$. Thus the deflection at any point is given by the equation

$$y = Px^2[(x/3) - L]/2EI$$

The maximum deflection occurs at $x = L$ and is

$$y_{max} = -PL^3/3EI \qquad\qquad Ans.$$

The negative sign indicates that the deflection is downward.

EXAMPLE 10.10

A simply supported steel beam (with $E = 30 \times 10^6$ psi) having the T-shaped cross section in Fig. E10.10a has a concentrated load of 1000 lb at its center. Determine (a) the maximum flexural stress, (b) the longitudinal shearing stress at the neutral surface and midway between the left support and the load, and (c) the deflection at the center of the beam.

SOLUTION

We must first locate the horizontal centroidal axis and compute the moment of inertia of the cross section with respect to this axis. This information is required before the flexural stresses can be determined. To locate the centroidal axis, we take the first moment of the area with respect to some convenient axis such as cc (Fig. E10.10b). Summing moments about cc, we have

$$y_c(A_1 + A_2) = (h/2)A_2 + [h + (a/2)]A_1$$

and

$$y_c = \frac{(6/2)(6) + [6 + (1/2)]8}{8 + 6} = 5 \text{ in.}$$

The moment of inertia of the cross section with respect to the centroidal axis can be obtained from the parallel axis theorem; i.e.,

$$\begin{aligned}
I &= I_{dd} + [y_c - (h/2)]^2 A_2 + I_{ee} + [h + (a/2) - y_c]^2 A_1 \\
&= (1/12)(1)(6)^3 + (5-3)^2(6) + (1/12)(8)(1)^3 + [6 + (1/2) - 5]^2(8) \\
&= 182/3 \text{ in.}^4
\end{aligned}$$

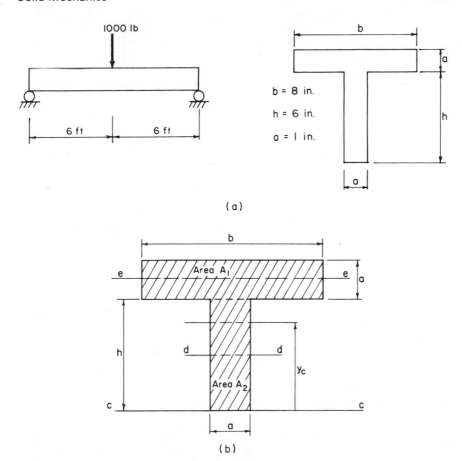

b = 8 in.

h = 6 in.

a = 1 in.

(a)

(b)

Fig. E10.10 (a), (b)

Since we also want to find the shearing stress at the neutral surface, we need to evaluate Q, the first moment with respect to the centroid of the area above the section at which the stress is to be evaluated. Thus

$$Q = (1)(1)(1/2) + (8)(1)(3/2) = 12.50 \text{ in.}^3$$

We next determine the moment $M(x)$ and shear $V(x)$ along the beam. Consider the free-body diagram of Fig. E10.10c where we have made use of the result that the left reaction $R_1 = P/2$. Equilibrium thus requires that $V = P/2$ and $M = Px/2$. The corresponding shear and moment diagrams are shown in Fig. E10.10d.

Since the maximum moment occurs at the center of the beam, this is the cross section at which the maximum flexural or normal stress will occur. The normal stress is obtained from the equation $\sigma_x = -M(x)y/I$; therefore at the top

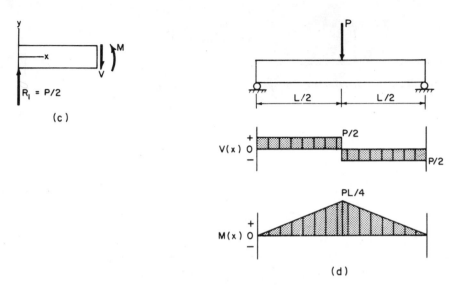

Fig. E10.10 (c), (d)

of the beam,

$$\sigma_x = - \frac{\{[(1000)(12)]/4\}\,[(12)(2)]}{182/3} = -1186 \text{ psi}$$

where the negative sign indicates a compressive stress. At the bottom of the beam

$$\sigma_x = - \frac{\{[(1000)(12)]/4\}\,[(12)(-5)]}{182/3} = 2967 \text{ psi} \qquad \textit{Ans.}$$

Thus the maximum flexural stress occurs in the bottom fibers.

The shearing stress is obtained from the equation $\tau = VQ/Ib$, where b is the width of the section at the location where τ is evaluated. Therefore

$$\tau = \frac{(1000/2)(12.5)}{(182/3)(1)} = 103 \text{ psi} \qquad \textit{Ans.}$$

Finally, we can obtain the deflection by means of the equation

$$\frac{d^2 y}{dx^2} = \frac{M(x)}{EI}$$

The first integration gives

$$EI \frac{dy}{dx} = \frac{Px^2}{4} + C_1$$

At the center of the beam $(x = L/2)$, $dy/dx = 0$ (due to symmetry); therefore

$C_1 = -PL^2/16$. The second integration yields

$$Ely = (Px^3/12) - (PL^2x/16) + C_2$$

However, $C_2 = 0$ since $y = 0$ at $x = 0$. The deflection equation can thus be written as

$$y = (P/EI)[(x^3/12) - (xL^2/16)]$$

The maximum deflection occurs at the center of the beam; thus

$$y_{max} = (P/EI)[(L^3/96) - (L^3/32)] = -PL^3/48EI$$

$$= -\frac{(1000)(12)^3(12)^3}{(48)(30)(10)^6(182/3)} = 0.0342 \text{ in.} \qquad Ans.$$

10.10   Buckling of Long Slender Beams

Special attention must be given to a long, slender, straight bar with a *compressive* axial load since the bar may "buckle," i.e., suddenly deform into some shape other than its initially straight configuration. A simple example of this phenomenon can be observed with a yardstick if it is subjected to an axial compressive load. If the bar is not long and slender but has cross-sectional dimensions of the same order as the length, this instability may not occur; and stresses beyond the proportional limit may be developed in accordance with the theory developed in Section 10.3. Long slender bars under compressive loads are usually referred to as *columns.*

To investigate the buckling problem, we consider the following question. If a column that is initially straight and subjected to an axial compressive load is given a small lateral displacement, will it return to its initial equilibrium position, or is it possible for it to remain in the deflected position? If two or more equilibrium configurations are possible, the column is said to be in *unstable equilibrium.* Experiments reveal that as the load applied to a column approaches that at which an unstable equilibrium condition may exist, the column may suddenly buckle with possible catastrophic consequences.

We again make simplifying assumptions which in some cases are not physically exact but allow us to analyze the problem. The solution is still of value in many situations since experimental results again indicate sufficiently close agreement with theory.

The assumptions made for an ideal column are: (1) the column is initially straight; (2) the compressive load is applied at the centroid of the cross section; (3) the cross section of the column remains constant, and the length of the column is much larger than the maximum

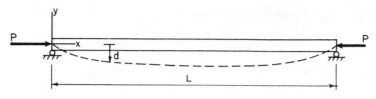

Fig. 10.22

cross-sectional dimension; and (4) the material being considered is linearly elastic, isotropic, and homogeneous.

Consider a simply supported beam with a compressive load P on the ends (Fig. 10.22) that is sufficiently large to hold the beam in a slightly deflected position after a small lateral load has been applied and removed. The deflection is assumed to be small so that the elastic deflection equation, Eq. (10.51), can be used. Again note we are considering only the neutral axis deflection. If we draw a free-body diagram of the deflected beam (Fig. 10.23) and use equilibrium con-

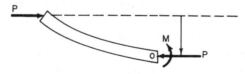

Fig. 10.23

siderations, we have $\Sigma M_0 = 0$, which gives $M = Pd$. But d is the distance from the undeformed neutral axis to the deformed neutral axis and for the coordinate system chosen is equal to $-y$. Thus we have

$$EI \frac{d^2 y}{dx^2} = M = -Py \qquad (10.52)$$

or

$$\frac{d^2 y}{dx^2} + k^2 y = 0 \qquad (10.53)$$

where $k^2 = P/EI$.

Equation (10.53) is a second-order, ordinary differential equation with constant coefficients. The solution to this equation is

$$y(x) = A \cos kx + B \sin kx$$

To evaluate the constants A and B, we must impose boundary conditions, which for this simply supported beam are

$$y = 0 @ x = 0, \qquad y = 0 @ x = L$$

Application of the first boundary condition gives

$$A = 0 \qquad (10.54)$$

while the second condition results in

$$B \sin kL = 0 \qquad (10.55)$$

From Eq. (10.55) we have two possible solutions. One is $B = 0$, which when coupled with Eq. (10.54) gives $y(x) = 0$. This solution simply indicates that one possible equilibrium configuration is the straight column merely being compressed by the load P. The other possible solution is $\sin kL = 0$. If we take this as the solution to Eq. (10.55), B is undetermined and kL must be an integer multiple of π. Thus

$$kL = n\pi, \qquad n = 0, 1, 2, 3, \ldots$$

With these values for k the possible values that the load P can attain are

$$P = n^2 \pi^2 EI/L^2, \qquad n = 0, 1, 2, 3, \ldots$$

The use of $n = 0$ gives $P = 0$, which is a beam that is not loaded and thus of little interest. The value $n = 1$ gives the smallest value of P that will satisfy Eq. 10.55. This value of the load P_E given by the equation

$$P_E = \pi^2 EI/L^2 \qquad (10.56)$$

is called the *Euler load* for a beam supported in this manner. Our deflection for this load is $y = B \sin (\pi x/L)$. We note here that B is undetermined so that the quantitative deflection is impossible to obtain.

The results of this analysis indicate that if a column is loaded slowly, starting with a zero load and given small lateral deflections, the equilibrium configuration remains the straight section until the critical load P_E is attained. At this load two equilibrium states are possible and the column is said to be in unstable equilibrium. This is commonly referred to as the *critical buckling load*.

Note that the buckling load depends on the moment of inertia of the cross section in addition to the modulus of elasticity and the length. Unless the column is somehow constrained, it will buckle around an axis for which the moment of inertia is a minimum. Thus in calculating the critical buckling load, this minimum moment of inertia is used.

Equation (10.56) has been found to agree closely with experimental data for long slender columns. The criterion for determining whether a beam is long and slender is usually taken as the ratio

$L\sqrt{A/I}$. For steel columns it has been determined that $L\sqrt{A/I} > 140$ for the Euler column theory to be valid.

The problem of determining critical buckling loads for columns is complicated and depends on many factors. The material presented in this section is given merely as an indication of the nature of buckling. The interested student should refer to the numerous excellent texts on strength of materials for a more thorough coverage of this important problem.

EXAMPLE 10.11

Find the Euler load for the cantilever beam of Fig. E10.11a.

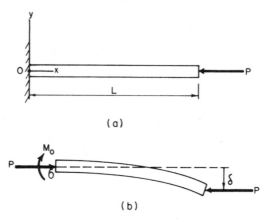

Fig. E10.11

SOLUTION

From overall equilibrium (Fig. E10.11b) we have $M_0 = -P\delta$ and the moment at any cross section is therefore

$$M(x) = M_0 - Py = -P(\delta + y)$$

The deflection curve thus can be written as

$$EI \frac{d^2 y}{dx^2} + Py = -P\delta$$

with the solution

$$y(x) = A \cos kx + B \sin kx - \delta$$

The boundary conditions for the cantilever beam are

$$y = 0, \quad \frac{dy}{dx} = 0 @ x = 0, \quad y = -\delta @ x = L$$

Application of these boundary conditions yields $A = \delta$, $B = 0$, and the condition $\delta \cos kL = 0$. The last equation gives again two solutions; $\delta = 0$ which implies no deflection, or $\cos kL = 0$ which implies $kL = [(2n + 1)\pi]/2$ where $n = 0, 1, 2, 3, \ldots$. Thus the Euler load for this beam occurs at $n = 0$ and is

$$P_E = \pi^2 EI/(2L)^2 \qquad\qquad Ans.$$

This result differs from that of the simply supported beam and demonstrates that the buckling loads will depend on the manner in which the column is constrained.

10.11   Combined Loading Problems

In this section we will consider problems in which the loading of the beam may experience a combination of tension, torsion, and flexural stresses. To best illustrate the method of solutions to these types of problems, consider the following examples.

EXAMPLE 10.12

A steel beam is loaded as in Fig. E10.12a. Find the maximum normal stress, maximum shear stress, and the points at which these stresses occur in the beam.

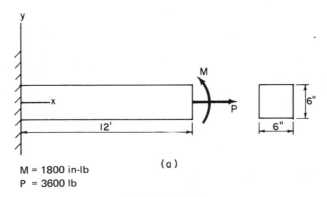

M = 1800 in-lb
P = 3600 lb

(a)

Fig. E10.12(a)

SOLUTION

It is easy to see in this case that the moment M is constant throughout the length of the beam. Hence every cross section behaves like any other. We further know that the maximum flexure stress occurs on the outer fibers of the beam and, for the case of a symmetric cross section, the neutral axis is at the center. Hence

$$(\sigma_x)_M = My/I = [(1800)(3)]/108 = 50 \text{ psi}$$

Fig. E10.12(b)

From Fig. E10.12a we see that due to the moment the fibers on top of the beam will be in compression while the bottom fibers are in tension. The maximum flexural stress is then

$$(\sigma_x)_M = 50 \text{ psi} \quad (C \text{ on top})$$

$$(\sigma_x)_M = 50 \text{ psi} \quad (T \text{ on bottom})$$

The load P gives a uniformly distributed normal stress

$$(\sigma_x)_P = P/A = 3600/36 = 100 \text{ psi } T$$

We now draw the stress picture for a point on top and bottom of the beam (Fig. E10.12b). No shearing stresses exist at these points on horizontal or vertical planes, therefore the stresses shown are the principal stresses. The maximum normal stress occurs at the bottom and is

$$\sigma_1 = 150 \text{ psi } T \qquad\qquad\qquad Ans.$$

The maximum shear stress also occurs at a point on the bottom fiber and is

$$\tau_{\max} = \sigma_1/2 = 75 \text{ psi} \qquad\qquad\qquad Ans.$$

and acts on a plane $45°$ to the principal plane.

EXAMPLE 10.13

For the beam loaded as in Fig. E10.13a, find the maximum normal and shearing stress.

SOLUTION

We have here a shear stress due to the torque T and a tensile stress due to the load P. Both stresses are independent of which cross section we examine. For a given cross section the shearing stress is maximum on the surface of the

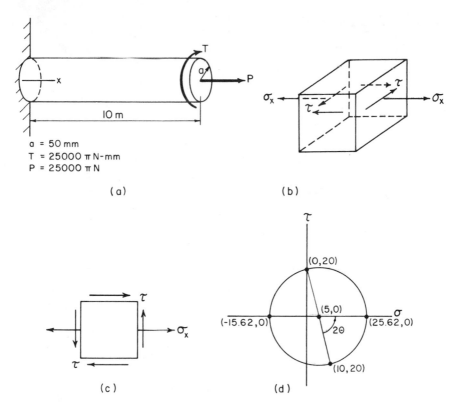

Fig. E10.13 (a), (b), (c), (d)

cylinder. Hence

$$\tau = \frac{Ta}{J} = \frac{(25000)\pi\,(50)}{[\pi(50)^3]/2} = 20 \text{ N/mm}^2$$

The tensile stress is $\sigma_x = P/A = 25000\pi/2500\pi = 10$ N/mm^2.

Since the normal stress is uniformly distributed over the cross section, the maximum normal stress occurs at the point of largest shear stress on the cross section. The state of stress at a point on the surface of the cylinder is shown in Fig. E10.13b where it is understood that the rectangular parallelepiped element is a graphical representation of the stresses acting on three orthogonal planes through the point. We see that one plane has no stress acting on it so that a two-dimensional representation can be drawn (Fig. E10.13c).

To find the maximum normal stress, we draw the Mohr's circle for the values of σ_x and τ given above (Fig. E10.13d). The radius of the circle is the maximum in-plane shear stress.

$$\tau_P = \sqrt{425} = 20.62 \text{ N/mm}^2$$

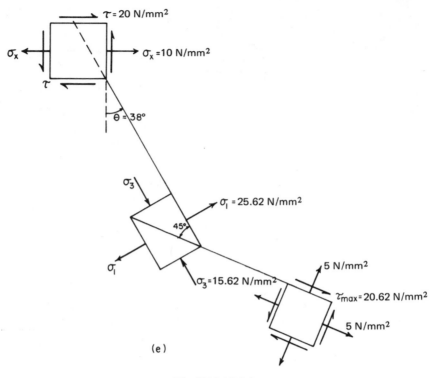

(e)

Fig. E10.13 (e)

The two principal stresses are

$$\sigma_3 = 15.62 \text{ N/mm}^2 \; C, \qquad \sigma_1 = 25.62 \text{ N/mm}^2 \; T$$

Thus the maximum normal stress is given by σ_1; and since $\sigma_2 = 0$ and σ_1 and σ_3 have different signs, τ_P is the maximum shear stress. The planes on which the principal stresses and maximum shear stress act are shown in Fig. E10.13e.

The angle θ is found from the Mohr's circle and is given by

$$\tan 2\theta = 2(20)/10 = 4$$

or $\theta = 38°$. Hence we have

$\sigma_1 = 25.62 \text{ N/mm}^2 \; T$ on a plane $38°$ counterclockwise from the
plane whose normal is in the positive x direction   *Ans.*
$\tau_{max} = 20.62 \text{ N/mm}^2$ on a plane $83°$ counterclockwise from the
plane whose normal is in the positive x direction   *Ans.*

EXAMPLE 10.14

For the I beam loaded as in Fig. E10.14a find the maximum normal stress and maximum in plane shear stress.

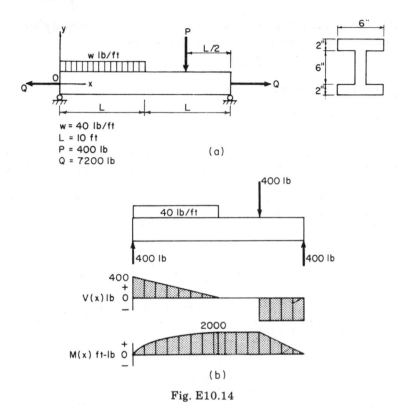

w = 40 lb/ft
L = I0 ft
P = 400 lb
Q = 7200 lb

(a)

Fig. E10.14

SOLUTION

The shear and moment diagram for the beam is shown in Fig. E10.14b. The details of finding the reactions on the beams due to the support is left as an exercise for the reader. The maximum value of the moment is seen to occur for $10 \leqslant x \leqslant 15$. In this range of x the shear force V is zero. Just to the right of the 400-lb concentrated load the shear force is 400 lb and the moment is almost equal to its maximum value. Hence we have

$$\sigma_x = [(2000)(12)y]/I = 24{,}000y/428$$

$$(\sigma_x)_{max} = [(24{,}000)(5)]/428 = 280 \text{ psi} \qquad C \text{ on top, } T \text{ on bottom}$$

The cross shear τ is zero at the outer surface of the beam where σ_x is a maximum. The stress due to the tensile load Q is

$$(\sigma_x)_Q = 7200/36 = 200 \text{ psi } T$$

Hence the normal stresses on top and bottom become

$$(\sigma_x)_{top} = -280 + 200 = -80 \text{ psi} = 80 \text{ psi } C$$

$$(\sigma_x)_{bottom} = 280 + 200 = 480 \text{ psi } T$$

Since the longitudinal shear stress is zero at the bottom, the normal stress is the principal stress and as such is the maximum normal stress.

$$\sigma_1 = 480 \text{ psi} \qquad T \text{ on bottom fiber} \qquad\qquad Ans.$$

The maximum in-plane shear is equal to $\sigma_1/2$ and thus

$$\tau_{max} = 240 \text{ psi on a plane } 45^\circ \text{ counterclockwise from the plane}$$
whose normal is in the positive x direction $\qquad\qquad Ans.$

We may feel that the value of the cross shear at the center line would be greater, but in fact $\tau = 26.6$ psi coupled with $\sigma_x = 200$ psi gives a τ_{max} of approximately 103 psi.

Another possible point in the cross section for large normal and shear stresses is where the flange and the web join. Here we find that $\tau_H = 22.4$ psi, $\sigma_x = 368$ psi T below the center line, and this leads to $\sigma_1 = 369$ psi, $\tau_{max} = 185$ psi.

PROBLEMS

10.1. Determine the maximum tensile stress in the skin of a basketball when it is inflated to a pressure of 15 psi. Assume the basketball has a diameter of 12 in. and a skin thickness of 1/8 in.

   Ans. $\sigma = 360$ psi

10.2. At a given location the pressure in a home water system is 100 psi. If copper tubing having an inside diameter of 0.75 in. and a maximum allowable tensile stress of 2500 psi is to be used for the system, determine the minimum allowable wall thickness for the tubing.

10.3. A standpipe 50 ft high and 6 ft in diameter is full of water (62.4 lb/ft^3). Determine the circumferential tensile stress at the bottom of the pipe if it is fabricated using 0.25 in. thick steel plates.

   Ans. $\sigma = 3125$ psi

10.4. A thin-walled cylindrical pressure vessel (Fig. P10.4) has a wall thickness of 0.3 in. and a diameter of 30 in. The normal stress on the plane of the weld is 4000 psi tension. Determine the pressure in the tank.

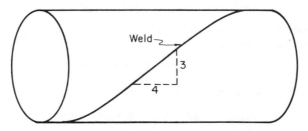

Fig. P10.4

10.5. A steel mine cable having a cross-sectional area $A = 1$ in.^{2} and an unstretched length $L = 2000$ ft is used to support an elevator cage weighing 5000 lb. If the cable weighs 0.3 lb/in.^{3} and has a modulus of elasticity $E = 30 \times 10^6$ psi, determine (a) the extension of the cable produced by the weight of the cage only and (b) the extension of the cable produced by the weight of the cable only, when the cage has been lowered to its maximum depth.

Ans. (a) $\delta_{\text{cage}} = 4$ in.,   (b) $\delta_{\text{cable wt.}} = 2.88$ in.

10.6. The round steel bar ($E = 30 \times 10^6$ psi) with fixed ends in Fig. P10.6 carries a total load of 9000 lbs. Determine the reactions at the ends and the maximum tensile stress in the bar.

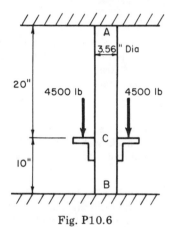

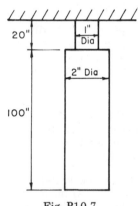

Fig. P10.6   Fig. P10.7

10.7. The long aluminum ($E = 10 \times 10^6$ psi, $w = 0.10$ lb/in.^{3}) bar in Fig. P10.7 hangs from a rigid support. Find the total elongation of the bar due to its own weight.

Ans. $\delta = 1.32 \times 10^{-4}$ in.

10.8. The rigid bar in Fig. P10.8 is supported by the aluminum rod AB ($E = 10 \times 10^6$ psi, $A_a = 1$ in.^{2}) and the steel rod CD ($E = 30 \times 10^6$ psi, $A_s = 0.5$ in.^{2}). When a 30,000 lb load is applied to the rigid bar, point A moves downward 0.1 in. more than point C. Determine the stress in the rods and the point of application of the 30,000-lb load.

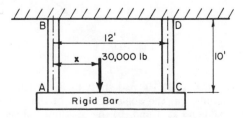

Fig. P10.8

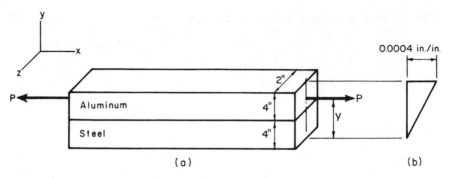

Fig. P10.9

10.9. A steel ($E = 30 \times 10^6$ psi) and an aluminum ($E = 10 \times 10^6$ psi) member are bonded together as in Fig. P10.9a. The bar is then loaded to produce the strain distribution in Fig. P10.9b (the strain does not vary in the z direction). Determine the magnitude and location of the load P.

Ans. $P = 48,000$ lb, $\quad y = 4.44$ in.

10.10. A solid steel shaft ($G = 12 \times 10^6$ psi) with a 4-in. diameter has an allowable shear stress of 5000 psi and an allowable angle of twist per foot of length of $0.1°$. Calculate the maximum torque T the shaft may transmit.

10.11. A 4-in. diameter steel shaft ($G = 12 \times 10^6$ psi) is loaded as in Fig. P10.11. Determine (a) the maximum shearing stress in the shaft and (b) the angle of twist of the free end of the shaft with respect to the fixed end.

Ans. $\tau_{\text{max}} = 10,000$ psi, $\quad \theta_{CA} = 0.0156$ rad ⤵

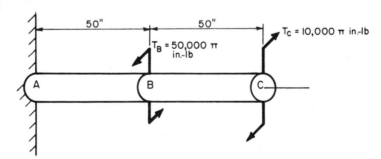

Fig. P10.11

10.12. The solid steel ($G = 12 \times 10^6$ psi) shaft in Fig. P10.12 is loaded in pure torsion. Determine (a) the angle of twist of the free end at E with respect to the fixed end at A and (b) the maximum shear stress in the shaft and the location where it occurs (neglect stress concentration effects).

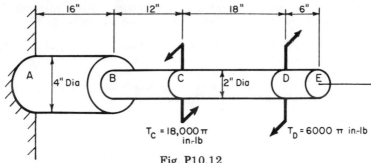

Fig. P10.12

10.13. A solid circular shaft is composed of two circular sections coupled together as in Fig. P10.13. Section AB is made of steel $(G = 12 \times 10^6$ psi), and section BC is made of brass $(G = 5 \times 10^6$ psi). Two torques are applied as shown. Determine (a) the maximum shearing stress and (b) the angle of twist of section C with respect to section A.

Ans. τ_{max} = 16,000 psi, θ_{CA} = 0.056 rad $\big)$

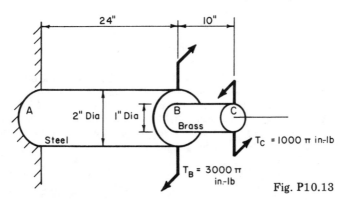

Fig. P10.13

10.14. A circular shaft is composed of two 6-in. diameter sections coupled together as in Fig. P10.14. Part AB is bronze $(G = 6 \times 10^6$ psi), and part BC is steel $(G = 12 \times 10^6$ psi). A torque of $100,000\pi$ in.-lb is applied at

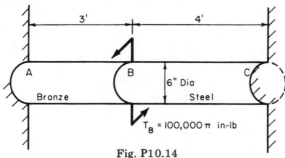

Fig. P10.14

section B. Determine (a) the maximum shearing stress in the shaft and
(b) the angle of twist of section B with respect to section A.

10.15. Shaft AB in Fig. P10.15 is rigidly attached to a collar bolted to shaft BC.
The bolts allow the collar to rotate 0.02 rad before engaging shaft BC.
Both shafts are made of steel ($G = 12 \times 10^6$ psi), and each has a diameter
of 4 in. Determine the maximum torque T that can be applied if the
shearing stress must not exceed 12 ksi in either shaft.

   Ans. $T = 271{,}000$ in.-lb

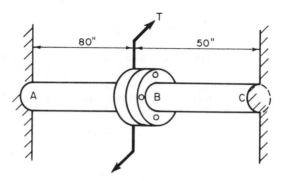

Fig. P10.15

10.16. Identical sheets of steel 12 in. wide by 0.125 in. thick by 50 in. long
were used to fabricate thin-walled tubes having the cross sections in
Fig. P10.16. The mean circumference of each tube is 12 in. Assume
$G = 12 \times 10^6$ psi. If a torque of 20,000 in.-lb is applied to each of the
tubes, determine (a) the average shear stress on a cross-section and (b) the
angle of twist in the 50-in. length.

Fig. P10.16

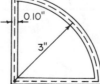

Fig. P10.17

10.17. A thin-walled aluminum ($G = 3.75 \times 10^6$ psi) tube has the cross section in Fig. P10.17. If the average shear stress in the wall is not to exceed 5000 psi and the angle of twist in a 60-in. length is not to exceed 0.05 rad, determine the maximum torque that can be transmitted.

Ans. $T = 5840$ in.-lb

10.18. An extruded aluminum ($G = 3.75 \times 10^6$ psi) tube has the cross section in Fig. P10.18. In a given design situation where this tube is to be used, the angle of twist in a 1-ft length must not exceed 0.001 rad. Determine the average shear stress that would be present on a cross section at this maximum angle of twist.

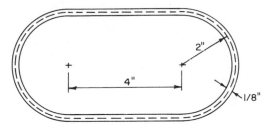

Fig. P10.18

10.19. A thin-walled steel ($G = 12 \times 10^6$ psi) tube having the cross section in Fig. P10.19 is subjected to a torque of 10,000 in.-lb. Determine (a) the maximum shearing stress on a cross section and (b) the angle of twist in a 36-in. length.

Ans. $\tau = 6250$ psi, $\theta = 0.012$ rad

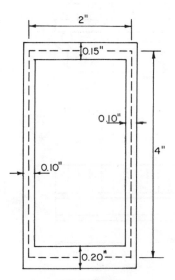

Fig. P10.19

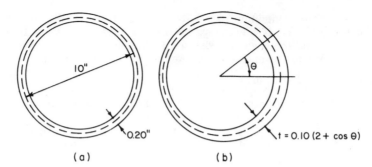

Fig. P10.20

10.20. The thin-walled steel ($G = 12 \times 10^6$ psi) tube having the cross section in Fig. P10.20a has been specified for a certain design application. If in the process of manufacture the wall thickness is not held constant but varies as in Fig. P10.20b, determine (a) the percent change in the maximum shear stress and (b) the percent change in the angle of twist per unit length.

10.21. Two timber beams are used to support a uniformly distributed load of 160 lb/ft as in Fig. P10.21. Determine the maximum axial tensile (flexure) stress on a cross section at the midspan.

    Ans. $\sigma = 1000$ psi

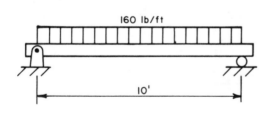
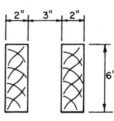

Fig. P10.21

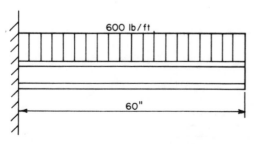

Fig. P10.22

10.22. The cantilever beam in Fig. P10.22 is a standard I beam having a 5-in. depth and a moment of inertia about the neutral axis of the cross section of 15 in^4. Determine the maximum axial tensile (flexure) stress on a section 6 in. from the wall when the beam is loaded as shown.

10.23. For the beam in Fig. P10.23 the maximum allowable axial (flexure) stresses on a cross section are 10,000 psi tension and 6000 psi compression. Determine the maximum allowable load P.

Ans. P_{max} = 3530 lb

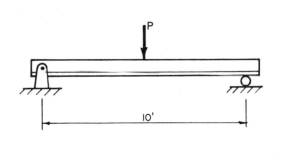

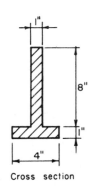

Cross section

Fig. P10.23

10.24. For the beam in Fig. P10.24 determine the maximum axial tensile (flexure) stress.

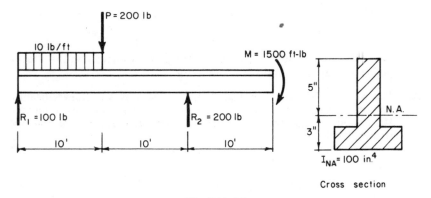

Cross section

Fig. P10.24

10.25. For the beam in Fig. P10.25 determine the maximum axial tensile and maximum axial compressive (flexure) stresses.

Ans. $\sigma_{max\ ten}$ = 4400 psi, $\sigma_{max\ comp}$ = 2640 psi

10.26. A wooden beam 12 ft long has the rectangular cross section in Fig. P10.26. The loads applied to the beam produce a moment about the x axis (M_x = 2400 ft-lb). During erection of the building the beam

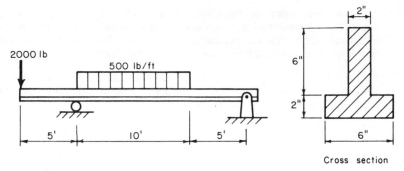

Fig. P10.25

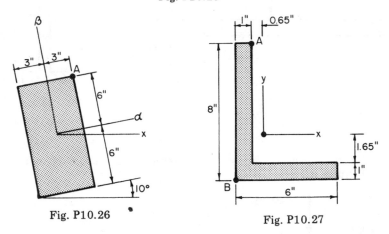

Fig. P10.26   Fig. P10.27

was tilted $10°$ as shown. Determine the percent increase in the stress at point A resulting from this erection error.

10.27. The $6'' \times 8'' \times 1''$ angle in Fig. P10.27 is used as a cantilever beam to support a large sign. The weight of the sign produces a moment $M_x = 5000$ ft-lb. at the cross section where the beam joins the wall. Determine the flexural stresses at points A and B of the beam at this critical section. For the cross section, use $I_x = 81$ in.^{4}., $I_y = 39$ in.^{4}., $I_{xy} = -32$ in.^{4}.

Ans. $\sigma_A = 5275$ psi, $\sigma_B = -4385$ psi

10.28. A simply supported beam has the Z cross section in Fig. P10.28. Loads on the beam produce a maximum moment $M_x = 12{,}500$ ft-lb. (a) Locate the neutral axis and (b) determine the maximum flexural stress and indicate the location on the cross section where it occurs. For the Z cross section use $I_x = 117$ in.^{4}, $I_y = 16$ in.^{4}, $I_{xy} = 9$ in.^{4}.

10.29. Find the maximum bending moment that can be carried by a structural member having the triangular cross section in Fig. P10.29. The moment is applied about the x axis to produce a tensile stress at point A. Stresses are to be limited to 10,000 psi compression and 8000 psi tension.

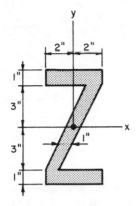

Fig. P10.28

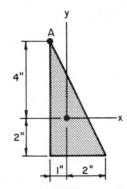

Fig. P10.29

Ans. $M_x = 36,100$ in.-lb

10.30. The timber cantilever beam in Fig. P10.30 carries a concentrated load P at its free end. Determine the maximum allowable load P if the maximum tensile stress is limited to 1000 psi and the maximum horizontal shearing stress is limited to 100 psi.

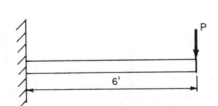
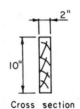

Fig. P10.30

10.31. A simply supported beam is loaded as in Fig. P10.31. Determine (a) the maximum axial tensile (flexure) stress and (b) the maximum vertical shearing stress.

Ans. $\sigma_{max} = 48,600$ psi, $\tau_{max} = 650$ psi

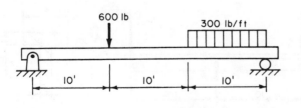

Fig. P10.31

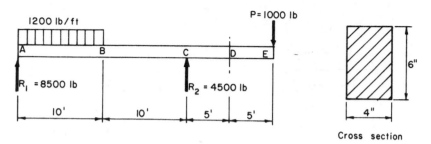

Fig. P.10.32

10.32. For the beam in Fig. P10.32 determine (a) the maximum axial tensile (flexure) stress and (b) the normal and shear stresses at a point on section D 1 in. from the top of the beam.

10.33. The beam in Fig. P10.33a has the cross section shown in Fig. P10.33b. For a point 7 ft from the left end and 2 in. below the top of the beam determine (a) the axial normal (flexure) stress, (b) the vertical shearing stress, and (c) the maximum shearing stress.

  Ans. σ_x = 643 psi (tension), τ_{xy} = 616 psi, τ_{max} = 694 psi

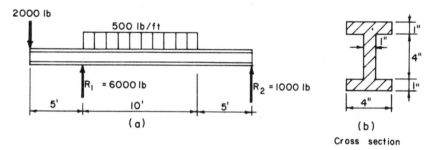

Fig. P10.33

10.34. For the beam in Fig. P10.34 determine (a) the maximum axial tensile (flexure) stress, (b) the maximum axial compressive (flexure) stress, (c) the maximum vertical shearing stress, and (d) the locations where they occur.

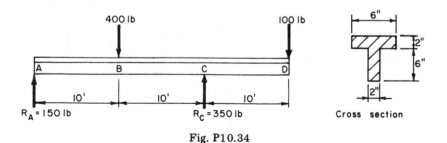

Fig. P10.34

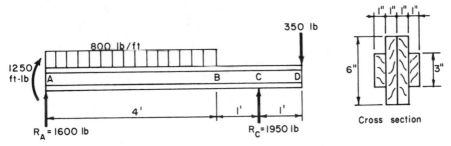

Fig. P10.35

10.35. For the built-up timber beam in Fig. P10.35 determine the maximum axial tensile (flexure) stress and the maximum horizontal shearing stress and indicate where they occur. Neglect stress concentrations.

Ans. σ = 2530 psi (ten) bottom of beam 2 ft from A
     τ = 134 psi at A and between B and C at change in section

10.36. For the beam in Fig. P10.36 determine (a) the equation of the deflection curve between A and B and (b) the angle of rotation of the beam at B.

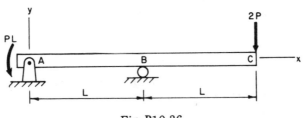

Fig. P10.36

10.37. Determine the deflection of point C in the beam in Fig. P10.37.

Ans. $y_c = wL^4/24EI$

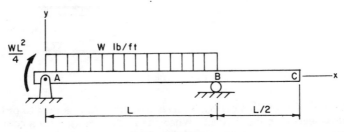

Fig. P10.37

10.38. For the beam in Fig. P10.38 determine (a) the equation of the deflection curve, (b) the magnitude and location of the maximum upward deflection, and (c) the magnitude and location of the maximum downward deflection.

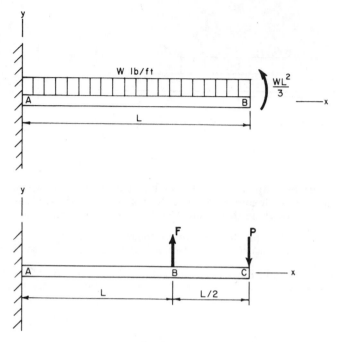

Fig. P10.39

10.39. For the beam in Fig. P10.39 determine the force F that must be applied at point B to produce zero deflection.

Ans. $F = 7P/4$

10.40. Determine the maximum axial compressive load that can be applied to the 1-in. diameter steel rod in Fig. P10.40.

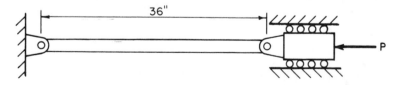

Fig. P10.40

10.41. Two $8'' \times 4'' \times 1''$ steel angles have been riveted together to produce the cross section in Fig. P10.41. When the angles are used for a column 20 ft long, fixed at the bottom and free to deflect at the top, determine the maximum compressive load that may be supported if a factor of safety of 2 with respect to buckling is to be maintained. For the cross section, use $A = 22$ in.^{2}, $I_x = 139$ in.^{4}, and $I_y = 48$ in.^{4}.

Ans. $P = 30,850$ lb

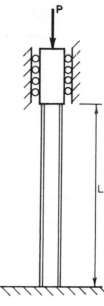

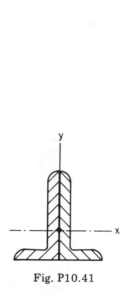

Fig. P10.41   Fig. P10.42

10.42. Determine the Euler load for the fixed-end beam in Fig. P10.42.

10.43. The linkage in Fig. P10.43 consists of two steel bars each with a 2-in. diameter. Determine the maximum load P that may be applied at pin B before failure by buckling or yielding occurs. $E = 30 \times 10^6$ psi, $\sigma_{yield} = 40{,}000$ psi.

   Ans. $P = 75{,}600$ lb

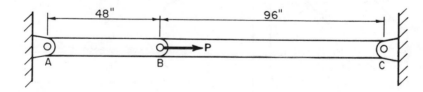

Fig. P10.43

10.44. A 4-in. diameter solid steel shaft is subjected to an axial load of $24{,}000\pi$ lb and a torque of $16{,}000\pi$ in.-lb (Fig. P10.44). Determine (a) the maximum normal stress in the shaft and (b) the maximum shear stress in the shaft and (c) locate the planes on which these stresses act.

10.45. An aluminum beam with a rectangular cross section carries a 30,000-lb load (Fig. P10.45). Determine the maximum normal and maximum shear stresses at point A on the cross section 10 in. from the wall.

   Ans. $\sigma_1 = 13{,}300$ psi, $\tau_{max} = 6800$ psi

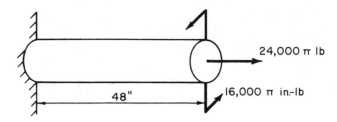

Fig. P10.44

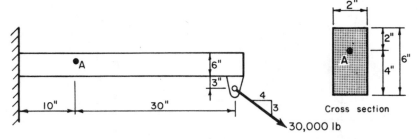

Fig. P10.45

10.46. For the steel block $6'' \times 8'' \times 30''$ long subjected to a 9600-lb load (Fig. P10.46) determine the maximum tensile and maximum compressive stresses that occur on plane *ABCD*.

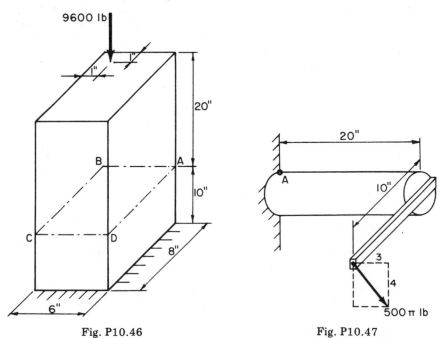

Fig. P10.46   Fig. P10.47

10.47. A 4-in. diameter solid steel shaft supports a load of 500π lb applied through a bar (Fig. P10.47). Determine the principal stresses and the maximum shear stress at point A of the cross section where the shaft joins the wall.

Ans. $\sigma_1 = 4308$ psi, $\sigma_2 = 0$, $\sigma_3 = -232$ psi, $\tau_{max} = 2270$ psi

chapter 11

Fluid Dynamics

11.1 Introduction

The subject of fluid dynamics is concerned with the motion of fluids, both liquids and gases. This field of mechanics obviously encompasses a vast collection of problems that may vary from a study of blood flow in the tiny capillaries of the circulatory system to a study of high-altitude flight of aircraft. One of the distinguishing features of a fluid as opposed to a solid is its great deformability; as a consequence very complex flow patterns are commonly encountered. Fluids in motion present a much more complicated situation than fluids at rest (Chapter 4) since we must be concerned with the kinematics of the flow field as well as the stress distribution throughout.

Fluid dynamics problems are broadly grouped into two categories: (1) incompressible flow (constant fluid density) and (2) compressible flow (variable fluid density). For most problems, liquids can be treated as incompressible fluids and, in general, gases must be considered as compressible. For low-speed gas flows, however, the compressibility of the gas can often be neglected; e.g., for air flows of less than approximately 300 fps the effect of compressibility is negligible (air at room temperature).

Flows in which the velocity at any point does not vary with time are called *steady flows*. If the velocity is time dependent, the flow is said to be *unsteady*. A *steady-state* flow is one for which none of the flow characteristics, such as the fluid density, are time dependent. The simplest problems usually involve steady, incompressible flows.

Shearing stresses are developed in moving real fluids; and to obtain a detailed description of the flow field, the relationship between the movement (deformation) and the stresses must be specified (i.e., the constitutive equation of the fluid must be known). Many common fluids such as air and water exhibit a linear relationship between stress and rate of deformation and are called Newtonian fluids. However, we will find that the analysis of problems involving Newtonian fluids is still very difficult and additional assumptions are frequently required. One such assumption is to treat the fluid as if it had no viscosity; i.e., even though the fluid is moving, no shearing stresses are developed. Such fluids are designated as *nonviscous*. *Perfect* or *ideal* fluids are incompressible and nonviscous and represent the simplest type of fluid that can be considered. Since real fluids are viscous, any analysis based on the assumption of zero viscosity is approximate. In certain types of problems this assumption may lead to useful results, whereas for others the results are completely in error. Thus, the concept of the nonviscous fluid must be applied with caution, and this point will be discussed more fully in Section 11.6. There is also a class of problems involving fluids that can be considered neither nonviscous nor Newtonian. We broadly classify such fluids as *non-Newtonian*.

Although for many problems the constitutive equations for the fluid must be specified, certain fundamental relationships hold for all materials. Three such basic relationships that are particularly useful in the study of fluid dynamics are

1. *Conservation of mass.*
2. *Newton's second law (momentum equation).*
3. *Conservation of energy.*

In the following sections these relationships will be developed in forms that are particularly convenient for the study of the flow of fluids; however, we will first consider a general method of describing the characteristics of a given mass of fluid.

11.2 Systems and Control Volumes

In the study of mechanics problems two viewpoints may be taken. We can focus attention on a particular particle or collection of matter and study the behavior of this given mass of material. Such a fixed quantity of matter is called a *system*. The system may change its shape and size, but by definition it always contains the same mass of material. For example, the gas contained in the closed cylinder in Fig. 11.1a may be considered a system. As the position of the piston changes, the properties of the gas (such as density and pressure) may change, but the same mass of material is always contained in the cylinder. The system viewpoint is particularly useful in rigid-body dynamics since for this case it is relatively simple to focus attention on a particular collection of matter.

In the study of moving fluids the system viewpoint is not usually

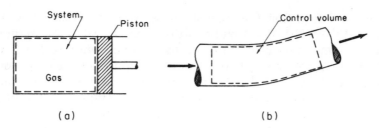

Fig. 11.1

convenient since it is difficult to label and follow a particular collection of fluid particles as they flow through some closed conduit or around some object immersed in the fluid. Thus for fluid dynamics we commonly focus our attention on a fixed volume (fixed relative to some coordinate system) called a *control volume* and analyze the fluid contained within the control volume at a particular instant in time. The boundary of the control volume is called the *control surface*. The identity and amount of material within the control volume may change with time, but the shape of the control volume is fixed. A typical example of a control volume is shown in Fig. 11.1b in which a fluid is flowing through a bend in a pipe. The system approach that focuses attention on particular particles of mass is sometimes referred to as the *Lagrangian* viewpoint, whereas the control volume approach in which we focus attention on particular points or volumes in space is called the *Eulerian* viewpoint. The Eulerian viewpoint has been discussed in Section 5.10 in connection with the kinematic description of fluid motion.

We will now show how the rate of change of certain fluid or flow characteristics for a system can be evaluated from a control volume viewpoint. Let B be some quantity associated with a system such that

$$B = \int_{\text{system}} \rho \beta(\mathbf{r}, t)\, dV \qquad (11.1)$$

where ρ is the density of the material, dV is a differential volume, and β represents the intensity of the quantity B per unit mass. Both the quantity β and density ρ will generally be functions of the position vector $\mathbf{r}$ and time t as measured relative to some coordinate system. For example, B may correspond to the momentum $\bar{\mathbf{P}}$ of the system so that

$$\bar{\mathbf{P}} = \int_{\text{system}} \rho \mathbf{q}\, dV$$

and in this case $\beta = \mathbf{q}$ the fluid velocity. We now wish to consider the

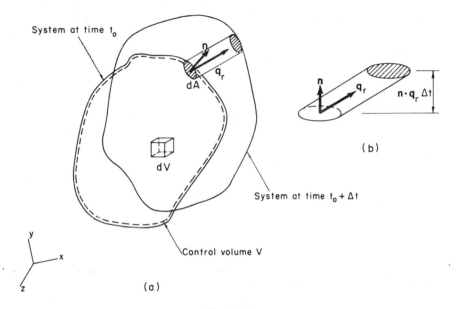

Fig. 11.2

rate of change of B for the system. As the system moves from one position to another in a time interval Δt, it may change shape and orientation and both β and ρ may change so that the quantity B will in general vary with time (Fig. 11.2a). We can express this time variation as

$$\left(\frac{dB}{dt}\right)_{\text{system}} = \frac{DB}{Dt} = \lim_{\Delta t \to 0} \left(\frac{\displaystyle\int_{\text{system @ } t_0 + \Delta t} \rho\beta\,dV \;-\; \displaystyle\int_{\text{system @ } t_0} \rho\beta\,dV}{\Delta t} \right)$$

(11.2)

where the special notation D/Dt will be used to emphasize the fact that we are considering the rate of change of B following the system. This derivative is sometimes called the *material derivative*.

At a time t_0 let the system coincide with a control volume V that is fixed relative to the xyz coordinate system in Fig. 11.2a. This coordinate system may be translating and rotating. In the time interval Δt the system may be displaced relative to the control volume and the system volume may change by an amount ΔV. Equation (11.2) can thus be written as

$$\frac{DB}{Dt} = \lim_{\Delta t \to 0} \left[\frac{\displaystyle\int_{V + \Delta V} \rho\beta(\mathbf{r}, t + \Delta t)\,dV - \int_{V} \rho\beta(\mathbf{r}, t)\,dV}{\Delta t} \right]$$

Also we can write

$$\int_{V+\Delta V} \rho\beta(\mathbf{r}, t + \Delta t)\, dV = \int_V \rho\beta(\mathbf{r}, t + \Delta t)\, dV + \int_{\Delta V} \rho\beta(\mathbf{r}, t + \Delta t)\, dV$$

so that

$$\frac{DB}{Dt} = \lim_{\Delta t \to 0} \left[\frac{\int_V \rho\beta(\mathbf{r}, t + \Delta t)\, dV - \int_V \rho\beta(\mathbf{r}, t)\, dV + \int_{\Delta V} \rho\beta(\mathbf{r}, t + \Delta t)\, dV}{\Delta t} \right]$$

$$\text{(a)}$$

Furthermore,

$$\rho\beta(\mathbf{r}, t + \Delta t) = \rho\beta(\mathbf{r}, t) + \frac{\partial\rho\beta(\mathbf{r}, t)}{\partial t} \Delta t \qquad \text{(b)}$$

where the higher order terms involving $(\Delta t)^2$, $(\Delta t)^3$, etc., are neglected. Equation (b) substituted into Eq. (a) yields

$$\frac{DB}{Dt} = \int_V \frac{\partial\rho\beta(\mathbf{r}, t)}{\partial t}\, dV + \lim_{\Delta t \to 0} \left[\frac{\int_{\Delta V} \rho\beta(\mathbf{r}, t)\, dV}{\Delta t} \right] + \lim_{\Delta t \to 0} \left[\int_{\Delta V} \frac{\partial\rho\beta(\mathbf{r}, t)}{\partial t}\, dV \right]$$

$$\text{(c)}$$

The last term in Eq. (c) vanishes, however, since $\Delta V \to 0$ as $\Delta t \to 0$.

The second term on the right-hand side of Eq. (c) can be evaluated in the following manner. The volume ΔV is the volume swept out by the fluid particles lying on the surface of the control volume in the time Δt. Thus a suitable differential volume dV for this integral is given by the expression (see Fig. 11.2b) $dV = \mathbf{n} \cdot \mathbf{q}_r \Delta t\, dA$, where $\mathbf{q}_r$ is the velocity of a fluid particle lying on the control surface measured relative to the control surface. The unit vector $\mathbf{n}$ is normal to the control surface (directed outward from the surface), and dA is a differential area on the control surface. Thus we can write

$$\lim_{\Delta t \to 0} \left[\frac{\int_{\Delta V} \rho\beta(\mathbf{r}, t)\, dV}{\Delta t} \right] = \int_A \rho\beta(\mathbf{r}, t)\mathbf{n} \cdot \mathbf{q}_r\, dA$$

where the surface integral is taken over the control surface A. The time rate of change of B can therefore be written as

$$\frac{DB}{Dt} = \int_V \frac{\partial\rho\beta}{\partial t}\, dV + \int_A \rho\beta\, \mathbf{n} \cdot \mathbf{q}_r\, dA \qquad (11.3)$$

where V and A refer to the control volume and control surface respectively. Since the control volume is independent of time, Eq. (11.3) can also be written in the form

$$\frac{DB}{Dt} = \frac{\partial}{\partial t} \int_V \rho\beta \, dV + \int_A \rho\beta \, \mathbf{n} \cdot \mathbf{q}_r \, dA \qquad (11.4)$$

The first integral represents the rate of change of B within the control volume, and the second integral represents the rate at which the quantity B is flowing across the boundary of the control surface.

11.3 Conservation of Mass

The mass M of a finite volume of fluid such as in Fig. 11.2a can be expressed in terms of the fluid density as

$$M = \int_{\text{system}} \rho \, dV$$

and since the mass is invariant for a system,

$$\frac{DM}{Dt} = 0$$

Thus for a control volume we can write

$$\frac{DM}{Dt} = \frac{\partial}{\partial t} \int_V \rho \, dV + \int_A \rho \mathbf{n} \cdot \mathbf{q}_r \, dA = 0 \qquad (11.5)$$

where we have made use of Eq. (11.4) with $\beta = 1$. Equation (11.5) can be written as

$$-\frac{\partial}{\partial t} \int_V \rho \, dV = \int_A \rho \mathbf{n} \cdot \mathbf{q}_r \, dA \qquad (11.6)$$

The left side of this equation represents the rate at which the mass within the control volume is decreasing, and the right side gives the rate at which mass is flowing out through the control surface. Thus Eq. (11.6) states that for a volume fixed relative to some reference frame the rate of decrease of mass within the volume is equal to the rate of flow of mass through the surface of the volume. This is a statement of the *conservation of mass*. Equation (11.6) is also commonly referred to as the *continuity equation*.

The continuity equation can be reduced to a simple and useful form for flow through closed conduits of the type illustrated in Fig. 11.3. We assume that a fluid is flowing into the rigid-walled conduit at section 1 and at this section the velocity and density may

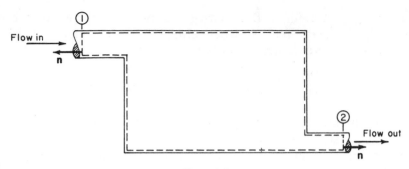

Fig. 11.3

vary over the cross section. The cross-sectional areas at the entrance and exits are A_1 and A_2 respectively. At the exit the fluid velocity and density may also vary. We restrict our attention to the case in which the density does not vary with time. Thus the time derivative on the left side of Eq. (11.6) is zero. A cross section of the fixed control volume to be considered is outlined by the dashed line in Fig. 11.3. From Eq. (11.6) it follows that

$$- \int_{A_1} \rho \mathbf{n} \cdot \mathbf{q} \, dA = \int_{A_2} \rho \mathbf{n} \cdot \mathbf{q} \, dA \qquad (11.7)$$

Since the control volume is fixed in space, $\mathbf{q}_r = \mathbf{q}$ where $\mathbf{q}$ is the fluid velocity measured relative to a fixed coordinate system. There is no contribution to the surface integral along the boundary walls since $\mathbf{n} \cdot \mathbf{q} = 0$ along any fixed, solid surface. Note that at section 1, $\mathbf{n}$ and $\mathbf{q}$ at all points on this cross section are in opposite directions so that at any point $(\mathbf{n} \cdot \mathbf{q})_1 = -q_1$. The integral

$$\int_A \rho \mathbf{n} \cdot \mathbf{q} \, dA = \dot{m}$$

represents the rate at which mass flows through the surface A. Thus for flow through a closed conduit we have

$$(\dot{m})_1 = (\dot{m})_2 \qquad (11.8)$$

where sections 1 and 2 can be any two cross sections. For an incompressible flow Eq. (11.7) reduces to

$$- \int_{A_1} \mathbf{n} \cdot \mathbf{q} \, dA = \int_{A_2} \mathbf{n} \cdot \mathbf{q} \, dA = Q \qquad (11.9)$$

where these integrals represent the volume rate of flow passing through the surface over which the integral is evaluated. The *average velocity* q_a over the surface is defined as $q_a = Q/A$. Thus Eq. (11.9) can be written in the useful form

$$(q_a)_1 A_1 = (q_a)_2 A_2 \qquad (11.10)$$

This equation is valid for both steady and unsteady incompressible flow through a closed conduit.

EXAMPLE 11.1

An incompressible fluid flows through the Y-section shown in Fig. E11.1. Determine the ratio of the mean velocity of the fluid at section 1 to the velocity at section 2. Assume that the flow is symmetrically split at the branch.

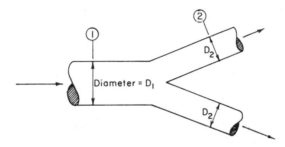

Fig. E11.1

SOLUTION

From Eq. (11.10) we can write

$$(\pi D_1^2/4)(q_a)_1 = (2)(\pi D_2^2/4)(q_a)_2$$

where $(q_a)_1$ and $(q_a)_2$ are the mean velocities preceding and following the branch respectively. Thus

$$(q_a)_1/(q_a)_2 = 2(D_2/D_1)^2 \qquad\qquad Ans.$$

The conservation of mass principle can be applied to a differential control volume (Fig. 11.4). We assume that the rectangular components of velocity at the center of the elements are u, v, and w and the fluid density is ρ. Thus the velocity and the density of fluid particles passing through the faces of the elements can be expressed in terms of the velocity and density at the center by means of a Taylor series expansion (Fig. 11.4). The rates of mass inflow and outflow through the two surfaces perpendicular to the x axis are shown on the figure. The net rate of outflow in the x direction is given by the relationship

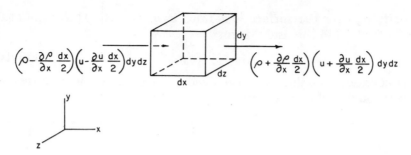

$$\left(\rho - \frac{\partial \rho}{\partial x}\frac{dx}{2}\right)\left(u - \frac{\partial u}{\partial x}\frac{dx}{2}\right)dy\,dz \qquad\qquad \left(\rho + \frac{\partial \rho}{\partial x}\frac{dx}{2}\right)\left(u + \frac{\partial u}{\partial x}\frac{dx}{2}\right)dy\,dz$$

Fig. 11.4

Net rate of outflow
in x direction
$$= \left[\left(\rho + \frac{\partial \rho}{\partial x}\frac{dx}{2}\right)\left(u + \frac{\partial u}{\partial x}\frac{dx}{2}\right)\right.$$
$$\left. - \left(\rho - \frac{\partial \rho}{\partial x}\frac{dx}{2}\right)\left(u - \frac{\partial u}{\partial x}\frac{dx}{2}\right)\right] dy\,dz$$
$$= \left[\rho\,\frac{\partial u}{\partial x} + u\,\frac{\partial \rho}{\partial x}\right] dx\,dy\,dz$$

Similar equations written for the y and z directions give the expression for the net rate of outflow; i.e.,

Net rate of
mass outflow
$$= \left[\rho\,\frac{\partial u}{\partial x} + u\,\frac{\partial \rho}{\partial x} + \rho\,\frac{\partial v}{\partial y} + v\,\frac{\partial \rho}{\partial y} + \rho\,\frac{\partial w}{\partial z} + w\,\frac{\partial \rho}{\partial z}\right] dx\,dy\,dz$$
$$= \left[\frac{\partial(\rho u)}{\partial x} + \frac{\partial(\rho v)}{\partial y} + \frac{\partial(\rho w)}{\partial z}\right] dx\,dy\,dz$$
$$= \left[\,\nabla \cdot (\rho \mathbf{q})\right] dx\,dy\,dz$$

From the conservation of mass principle it follows that the net rate of mass outflow must equal the rate of decrease of mass within the volume; i.e.,

$$[\,\nabla \cdot (\rho \mathbf{q})]\,dx\,dy\,dz = -\left[\frac{\partial \rho}{\partial t}\right] dx\,dy\,dz$$

or

$$\nabla \cdot (\rho \mathbf{q}) = -\frac{\partial \rho}{\partial t} \tag{11.11}$$

Equation (11.11) is the differential form of the continuity equation that applies at a point. For incompressible fluids

$$\nabla \cdot \mathbf{q} = 0 \tag{11.12}$$

or

$$\frac{\partial u}{\partial x} + \frac{\partial v}{\partial y} + \frac{\partial w}{\partial z} = 0 \qquad (11.13)$$

Integration of Eq. (11.11) over a fixed finite volume V gives

$$-\frac{\partial}{\partial t} \int_V \rho \, dV = \int_V \nabla \cdot (\rho \mathbf{q}) \, dV \qquad (11.14)$$

A comparison of Eqs. (11.6) and (11.14) reveals that

$$\int_A \rho \mathbf{n} \cdot \mathbf{q} \, dA = \int_V \nabla \cdot (\rho \mathbf{q}) \, dV \qquad (11.15)$$

Equation (11.15), which relates a surface integral and a volume integral, is a special application of the *divergence theorem* that states

$$\int_A \mathbf{n} \cdot \mathbf{T} \, dA = \int_V \nabla \cdot \mathbf{T} \, dV \qquad (11.16)$$

where $\mathbf{T}$ is a vector, A the surface bounding some volume V, and $\mathbf{n}$ the unit outward normal of the surface.

The continuity equation, either for a finite volume or in differential form, is one of the basic equations of fluid mechanics. All physically possible flow fields regardless of the type of fluid must satisfy the continuity equation.

EXAMPLE 11.2

The velocity components in an incompressible flow field are given by the equations

$$u = x^2 - y^2, \qquad v = -2xy + f(y), \qquad w = 0$$

Determine the permissible form for the function $f(y)$.

SOLUTION

Since all incompressible flow fields must satisfy the continuity equation $\nabla \cdot \mathbf{q} = 0$, we can write

$$\frac{\partial u}{\partial x} + \frac{\partial v}{\partial y} = 0$$

For this example

$$\frac{\partial u}{\partial x} = 2x, \qquad \frac{\partial v}{\partial y} = -2x + \frac{\partial f}{\partial y}$$

so that

$$2x - 2x + \frac{\partial f}{\partial y} = 0$$

Thus

$$\frac{\partial f}{\partial y} = 0$$

therefore

$$f(y) = \text{constant} \qquad\qquad Ans.$$

The specific value of the constant could be determined if the velocity component v were specified at some point.

11.4 Linear Momentum Equation

In Chapter 7 we have shown that the resultant force acting on a collection of particles is related to the motion of the particles through the equation

$$\mathbf{F} = \frac{d}{dt} \sum_{i=1}^{n} \mathbf{q}_i m_i \qquad\qquad (11.17)$$

where m_i is the mass and $\mathbf{q}_i$ the velocity of each particle. If we consider a fluid mass as consisting of a collection of particles (Fig. 11.5), each having a mass dm, Eq. (11.17) can be written in the form

$$\mathbf{F} = \frac{D}{Dt} \int_{\text{system}} \mathbf{q}\, dm = \frac{D}{Dt} \int_{\text{system}} \rho \mathbf{q}\, dV \qquad\qquad (11.18)$$

The special notation D/Dt has been employed to emphasize the fact that we are considering the time rate of change of the integral for a given mass. For Eq. (11.18) to be valid, the velocity must be measured relative to an inertial reference frame, i.e., one that is fixed relative to the earth or moving with a constant velocity. The force $\mathbf{F}$ is the resultant force acting on the system of particles and arises from the surface forces acting on the surface of the system and the body forces distributed throughout. The force arising from the distribution of pressure over the surface is an example of a surface force, whereas the weight of the system is an example of a body force.

We now consider a control volume that coincides with the system at $t = t_0$ and is fixed relative to the inertial reference frame. From

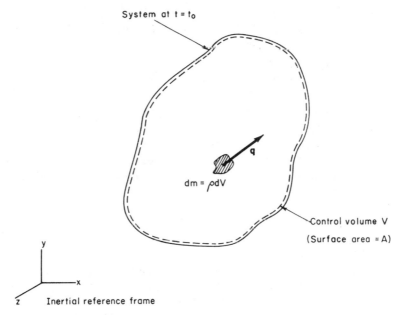

Fig. 11.5

Eq. (11.4) we can write

$$\mathbf{F} = \frac{D}{Dt} \int_{\text{system}} \rho\mathbf{q} \, dV = \frac{\partial}{\partial t} \int_{V} \rho\mathbf{q} \, dV + \int_{A} \rho\mathbf{q} \, \mathbf{n} \cdot \mathbf{q} \, dA \quad (11.19)$$

Since the quantity $\rho\mathbf{q}$ represents the linear momentum per unit volume of a differential element of fluid, the last term in Eq. (11.19) represents the rate at which momentum is flowing through the control surface. The term involving the volume integral represents the rate at which the momentum within the control volume is changing with respect to time. Thus we can state that the resultant force acting on the fluid particles contained within a control volume is equal to the rate of increase of the linear momentum within the control volume plus the rate of efflux of momentum through the control surface.

Equation (11.19) is called the *linear momentum equation* and in terms of a rectangular coordinate system can be written as three scalar equations:

$$F_x = \frac{\partial}{\partial t} \int_{V} \rho u \, dV + \int_{A} \rho u \, \mathbf{n} \cdot \mathbf{q} \, dA$$

$$F_y = \frac{\partial}{\partial t} \int_V \rho v \, dV + \int_A \rho v \, \mathbf{n} \cdot \mathbf{q} \, dA \qquad (11.20)$$

$$F_z = \frac{\partial}{\partial t} \int_V \rho w \, dV + \int_A \rho w \, \mathbf{n} \cdot \mathbf{q} \, dA$$

In the derivation of the momentum equation no restrictions were placed on the type of fluid under consideration. The momentum equation is particularly useful for problems in which $\rho \mathbf{q}$ does not depend on time. In this case the volume integral in Eq. (11.19) is zero, and only velocities at the surface of the control volume need be considered.

EXAMPLE 11.3

An incompressible fluid flows steadily through the converging circular tube in Fig. E11.3a. The pressures at sections 1 and 2 are p_1 and p_2 respectively and may be assumed to be constant over the cross section. Determine an expression for the resultant force the fluid exerts on the tube for the following two cases:

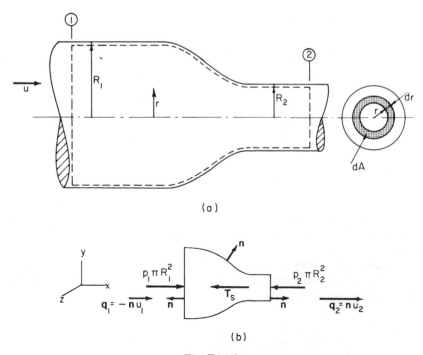

(a)

(b)

Fig. E11.3

(a) the velocity is constant over the cross section and (b) the velocity distribution is parabolic; i.e.,

$$u = 2U[1 - (r/R_i)^2]$$

where U = mean velocity
      r = radius from tube center line
      R_i = tube radius

SOLUTION

Since we are looking for a force acting on a surface, it is apparent that the momentum equation can be used. We select as the control volume the dashed section in Fig. E11.3a. The forces acting on this volume are shown in Fig. E11.3b. On the two ends of the control volume the pressure forces are acting, and the resultant force exerted on the fluid by the wall is T_S. Thus the force the fluid exerts on the tube is equal and opposite to the force shown in Fig. E11.3b. The x component of the momentum equation for steady incompressible flow is

$$F_x = \rho \int_A u \, \mathbf{n} \cdot \mathbf{q} \, dA$$

When using the scalar components of the momentum equation, a sign convention must be established. We will take the positive x direction as our positive direction. Thus

$$p_1 \pi R_1^2 - p_2 \pi R_2^2 - T_S = \rho \int_A u \, \mathbf{n} \cdot \mathbf{q} \, dA$$

The surface integral can be considered in two parts: the contribution over the curved surface and the contribution over the two ends. However, the contribution over the curved portion of the surface is zero since $\mathbf{n} \cdot \mathbf{q}$ is everywhere zero on this surface. Thus

$$\int_A u \, \mathbf{n} \cdot \mathbf{q} \, dA = \int_{A_1} u_1 \mathbf{n} \cdot \mathbf{q}_1 \, dA + \int_{A_2} u_2 \mathbf{n} \cdot \mathbf{q}_2 \, dA$$

$$= -2\pi \int_0^{R_1} u_1^2 r \, dr + 2\pi \int_0^{R_2} u_2^2 r \, dr$$

where $dA = 2\pi r \, dr$ as in Fig. E11.3a.

(a) If u_1 and u_2 are constant over the cross section, then

$$\int_A u \, \mathbf{n} \cdot \mathbf{q} \, dA = \pi(u_2^2 R_2^2 - u_1^2 R_1^2)$$

From the continuity equation we have

$$\pi R_2^2 u_2 = \pi R_1^2 u_1 = Q$$

where Q is the discharge. Thus we can write

$$p_1 \pi R_1^2 - p_2 \pi R_2^2 - T_S = \rho Q(u_2 - u_1)$$

and

$$\mathbf{T}_S = p_1 \pi R_1^2 - p_2 \pi R_2^2 - \rho Q(u_2 - u_1) \rightarrow \qquad\qquad Ans.$$

Recall that the force $\mathbf{T}_S$ the fluid exerts on the tube is in a direction opposite to that shown in Fig. E11.3b.

(b) If the velocity varies over the cross section, the momentum integral becomes

$$2\pi \int_0^R u^2 r \, dr = 2\pi \int_0^R (2U)^2 \left[1 - \left(\frac{r}{R} \right)^2 \right]^2 r \, dr = \frac{4}{3} \pi R^2 U^2$$

and thus

$$\int_A u \, \mathbf{n} \cdot \mathbf{q} \, dA = \frac{4}{3} [(\pi R_2^2 U_2)U_2 - (\pi R_1^2 U_1)U_1] = \frac{4}{3} Q(U_2 - U_1)$$

since $Q = \pi R_1^2 U_1 = \pi R_2^2 U_2$. The expression for $\mathbf{T}_S$ becomes

$$\mathbf{T}_S = p_1 \pi R_1^2 - p_2 \pi R_2^2 - (4/3)\rho Q(U_2 - U_1) \rightarrow \qquad\qquad Ans.$$

We must specify p_1, p_2, and Q to evaluate $\mathbf{T}_S$.

EXAMPLE 11.4

A jet of water ($\rho = 10^3$ kg/m^3) having an area of 10^{-3} m^3 and moving with a steady velocity (uniform over cross section) of 30 m/s is deflected through a $30°$ angle by a moving vane (Fig. E11.4). Determine the horizontal component

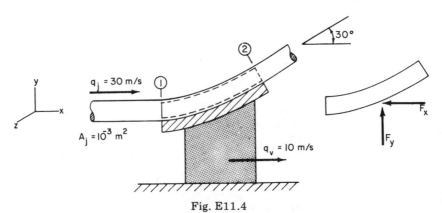

Fig. E11.4

of the force the jet exerts on the vane. The velocity of the vane is constant and equal to 10 m/s as shown, and friction and gravity can be neglected.

SOLUTION

In this example it is convenient to select a control volume that is fixed relative to a coordinate system attached to the vane. Since this coordinate system is moving with a constant velocity, it can also be considered as an inertial frame; therefore Eq. 11.19 is applicable with the velocities measured relative to the vane.

Since the flow is steady, we can write

$$-F_x = \int_A \rho(u_j - u_v) \, \mathbf{n} \cdot (\mathbf{q}_j - \mathbf{q}_v) \, dA$$

where F_x is the force the vane exerts on the jet (Fig. E11.4). Since momentum flows only through the control surface at the end sections 1 and 2, we can write

$$-F_x = -\int_{A_1} \rho(u_j - u_v)_1^2 \, dA + \int_{A_2} \rho(u_j - u_v)_2 \, \mathbf{n} \cdot (\mathbf{q}_j - \mathbf{q}_v)_2 \, dA$$

The magnitude of the velocity of jet relative to vane does not change along the vane, so that

$$(u_j - u_v)_2 = (u_j - u_v)_1 \cos \theta$$

Thus

$$-F_x = -\rho(u_j - u_v)_1^2 A_1 + \rho(u_j - u_v)_1^2 (\cos \theta) A_2$$

From the continuity equation applied to the moving control volume, we know that $A_1 = A_2 = A_j$ since the velocity does not change along the vane. Also $(u_j)_1 = q_j$ and $(u_v)_1 = q_v$; therefore

$$F_x = \rho(q_j - q_v)^2 A_j (1 - \cos \theta)$$
$$= 10^3 (20)^2 \, 10^{-3} (1 - (0.866)) = 53.6 \text{ N}$$

Thus the horizontal component of the force on the vane is

$$(\mathbf{F}_x)_{\text{vane}} = 53.6 \text{ N} \rightarrow \qquad\qquad Ans.$$

11.5 Energy Equation

Another fundamental equation of fluid mechanics can be developed from the *first law of thermodynamics*. This law states that the rate at which heat $\dot{Q}_h$ is added to a system minus the rate at which the system does work $\dot{W}$ on its surroundings is equal to the rate of increase of the energy $\dot{E}$ of the system; i.e.,

$$\dot{Q}_h - \dot{W} = \dot{E} \qquad\qquad (11.21)$$

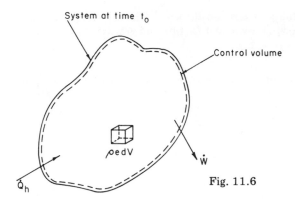

Fig. 11.6

At a time t_0 we let the system under consideration occupy the control volume shown in Fig. 11.6. The total energy of the system at this instant can be expressed in terms of the "stored" energy per unit mass e integrated over the volume. Thus

$$E = \int_V \rho e \, dV$$

and the time rate of change of this integral can be expressed in terms of a fixed control volume as

$$\frac{DE}{Dt} = \int_A \rho e \mathbf{n} \cdot \mathbf{q} \, dA + \frac{\partial}{\partial t} \int_V \rho e \, dV$$

We know from elementary physics that the energy of a particle consists of energy due to its motion (kinetic energy), position (potential energy), and molecular state (internal energy). The kinetic energy per unit mass is equal to $q^2/2$, and the potential energy per unit mass due to gravitational effects is gy where y is a vertical coordinate measured from some arbitrary datum. The internal energy per unit mass will be denoted as $\hat{u}$ and in the absence of chemical, electrical, and magnetic effects is due to random atomic motion and is a function of temperature for most common fluids. Thus we can express e in the form $e = (q^2/2) + gy + \hat{u}$.

It is convenient to consider in two parts the rate at which the system is doing work on its surroundings.

1. Shaft work $\dot{W}_s$ is transferred to or from the system through some mechanical device such as the shaft of a pump or turbine. Work transferred from the system via a turbine would be positive, whereas work done on the system via a pump would be negative.
2. Work due to surface forces $\dot{W}_f$ is done by the forces developed on the surface of the system. As illustrated in Fig. 11.7, at the surface of the system (which coincides with the surface of the con-

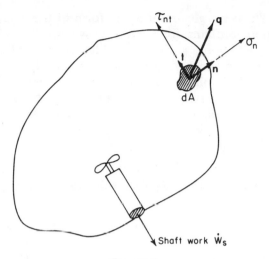

Fig. 11.7

trol volume at the instant we are considering) the surface forces
are due to a normal stress σ_n in the n direction (normal to surface) and a shearing stress τ_{nt} in the t direction (tangent to surface). Since the rate at which a given force **F** moving with a
velocity **q** does work is **F** · **q**, the rate at which these stresses are
doing work on the surroundings is

$$\dot{W}_f = -\int_A \sigma_n \, \mathbf{n} \cdot \mathbf{q} \, dA - \int_A \tau_{nt} \, \mathbf{t} \cdot \mathbf{q} \, dA$$

where **n** and **t** are unit vectors in the n and t directions respectively. Here we have presumed that the stresses in Fig. 11.7 are
due to forces exerted by the surroundings on the system so that
the work done by the system is equal to but opposite in sign from
the rate at which work is done by the surface stresses shown. Unless **q** has a component in either the n or t direction (i.e., unless
there is flow parallel or perpendicular to the control surface), $\dot{W}_f$
will be zero. Work due to flow across the control surface is sometimes called *flow work*.

The foregoing relationships for energy and work can be substituted into Eq. (11.21) to give

$$\dot{Q}_h - \dot{W}_s = -\int_A \sigma_n \, \mathbf{n} \cdot \mathbf{q} \, dA - \int_A \tau_{nt} \, \mathbf{t} \cdot \mathbf{q} \, dA$$

$$+ \int_A \rho \left(\frac{q^2}{2} + gy + \hat{u} \right) \mathbf{n} \cdot \mathbf{q} \, dA + \frac{\partial}{\partial t} \int_V \rho e \, dV \qquad (11.22)$$

Equation (11.22) is the general integral form of the energy equation.

In a common configuration for a flow system (Fig. 11.8) the fluid enters the device at some section 1 and leaves at some section 2, and there is no flow through the intervening portion of the system. If we choose as our control volume the dashed surface, the surface integrals involving the shearing stress will be zero since $\mathbf{t} \cdot \mathbf{q}$ is everywhere zero. Also the integrals involving $\mathbf{n} \cdot \mathbf{q}$ will be zero over any surface in which there is no flow. Thus for steady-state operation of such a device the energy equation can be written as

$$\dot{Q}_h - \dot{W}_s = \int_A \rho[(q^2/2) + gy + \hat{u} - (\sigma_n/\rho)] \, \mathbf{n} \cdot \mathbf{q} \, dA \quad (11.23)$$

where the integral is evaluated over the area of the inlet and outlet only.

In dealing with fluids, it is common practice to define the pressure p as the negative of the average normal stress at a point; i.e.,

$$p = -[(\sigma_x + \sigma_y + \sigma_z)/3]$$

As noted in Chapter 3, the sum of these normal stresses is invariant and not dependent on the orientation of the xyz coordinate system. The negative sign simply indicates that a positive pressure indicates a compressive stress, whereas a positive normal stress has been taken as a tensile stress. It can be shown from a more detailed study of constitutive relationships for Newtonian fluids that in those cases in which the fluid particles are following all straight and parallel paths (such as at the inlet and exit sections in Fig. 11.8), the normal stress

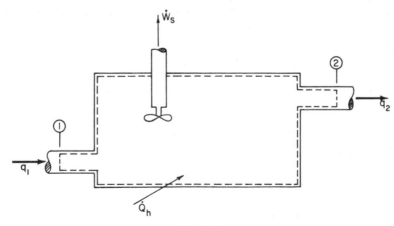

Fig. 11.8

on a plane perpendicular to the direction of flow is equal to the negative of the pressure. Thus σ_n in Eq. (11.23) can be replaced with $-p$.

If the flow at the inlet and exit sections is *one dimensional* so
that all characteristics are constant over the cross section, Eq. (11.23)
can be expressed as

$$\dot{Q}_h - \dot{W}_s = [(q_2^2/2) + gy_2 + \hat{u}_2 + (p_2/\rho_2)] \, \dot{m}_2$$
$$- [(q_1^2/2) + gy_1 + \hat{u}_1 + (p_1/\rho_1)] \, \dot{m}_1 \qquad (11.24)$$

where the subscripts 1 and 2 refer to the inlet and exit sections respectively, y_1 and y_2 are vertical coordinates to the centroid of the inlet and exit cross sections, and $\dot{m}_1$ and $\dot{m}_2$ are mass rates of flow.
From the conservation of mass equation we know that $\dot{m}_1 = \dot{m}_2$;
therefore Eq. (11.24) can be written as

$$\frac{\dot{Q}_h}{\dot{m}} + \frac{q_1^2}{2} + gy_1 + \hat{u}_1 + \frac{p_1}{\rho_1} = \frac{\dot{W}_s}{\dot{m}} + \frac{q_2^2}{2} + gy_2 + \hat{u}_2 + \frac{p_2}{\rho_2} \qquad (11.25)$$

Each term in Eq. (11.25) represents energy per unit mass of fluid
flowing through the control volume, and this equation states that the
energy per unit mass available at section 1 plus the energy per unit
mass added in the form of heat must equal the shaft work done by
the system per unit mass plus energy available at section 2 per unit
mass. Equation (11.25) is called the *one-dimensional steady-flow
energy equation.*

EXAMPLE 11.5

An incompressible fluid flows steadily at a rate of 0.016 m³/s through a
horizontal 0.2-m diameter pipe. The pressure drop over a 10-m length of pipe
(Fig. E11.5) is 200 N/m². Determine the rate of heat flow in watts through the
walls of the pipe, assuming there is no heat flow in the direction of flow and the

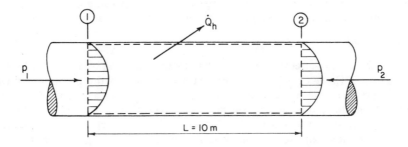

Fig. E11.5

internal energy is constant. The velocity distributions at sections 1 and 2 are the same, and the pressure may be assumed constant over a given cross section.

SOLUTION

Steady-state conditions exist and Eq. 11.23 is applicable. For the control volume, indicated by the dashed line in the figure, this equation reduces to

$$\dot{Q}_h = \int_A p\,\mathbf{n}\cdot\mathbf{q}\,dA$$

since there is no variation in the potential, kinetic, or internal energy crossing the control surfaces at sections 1 and 2. There is no flow across the surface parallel to the pipe walls so that $\mathbf{n}\cdot\mathbf{q} = 0$, and there is no shaft work. The surface integral can be written

$$\int_A p\,\mathbf{n}\cdot\mathbf{q}\,dA = p_1\int_{A_1}\mathbf{n}\cdot\mathbf{q}\,dA + p_2\int_{A_2}\mathbf{n}\cdot\mathbf{q}\,dA = (p_2 - p_1)Q$$

where we have made use of the continuity equation

$$-\int_{A_1}\mathbf{n}\cdot\mathbf{q}\,dA = \int_{A_2}\mathbf{n}\cdot\mathbf{q}\,dA = Q$$

Thus

$$\dot{Q}_h = -\Delta p\,Q = -(200)(0.016) = -3.20\ \text{N-m/s}$$

Since $1\ \text{W} = 1\ \text{N-m/s}$, the rate of heat flow out of the pipe (as indicated by the negative sign) is

$$(\dot{Q}_h)_{\text{out}} = 3.20\ \text{W} \qquad\qquad\qquad Ans.$$

EXAMPLE 11.6

A pump draws water ($\gamma = 62.4$ pcf) from a sump and discharges it through a nozzle into the atmosphere (Fig. E11.6). Determine the power the pump adds to the fluid to maintain a steady discharge of 0.1 cfs. Assume there is negligible heat transfer from the fluid and negligible changes in internal energy.

SOLUTION

We choose as our control volume the volume indicated by the dashed line in the figure. Although the local velocity at a point within the pump would vary with respect to time, we assume that when averaged over a period of time the flow within the pump and throughout the system is steady (if the pump is operating at a constant speed). We further assume that the flow is one dimensional at sections 1 and 2, so the appropriate form of the energy equation is

$$(q_1^2/2) + gy_1 + (p_1/\rho_1) = (\dot{W}_s/\dot{m}) + (q_2^2/2) + gy_2 + (p_2/\rho_2)$$

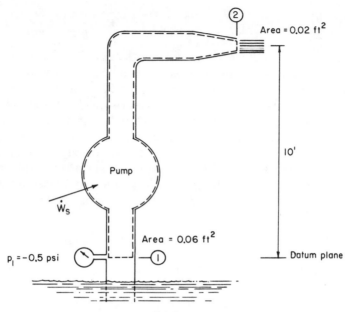

Fig. E11.6

We pass the datum for the elevation terms through section 1 and recognize that $p_2 = 0$ since the water discharges freely into the atmosphere. Thus

$$(\dot{W}_s/\dot{m}) = [(q_1^2 - q_2^2)/2] - gy_2 + (p_1/\rho)$$

From the continuity equation $q_1 A_1 = q_2 A_2 = 0.1$ so that

$$\frac{\dot{W}_s}{\dot{m}} = \frac{q_1^2 \lfloor 1 - (A_1/A_2)^2 \rfloor}{2} - gy_2 + \frac{p_1}{\rho}$$

and

$$\frac{\dot{W}_s}{\dot{m}} = \left(\frac{0.1}{0.06}\right)^2 \frac{[1 - (3)^2]}{2} - (32.2)(10) - \frac{(0.5)(144)}{62.4/32.2} = -705 \frac{\text{ft-lb}}{\text{slug}}$$

This result gives the rate at which work is being done on the fluid per unit of mass flowing. Since power is the rate of work done per unit time, we can write

$$\text{Power added by pump} = (705)\dot{m} = (705)(0.1)\frac{(62.4)}{(32.2)} = 136.5 \text{ ft-lb/s} \quad Ans.$$

Since 1 hp is equivalent to 550 ft-lb/s, the pump is adding 0.248 hp to the fluid.

For incompressible flow through conduits (such as pipe flow) Eq. 11.25 can be written in a particularly convenient form by dividing by g and regrouping the terms to give

$$\frac{q_1^2}{2g} + y_1 + \frac{p_1}{\gamma} - \frac{\dot{W}_s}{g\dot{m}} = \frac{q_2^2}{2g} + y_2 + \frac{p_2}{\gamma} + \frac{\hat{u}_2 - \hat{u}_1}{g} - \frac{\dot{Q}_h}{g\dot{m}} \qquad (11.26)$$

Each term in the equation now represents energy per unit weight of fluid flowing, and each term thus has the dimensions of a length $(FL/F = L)$. The various terms such as $q^2/2g$, y, p/γ are called the velocity head, elevation head, and pressure head respectively. We also let $-\dot{W}_s/g\dot{m}$ represent the head added to the fluid H_a via shaft work. For an incompressible fluid any increase in the internal energy and heat flow from the system represents a conversion from mechanical energy (as represented by the velocity, elevation, and pressure head) to thermal energy. In this context such increases in internal energy and heat flow from the system are considered as "losses." Thus we write

$$\text{head loss} = H_f = [(\hat{u}_2 - \hat{u}_1)/g] - (\dot{Q}_h/g\dot{m})$$

Equation (11.26) can now be written as

$$\frac{q_1^2}{2g} + y_1 + \frac{p_1}{\gamma} + H_a = \frac{q_2^2}{2g} + y_2 + \frac{p_2}{\gamma} + H_f \qquad (11.27)$$

and states that the sum of the velocity, elevation, and pressure head as some section 1 plus any head added must equal the sum of the velocity, elevation, and pressure head at some section 2 plus any head lost. A further discussion of the evaluation of the head-lost term will be given in Section 11.8.

11.6 Bernoulli Equation

The integral form of the momentum equation does not provide information with regard to the behavior of a fluid at any particular point. To obtain this more detailed picture, we consider a small element of fluid (Fig. 11.9). We assume there are no shearing stresses in the fluid, a so-called *nonviscous* fluid, so the only surface forces acting on the element are due to the pressure. As in Chapter 4 the pressure on each face of the element can be expressed in terms of the pressure at the center of the element (Fig. 11.9). In general, a body force may act on the fluid, and we let **B** represent the body force per unit mass with Cartesian components B_x, B_y, and B_z. If we now sum forces in the x direction and employ Newton's second law,

$$-\frac{\partial p}{\partial x} dx\, dy\, dz + B_x \rho\, dx\, dy\, dz = \rho\, dx\, dy\, dz\, a_x \qquad (11.28)$$

where a_x is the acceleration of the fluid element in the x direction.

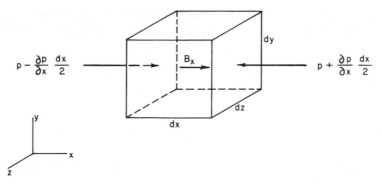

Fig. 11.9

Recall from Chapter 5 that the acceleration expressed from the Eulerian viewpoint is given by the equation

$$a = q \cdot \nabla q + \frac{\partial q}{\partial t}$$

so that

$$a_x = q \cdot \nabla u + \frac{\partial u}{\partial t} = u\frac{\partial u}{\partial x} + v\frac{\partial u}{\partial y} + w\frac{\partial u}{\partial z} + \frac{\partial u}{\partial t}$$

Equation (11.28) can thus be written in the form

$$u\frac{\partial u}{\partial x} + v\frac{\partial u}{\partial y} + w\frac{\partial u}{\partial z} + \frac{\partial u}{\partial t} = -\frac{1}{\rho}\frac{\partial p}{\partial x} + B_x \qquad (11.29a)$$

Similarly, for the y and z directions we can write

$$u\frac{\partial v}{\partial x} + v\frac{\partial v}{\partial y} + w\frac{\partial v}{\partial z} + \frac{\partial v}{\partial t} = -\frac{1}{\rho}\frac{\partial p}{\partial y} + B_y \qquad (11.29b)$$

$$u\frac{\partial w}{\partial x} + v\frac{\partial w}{\partial y} + w\frac{\partial w}{\partial z} + \frac{\partial w}{\partial t} = -\frac{1}{\rho}\frac{\partial p}{\partial z} + B_z \qquad (11.29c)$$

Equations (11.29) are the partial differential equations governing the flow of nonviscous fluids and are known as the *Euler equations*. In vector form Eqs. (11.29) can be expressed as

$$q \cdot \nabla q + \frac{\partial q}{\partial t} = -\frac{\nabla p}{\rho} + B \qquad (11.30)$$

It should be noted that the Euler equations apply to both compressible and incompressible fluids.

The body force is commonly assumed to be expressible in the form $B = -\nabla\Omega$ where Ω is a scalar function. A body force expressible in this form is said to be *conservative* and if the fluid weight is the only body force, then $\Omega = gy$ where y is a coordinate in the vertical direction. If we now make use of the vector identity

$$\mathbf{q} \times (\nabla \times \mathbf{q}) = (1/2) \nabla (\mathbf{q} \cdot \mathbf{q}) - (\mathbf{q} \cdot \nabla)\mathbf{q} \qquad (11.31)$$

Equation (11.30) can be written in the form

$$\frac{\nabla p}{\rho} + \nabla \left(\frac{q^2}{2} + \Omega \right) + \frac{\partial \mathbf{q}}{\partial t} = \mathbf{q} \times (\nabla \times \mathbf{q}) \qquad (11.32)$$

Under certain special conditions Eq. (11.32) can be integrated to obtain a very useful result.

Consider a differential length $d\mathbf{r}$ measured along a streamline in a flow field (Fig. 11.10). We now take the dot product of this vector

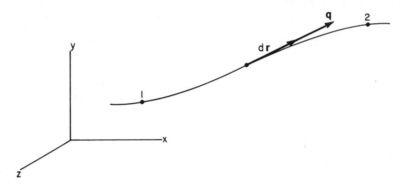

Fig. 11.10

of differential length with each term in Eq. (11.32) to obtain

$$\frac{\nabla p}{\rho} \cdot d\mathbf{r} + \left[\nabla \left(\frac{q^2}{2} + \Omega \right) \right] \cdot d\mathbf{r} + \frac{\partial \mathbf{q}}{\partial t} \cdot d\mathbf{r} = [\mathbf{q} \times \nabla \times \mathbf{q}] \cdot d\mathbf{r} \qquad (11.33)$$

Since we have selected $d\mathbf{r}$ to have a direction along the streamline, it follows that $d\mathbf{r}$ and $\mathbf{q}$ are parallel. However, the vector $\mathbf{q} \times (\nabla \times \mathbf{q})$ is perpendicular to $\mathbf{q}$ (from the definition of the cross product); therefore it is also perpendicular to $d\mathbf{r}$ so that $\mathbf{q} \times (\nabla \times \mathbf{q}) \cdot d\mathbf{r} = 0$. We now integrate Eq. (11.33) along the streamline to obtain

$$\int \frac{dp}{\rho} + \int d \left(\frac{q^2}{2} + \Omega \right) + \int \frac{\partial \mathbf{q}}{\partial t} \cdot d\mathbf{r} = F(t) \qquad (11.34)$$

where $F(t)$ is either a constant or a function of time. For steady, incompressible flow Eq. (11.34) reduces to

$$(p/\rho) + (q^2/2) + \Omega = \text{constant}[1] \qquad (11.35)$$

Equation (11.35) is the well-known *Bernoulli equation*, named in honor of Daniel Bernoulli (1700–1782). This equation applies between any two points (Fig. 11.10) along the *same streamline*, and we can write the equation as

$$(p_1/\rho) + (q_1^2/2) + gy_1 = (p_2/\rho) + (q_2^2/2) + gy_2 \qquad (11.36)$$

or

$$(p_1/\rho g) + (q_1^2/2g) + y_1 = (p_2/\rho g) + (q_2^2/2g) + y_2 \qquad (11.37)$$

where we have used the relation $\Omega = gy$. The Bernoulli equation relates the velocity, pressure, and elevation at one point to another point on the same streamline. However, when applying Eq. (11.36) or (11.37), the several conditions used in the derivation should be recalled. These are: (1) the fluid is nonviscous, (2) the equation applies along a streamline, (3) the flow is steady, (4) the fluid is incompressible, and (5) the only body force is the weight of the fluid.

We note that if the vector $\nabla \times q$ is everywhere zero in the flow field, the right side of Eq. (11.33) is zero regardless of the direction of dr. It follows immediately that Eq. (11.33) can be integrated as before. The resulting equation will be the same as Eq. (11.36) or (11.37), except that points 1 and 2 may be anywhere in the flow field since dr is arbitrary. It can be shown that $(1/2)(\nabla \times q)$ is the vector representation of angular velocity of a fluid particle so that the condition $\nabla \times q = 0$ corresponds to "no rigid body rotation" of a fluid particle. This type of flow is called *irrotational* and is sometimes found to be a reasonable approximation for some flow fields. It also follows that if $\nabla \times q = 0$, then q can be expressed as the gradient of a scalar function ϕ so that for irrotational flow $q = \nabla \phi$. For an incompressible fluid we have shown that $\nabla \cdot q = 0$ so that

$$\nabla^2 \phi = 0 \qquad (11.38)$$

for the irrotational flow of an incompressible fluid. The function ϕ is commonly called the *velocity potential*. We recognize Eq. (11.38) as the well-known Laplace equation, and the solution of "potential flow" problems reduces to finding a solution to this linear partial differential equation subject to the appropriate initial and boundary conditions. Much attention has been given to potential flow theory, and it is a highly developed branch of fluid mechanics. However, it

---

1. Actually, this "constant" could depend on time if, for example, the overall hydrostatic pressure of the system were changing with time. However, this would be a rather unusual situation.

should be reemphasized that the use of the irrotational flow condition imposes a severe restriction on the flow that may be reasonable for some flow fields but completely erroneous for others.

EXAMPLE 11.7

Show how the converging-diverging tube in Fig. E11.7a, when combined with a manometer, can be used as a flowmetering device. Assume the fluid is incompressible and the velocities and pressures at the inlet and throat sections are constant across the cross section of the tube. The flow is steady.

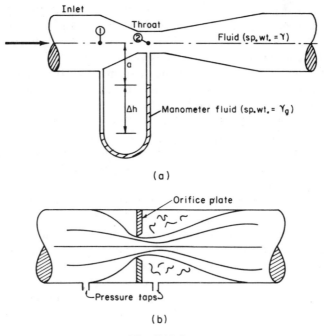

(a)

(b)

Fig. E11.7

SOLUTION

We first apply the Bernoulli equation, Eq. (11.36), between the inlet and throat section; i.e., $(p_1/\rho) + (q_1^2/2) = (p_2/\rho) + (q_2^2/2)$. Since there is no difference in elevation between points 1 and 2, it follows that $y_1 = y_2$. Next apply the continuity equation between the inlet and throat sections to obtain the relation $q_1(\pi/4)D_1^2 = q_2(\pi/4)D_2^2$ or $q_2 = q_1(D_1/D_2)^2$ where D_1 and D_2 are the inlet and throat tube diameters respectively. The Bernoulli equation can now be written as

$$p_1 + (\rho q_1^2/2) = p_2 + (\rho q_1^2/2)(D_1/D_2)^4$$

or

$$q_1 = \sqrt{\frac{2(p_1 - p_2)}{\rho} \frac{1}{(D_1/D_2)^4 - 1}}$$

Thus we see that the velocity at the inlet can be expressed in terms of the pressure differential between the inlet and throat. This is actually an ideal velocity since we have neglected viscous effects. The ideal discharge is $Q_{\text{ideal}} = q_1 (\pi/4) D_1^2$. To correct for viscous effects, we define a *meter coefficient* C_D such that $Q_{\text{actual}} = C_D Q_{\text{ideal}}$. The value of the meter coefficient must be determined experimentally. For a well-designed flowmeter of the type in Fig. E11.7a (commonly called a *Venturi meter*) C_D is of the order of 0.98.

The pressure difference can be obtained from the U-tube manometer, and the manometer equation is

$$p_1 + a\gamma + \Delta h \, \gamma = \gamma_g \, \Delta h + a\gamma + p_2$$

or

$$p_1 - p_2 = \Delta h \, (\gamma_g - \gamma)$$

Thus a convenient expression for the discharge through this type of flowmeter is

$$Q_{\text{actual}} = C_D (\pi/4) D_1^2 \sqrt{\frac{2g \, \Delta h \, [(\rho_g/\rho) - 1)]}{(D_1/D_2)^4 - 1}} \qquad Ans.$$

where g is the acceleration of gravity and ρ_g is the density of the manometer fluid.

Essentially any geometrical constriction that can be placed in a pipe or tube through which a fluid is flowing can be used as a flowmeter. The important characteristic is simply a change in velocity which gives rise to a pressure differential. For example, another common flowmeter of this type is the *flat-plate orifice* (Fig. E11.7b). As the fluid stream contracts to flow through the orifice, the pressure drops and this can be related to the discharge by the same equations as those for the Venturi meter. However, due to the abrupt contraction the losses in the orifice are higher than those for the Venturi, and orifice meter coefficients are of the order of 0.6.

EXAMPLE 11.8

Point velocities in a moving stream can be measured by means of a small tube placed in the stream with its axis oriented parallel to the streamlines, (Fig. E11.8). The tube is designed so that the velocity at its tip is zero; i.e., at the tip (point 2) there is a *stagnation point*. Derive an equation that relates the velocity q at some distance from the wall to the differential height of fluid Δh. Assume that the flow is nonviscous, incompressible, and steady.

SOLUTION

We apply the Bernoulli equation between points 1 and 2 to obtain

$$p_1 + (\rho q_1^2/2) = p_2$$

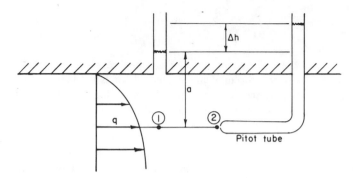

Fig. E11.8

Note that we have set q_2 equal to zero since a stagnation point exists at point 2. We can solve for q_1 to obtain

$$q_1 = \sqrt{2(p_2 - p_1)/\rho}$$

Since $p_1 = \rho g a$ and $p_2 = \rho g(a + \Delta h)$, it follows that $p_2 - p_1 = \rho g \, \Delta h$. Thus

$$q_1 = \sqrt{2g \, \Delta h} \qquad\qquad Ans.$$

which is the desired relationship.

This is actually an ideal velocity since the fluid has been assumed to be nonviscous. However, over the short distance between the points 1 and 2 the losses are negligibly small so that the derived relationship is not commonly corrected. It should be emphasized that the Pitot tube must be aligned with the streamlines so that a stagnation point exists at its tip. Misalignment of the Pitot tube can introduce an error far more serious than that caused by viscous effects.

11.7   Viscous Flow

Shearing stresses are developed whenever a real fluid is set in motion. Thus any analysis of moving fluids based on the assumption of zero shearing stresses (such as in the development of the Bernoulli equation) must be approximate. In this section we will consider the influence of shearing stresses on the flow in two relatively simple geometric configurations.

Couette Flow—Consider the case of an incompressible Newtonian fluid located between two infinite parallel plates (Fig. 11.11). Let the bottom plate be fixed and the upper plate move with a constant velocity U. We recognize that the fluid will be in motion since the fluid in contact with the upper plate will move with a velocity U, whereas the fluid in contact with the lower plate will be at rest. We assume that since the plates are very large and parallel, there will

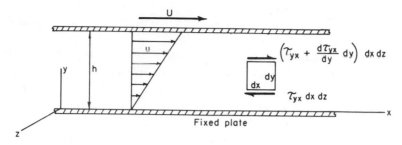

Fig. 11.11

only be a velocity component in the x direction; i.e., $v = w = 0$. It follows from the continuity equation that

$$\frac{\partial u}{\partial x} = 0$$

therefore $u = f(y, z, t)$. A flow field of this type in which there is only one velocity component is commonly referred to as a *uniform* flow field. However, if the flow is steady, and since we are dealing with infinite plates so there is no variation in velocity in the z direction, then $u = f(y)$.

We now consider a differential element of fluid located between the two plates as in Fig. 11.11. The motion of the fluid is due entirely to the movement of the upper plate; i.e., there is no pressure gradient in the x direction so the only forces acting on the element are due to the shearing stresses. On the bottom surface of the element the shearing stress is τ_{yx} and on the upper surface the shearing stress can be expressed in the form $\tau_{yx} + (d\tau_{yx}/dy)dy$. If the fluid element is not accelerating we can write

$$\Sigma F_x = 0$$

$$\left(\tau_{yx} + \frac{d\tau_{yx}}{dy} dy\right) dx\, dz - \tau_{yx}\, dx\, dz = 0$$

or

$$\frac{d\tau_{yx}}{dy} = 0 \qquad (11.39)$$

This result shows that the shearing stress τ_{yx} is a constant in this type of parallel-plate device. Experiments performed with this type of device (see Section 9.6) reveal that the shearing stress is proportional to the mean velocity gradient U/h, and we assume that

$$\tau_{yx} = \mu \frac{du}{dy} \qquad\qquad (11.40)$$

where μ is the absolute viscosity of the fluid. Substitution of Eq. (11.40) into Eq. (11.39) gives the simple differential equation

$$\frac{d^2 u}{dy^2} = 0$$

which can be integrated to give $u = Ay + B$ where A and B are constants of integration. We now apply the boundary conditions

$$u = 0 @ y = 0$$

$$u = U @ y = h$$

to obtain the linear velocity distribution

$$u = U(y/h) \qquad\qquad (11.41)$$

This simple flow is called *Couette flow*, named in honor of M. Couette. In practice this type of flow cannot be produced exactly since we cannot build an apparatus consisting of infinite parallel plates. However, a very useful device which develops a flow field that may closely approximate Couette flow consists of two long, concentric cylinders (Fig. 11.12). The inner cylinder is fixed and the outer cylinder rotates with a constant angular velocity ω. We see that if the gap width $r_2 - r_1$ is small and the radius r_1 large, the fluid in the gap would behave approximately as if it were contained between two infinite parallel plates. The shearing stress at the inner wall is

$$\tau_{r=r_1} = \mu\, r_2\, \omega / (r_2 - r_1)$$

and the torque (force times distance) on the inner cylinder due to this stress is

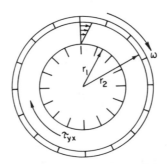

Fig. 11.12

$$T = (2\pi r_1 L)(r_1)\tau = 2\pi\mu L r_2\,\omega[r_1^2/(r_2 - r_1)] \qquad (11.42)$$

where L is the length of the cylinder. If the torque and angular velocity can be measured for a given geometry (r_1, r_2, L), then Eq. (11.42) can be used to determine the viscosity of the fluid. The torque is usually measured by suspending the inner cylinder from an elastic wire and measuring the angle through which the wire is twisted. The wire twists until an equilibrium condition is achieved between the torque developed by the wire and the torque developed by viscous action on the inner cylinder. As shown in Section 10.4, the torque and angle of twist are uniquely related. This type of device is called a *rotating cylinder viscometer* and is commonly used for the measurement of viscosity.

An exact solution to the problem of flow between concentric cylinders can be obtained and gives for the torque

$$T = 4\pi\mu L r_2\,\omega[r_1^2 r_2/(r_2^2 - r_1^2)] \qquad (11.43)$$

which will reduce to Eq. (11.42) for small gaps.

Poiseuille Flow—Consider the steady flow of an incompressible fluid through a straight horizontal circular tube of constant cross section. This type of flow is commonly found in hydraulic systems, and a knowledge of the pertinent flow characteristics in pipe flow is of practical importance.

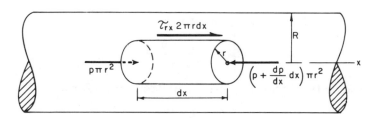

Fig. 11.13

We assume for the tube in Fig. 11.13 that the fluid velocity is in the x direction and the velocity components in the y and z directions are everywhere zero. It again follows from the continuity equation that the rate of change of the velocity in the x direction must be zero; therefore for axisymmetric steady flow the velocity u is a function only of the radial coordinate r. The forces acting on a cylindrical element of fluid of length dx are shown in Fig. 11.13 where p is the average pressure acting at the left end of the element and τ_{rx} is the shearing stress acting on the curved surface. For equilibrium to exist

$$\Sigma F_x = 0$$

$$p\pi r^2 - \left(p + \frac{dp}{dx}\,dx\right)\pi r^2 + \tau_{rx}\,2\pi r\,dx = 0$$

and upon simplification

$$\tau_{rx} = \frac{r}{2}\frac{dp}{dx} \tag{11.44}$$

We assume that the fluid is Newtonian and the shear stress is proportional to the velocity gradient; i.e.,

$$\tau_{rx} = \mu\frac{du}{dr}$$

Thus Eq. (11.44) can be written in the form

$$\frac{du}{dr} = \frac{1}{2\mu}\frac{dp}{dx}\,r \tag{11.45}$$

Since there is no flow in a direction perpendicular to the axis of the tube, the pressure distribution over any cross section must be hydrostatic so that dp/dx is not a function of r and Eq. (11.45) can be integrated to yield

$$u = \frac{1}{4\mu}\frac{dp}{dx}\,r^2 + C_1$$

At the wall $r = R$ the velocity is zero so

$$C_1 = -\frac{1}{4\mu}\frac{dp}{dx}R^2$$

and the velocity distribution can be written in the form

$$u = -\frac{R^2}{4\mu}\frac{dp}{dx}\left(1 - \frac{r^2}{R^2}\right) \tag{11.46}$$

Thus at a given cross section the velocity distribution is parabolic. At the center of the tube the velocity is a maximum u_{max} so that

$$u_{max} = -\frac{R^2}{4\mu}\frac{dp}{dx}$$

and Eq. (11.46) can also be written as

$$u = u_{max}\left[1 - (r^2/R^2)\right] \tag{11.47}$$

The discharge of fluid passing through a ring of differential width (Fig. 11.14) is $dQ = u2\pi r\,dr$, and the total discharge passing

Fig. 11.14

any cross section is

$$Q = 2\pi \int_0^R ur\,dr \qquad (11.48)$$

Equation (11.47) can be substituted into Eq. (11.48) and the resulting equation integrated to give $Q = (u_{max}/2)\pi R^2$. We define the average velocity U so that $Q = \pi R^2 U$ and thus $u_{max} = 2U$.

The expression relating the discharge and the pressure gradient is

$$Q = -\frac{\pi R^4}{8\mu}\frac{dp}{dx} \qquad (11.49)$$

where we have replaced u_{max} by its equivalent in terms of the pressure gradient. Equation (11.49) is one form of *Poiseuille's law*. This type of flow is known as Poiseuille flow, named in honor of Jean Louis Poiseuille (1799–1869), a French physician. Poiseuille was interested in blood flow through capillaries and experimentally deduced the resistance laws for such flows. It was actually after his work that the theoretical results presented in this section were determined, but his name is nevertheless associated with the solution to this problem.

From Eq. (11.49) we see that the pressure gradient is a constant for a given discharge Q and thus

$$-\frac{dp}{dx} = \frac{\Delta p}{L}$$

where Δp is the pressure drop between any two points along the tube a distance L apart. Equation (11.49) can therefore also be written in the form

$$Q = \frac{\pi R^4}{8\mu}\frac{\Delta p}{L} \qquad (11.50)$$

or in terms of the average velocity

$$U = \frac{R^2}{8\mu} \frac{\Delta p}{L} \qquad (11.51)$$

11.8 Energy Loss in Pipe Flow

Application of the energy equation, Eq. (11.23), to a control volume of length L and bounded by the pipe wall (see Example 11.5) yields

$$\dot{Q}_h = - Q\Delta p + \rho Q (\hat{u}_2 - \hat{u}_1)$$

or

$$\frac{\Delta p}{\rho g} = \frac{(\hat{u}_2 - \hat{u}_1)}{g} - \frac{\dot{Q}_h}{g\dot{m}} \qquad (11.52)$$

where Δp is the pressure drop, and $\hat{u}_2 - \hat{u}_1$ represents the difference between the average internal energies at sections 1 and 2. Thus we see that the pressure drop is reflected in an increase in the internal energy and heat flow from the pipe. As discussed in Section 11.5 the terms on the right side of Eq. (11.52) are frequently lumped into a single term called the head loss H_f. Thus

$$H_f = \frac{\Delta p}{\rho g}$$

and if the expression for the pressure drop, as obtained from Eq. (11.51), is utilized we have

$$H_f = 8\mu L U / \rho g R^2 \qquad (11.53)$$

It is common practice in engineering literature to express Eq. (11.53) in the form

$$H_f = f \frac{L}{D} \frac{U^2}{2g} \qquad (11.54)$$

where f = friction factor = $64/R_e$
          R_e = Reynolds number = $\rho D U/\mu$
          D = tube diameter

The Reynolds number is a dimensionless quantity that plays an important role in many fluid mechanics problems. Its importance to this particular problem is discussed in Section 11.9.

Application of the one-dimensional energy equation, Eq. (11.27), to a small control volume surrounding a streamline in a steady in-

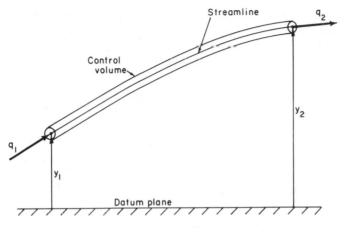

Fig. 11.15

compressible flow (Fig. 11.15) gives

$$(q_1^2/2g) + y_1 + (p_1/\gamma) = (q_2^2/2g) + y_2 + (p_2/\gamma) + H_f$$

with no head added. If the fluid is nonviscous, the Bernoulli equation written along the streamline is

$$(q_1^2/2g) + y_1 + (p_1/\gamma) = (q_2^2/2g) + y_2 + (p_2/\gamma)$$

Since both equations are valid for nonviscous fluids, it follows that any heat transfer into or out of the control volume must be reflected in a change in the internal energy so that

$$H_f = [(u_2 - u_1)/g] - (\dot{Q}_h/g\dot{m}) = 0$$

However, as illustrated by Eq. (11.53) for Poiseuille flow, the head loss is nonzero when viscosity is considered. Thus for steady incompressible flow along a streamline or for one-dimensional flow, the energy and Bernoulli equations reduce to the same equation for nonviscous fluids with no head added. But for viscous flow the one-dimensional energy equation is required. Equation (11.27) is sometimes called the *modified Bernoulli equation*.

11.9 Turbulence

Numerous experiments have been performed to check the validity of Poiseuille's law, and excellent agreement is found between theory and experiment as long as the Reynolds number is below a certain critical value, approximately 2000. As the critical Reynolds number is exceeded, the pressure drop is no longer linearly related to the average velocity, as indicated by Eq. (11.51), and the nature of the flow is drastically changed. Filaments of dye injected into the stream follow smooth paths parallel to the axis of the tube for Reynolds numbers less than critical. In this case, the flow is said to be *laminar*. However, as the Reynolds number is increased beyond the critical value, the dye filaments begin to waver; upon further increase the dye rapidly disperses throughout the tube. The paths of individual fluid particles are no longer smooth and parallel to the tube axis but are tortuous and randomly oriented. The flow is now said to be *turbulent*.

Transition from laminar to turbulent flow appears to be due to inherent instabilities in the flowing stream. Once turbulence is initiated, it is maintained by energy supplied by the main stream. The external source creating the flow may for all practical purposes be completely free of random fluctuations. The Reynolds number, named in honor of Osborne Reynolds (1842–1912) who first reported this phenomenon in detail, is a dimensionless combination of four parameters. Thus turbulence can be induced by increasing the fluid density, tube diameter, or fluid velocity or by decreasing the viscosity.

For the rotating cylinder apparatus described in Section 11.7 the critical Reynolds number, defined as

$$R_e = [\rho(r_2 - r_1)r_1 \omega]/\mu$$

is much higher than 2000. It should be emphasized that the number 2000 is not some form of "universal constant," and the specific value of the critical Reynolds number depends on the geometry of the flow system and possibly on other factors. For example, the critical Reynolds number for flow in a pipe bend is not the same as that for flow in a straight pipe (critical Reynolds numbers appear to be higher in the bend). Even for flow through straight tubes the transition may not occur until the Reynolds number has a value in the tens of thousands. However, these very large values occur only under carefully controlled laboratory conditions in which external sources of disturbances from noise, vibration, etc., are greatly reduced.

Turbulent flow is extremely complex. Fluctuations in a velocity component, typical for a turbulent stream, are shown in Fig. 11.16. The fluctuations are random, and it is necessary to utilize statistical concepts to characterize turbulent flow. The time average value of a given velocity component, say $\tilde{v}$, in a turbulent flow is defined as

$$\tilde{v} = \frac{1}{T} \int_0^T v \, dt$$

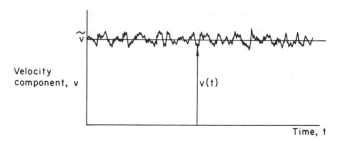

Fig. 11.16

where T is some time interval over which the averaging takes place. We can thus express the instantaneous velocity in the form $v = \tilde{v} + v'$ where v' represents the fluctuating component.

A common measure of the intensity of a given velocity component in a turbulent flow field is the root mean square (rms) value of the fluctuating component; i.e.,

$$v'_{\text{rms}} = \left[\frac{1}{T} \int_0^T (v')^2 \, dt \right]^{1/2}$$

Typically, v'_{rms} may be of the order of 5–10% of the average velocity. Similar relationships can be used for the other velocity components, and other statistical characteristics are commonly used to characterize a turbulent flow field. Although much theoretical work has been done in this field, using statistical concepts, all practical problems involving turbulence rely on experimental data for their solution.

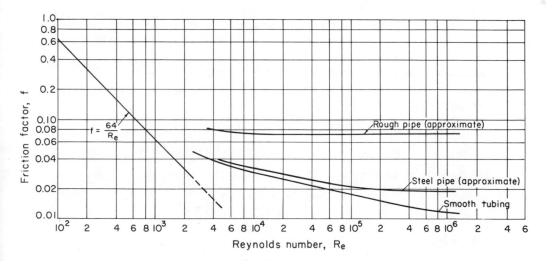

Fig. 11.17

For example, the pressure drop Δp for turbulent flow through a circular tube is commonly obtained through the use of Eq. (11.54) with the friction factor determined experimentally. Typical values of f are shown in Fig. 11.17, and we observe that for turbulent flow the friction factor is a function of both the Reynolds number and the surface characteristics of the tube through which the fluid is flowing.

EXAMPLE 11.9

A viscous incompressible fluid ($\mu = 5 \times 10^{-3}$ N-s/m^2; $\rho = 1.2 \times 10^3$ kg/m^3) is flowing steadily through a long horizontal glass tube having a diameter of 0.02 m (Fig. E11.9). Determine the pressure drop between two points along the tube separated by a distance of 0.5 m if the rate of flow through the tube is: (a) 10^{-6} m^3/s and (b) 10^{-3} m^3/s. Express the answer in N/m^2.

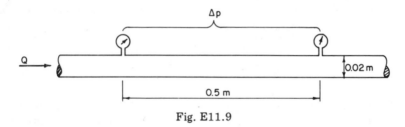

Fig. E11.9

SOLUTION

The pressure drop through a circular horizontal tube can be obtained from Eq. (11.54),

$$\frac{\Delta p}{\rho g} = f \frac{L}{D} \frac{U^2}{2g}$$

where $f = 64/R_e$ for laminar flow but must be determined empirically from a diagram such as Fig. 11.17 for turbulent flow. Thus we must first obtain the Reynolds number to determine whether the flow is laminar or turbulent. The average velocity is

$$U = Q/[\pi(10^{-4})] = 10^{-2}/\pi \text{ m/s for (a)}$$

$$= 10/\pi \text{ m/s for (b)}$$

The Reynolds number for (a) is

$$R_e = \frac{(1.2 \times 10^3)(2 \times 10^{-2})(10^{-2})}{(5 \times 10^{-3})(\pi)} = 15.28$$

and for (b)

$$R_e = 15{,}280$$

For (a) the Reynolds number is less than 2000 so for this case the flow is laminar and $f = 64/15.28 = 4.18$. For (b) the flow is turbulent and the friction factor can be obtained from Fig. 11.17. Since a glass tube is considered to be "smooth", we find that $f \approx 0.027$ for $R_e = 15,280$. Thus for (a)

$$\Delta p = (4.18) \frac{(5 \times 10^{-1})(10^{-4})(1.2 \times 10^{3})}{(2 \times 10^{-2})(\pi)^2 (2)} = 0.637 \text{ N/m}^2 \qquad Ans.$$

and for (b)

$$\Delta p = (0.027) \frac{(5 \times 10^{-1})(10^{2})(1.2 \times 10^{3})}{(2 \times 10^{-2})(\pi)^2 (2)} = 4.11 \times 10^{3} \text{ N/m}^2 \qquad Ans.$$

EXAMPLE 11.10

An incompressible viscous fluid ($\mu = 24 \times 10^{-6}$ lb-s/ft^2; sp. wt. = 64 pcf) flows under the action of gravity from a large tank (Fig. E11.10). The fluid discharges into the atmosphere and the free surface of the tank is open to the atmosphere. What is the required change in elevation a between the free surface and the discharge end of the tube for the mean velocity in the tube to be 3 fps? Assume the fluid level in the tank remains constant.

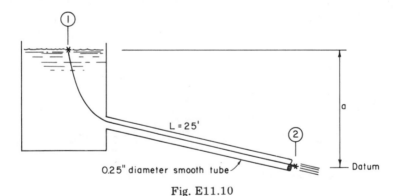

Fig. E11.10

SOLUTION

We apply the one-dimensional energy equation, Eq. (11.27), between the fluid surface at 1 and the end of the tube at 2. Thus

$$(q_1^2/2g) + y_1 + (p_1/\gamma) = (q_2^2/2g) + y_2 + (p_2/\gamma) + H_f$$

This equation indicates that the energy at some point 1 (due to velocity, elevation, and pressure) is equal to the energy at some point 2 plus any energy loss that has taken place between points 1 and 2. The head-loss term is given by Eq. (11.54); i.e., $H_f = f(L/D)(U^2/2g)$. This formula accounts for the loss in the straight section of the tube. So-called *minor losses* at the entrance to the pipe also exist, but these are usually negligible if the tube length–diameter ratio is large, say of the order of 1000.

In this particular example, $q_1 \approx 0$ if the tank is large; and $p_1 = p_2 = 0$ since both points 1 and 2 are at atmospheric pressure. Also, if we select point 2 as our reference or datum for elevation changes, then $y_2 = 0$ and $y_1 = a$. Thus the energy equation becomes

$$a = (q_2^2/2g) + f(L/D)(q_2^2/2g)$$

To calculate the friction factor, the Reynolds number is determined:

$$R_e = \frac{(64)(1)(3)(10)^6}{(32.2)(48)(24)} = 5.18 \times 10^3$$

From Fig. 11.17 we obtain $f \approx 0.036$, and the required distance a is

$$a = \frac{(3)^2}{2(32.2)} + (0.036)\frac{(25)}{(1/48)}\frac{(3)^2}{2(32.2)}$$

$$= 0.014 \, (1 + 43) = 0.615 \text{ ft} \qquad\qquad\qquad Ans.$$

The distinction between laminar and turbulent flow is not always easily made. For example, a laminar flow may be unsteady, and in addition the streamlines are not necessarily straight and parallel (Fig. 11.18). Thus the common definition of laminar flow as "one for

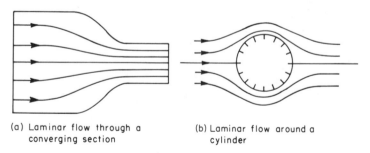

(a) Laminar flow through a   (b) Laminar flow around a
    converging section   cylinder

Fig. 11.18

which the paths of the fluid particles are parallel to the boundaries" is not adequate.

As a matter of practice we usually classify various flows as steady laminar flow, unsteady laminar flow, steady turbulent flow, or unsteady turbulent flow. A *steady laminar flow field* is one in which at any given point the velocity is independent of time and for which the flow can be reproduced precisely from experiment to experiment. This type of flow is completely determinate. An *unsteady laminar flow field* is one in which the velocity at a point varies with respect to time, but the flow field is still determinate in the sense that it can be reproduced precisely from experiment to experiment. A *turbulent flow field* is one which cannot be reproduced precisely from experiment to experiment even though all factors under the control of the experimenter are invariant between experiments. Thus the distinguishing feature of a turbulent flow field is that the characteristics of the field, such as pressure and velocity, are random functions of time; it is indeterminate in the sense that two experiments performed under the same conditions will not produce identical results at any particular instant in time. The terms steady and unsteady as they apply to turbulent flows refer to the mean velocity [e.g., for a steady turbulent flow $\tilde{v} \neq f(t)$], whereas for an unsteady turbulent flow $\tilde{v} = f(t)$.

11.10   Boundary Layer Concept

A serious defect of the nonviscous fluid model is the fact that the fluid is allowed to "slip" past a fixed rigid surface. From a physical standpoint this follows since it is the property of viscosity that is presumed to give a fluid its "stickiness." It also follows from the mathematics of nonviscous fluid theory that only the component of fluid velocity normal to a surface can be specified. Nevertheless, from many experimental observations it is known that even for fluids with relatively low viscosities such as air, the "no-slip" condition still applies. However, experiments also reveal that under certain conditions the effects of viscosity appear to be confined to a very thin layer near the solid surface and outside this layer the fluid behaves as if it were nonviscous.

In a paper presented in 1904, Ludwig Prandtl (1875–1953) proposed a method of solution for certain types of fluid mechanics problems whereby viscosity is taken into account in the thin *boundary layer* near the surface and neglected outside. Since the formulation of the boundary layer concept this technique has met with considerable success. The classical problem, which illustrates many important characteristics of the boundary layer, is that of flow past a wide thin plate (Fig. 11.19) fixed in a uniform stream. Experimental measurements of the velocity profile near the plate reveal that the velocity is zero at the plate surface, increases rapidly as we move away along a normal to the surface, and for all practical purposes attains the free-stream velocity at a short distance δ away from the surface. The distance δ is denoted the *boundary layer thickness*.

Boundary layer theory is based on the fact that the layer is very thin compared with the overall size of the system. However, it

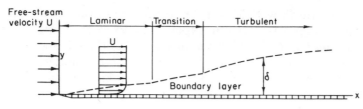

Fig. 11.19

should be recognized that a boundary layer will exist only for relatively high-velocity flows of low-viscosity fluids. Actually the proper criterion is again the Reynolds number based on some characteristic length and velocity, and boundary layer theory is applicable for high Reynolds number flows. For the flat-plate problem, theory shows that if the flow within the boundary layer is laminar, the thickness of the layer varies according to the equation

$$\delta/x \approx 5 \, (\rho U x/\mu)^{-1/2} \tag{11.55}$$

where x is measured from the leading edge of the plate. A laminar-turbulent transition similar to that occurring in pipe flow can also take place within the boundary layer (Fig. 11.19). When the local Reynolds number, defined as $R_e = \rho U x/\mu$, exceeds a value of the order of 3×10^5, transition normally takes place. A turbulent boundary layer grows at a faster rate and a good approximation for the thickness of a turbulent boundary layer on a flat plate is given by the equation

$$\delta/x \approx 0.4 \, (\rho U x/\mu)^{-1/5} \tag{11.56}$$

Although it is beyond the scope of this book to consider the characteristics of boundary layers in greater detail, it is clear that the presence of a boundary layer plays a major role in any heat and mass transfer processes that may occur at a fluid-solid interface. In addition, in many instances the boundary layer is responsible for the phenomenon of separation whereby the main stream "separates" from a solid boundary. For example, the flow around a long circular cylinder fixed in a moving stream of fluid will "separate" if the Reynolds number $(\rho U D/\mu)$ is large. For such a cylinder a boundary layer will build up (Fig. 11.20) and eventually leave the surface at the *separation point*. Beyond the separation point, the flow near the cylinder surface will be reversed as shown and a *wake* will form behind the cylinder. The wake is observed to contain a mass of fluid moving very randomly. The separation phenomenon is due to the fact that viscous forces in the boundary layer simply retard the flow to such an extent that the fluid cannot follow along the surface.

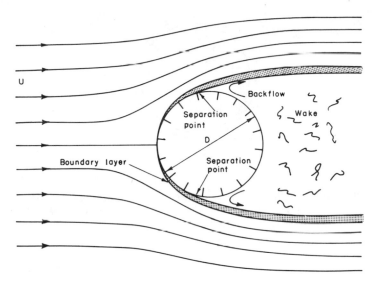

Fig. 11.20

PROBLEMS

11.1. For the conduit in Fig. P11.1 the average velocity at section 1 is 10 fps and at section 2 it is 3 fps. What is the average velocity at section 3? Assume that the cross-sectional areas at all three sections are the same and the fluid is incompressible.

Ans. 7 fps

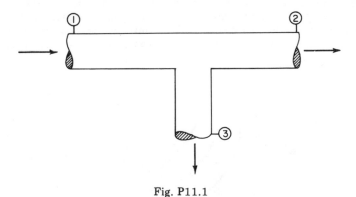

Fig. P11.1

11.2. At a certain cross section (Fig. P11.2) in a pipeline carrying a liquid the velocity profile is given by the equation $u = U[1 - (r/R)^2]$. Determine an expression for the average velocity.

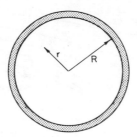

Fig. P11.2

11.3. Oil flows through a 0.2-m diameter pipe at an average velocity of 2 m/s. At a certain point in the pipe the diameter is reduced to 0.1 m. What is the average velocity in this smaller section?

Ans. 8 m/s

11.4. The velocity in a certain incompressible flow field is given by the expression $q = 3yz^2 i + xz j + y k$. Show that the continuity equation is satisfied.

11.5. The velocity of a fluid particle moving along a streamline that coincides with the x axis in a plane two-dimensional flow field was experimentally found to be given by the equation $u = x^2$. Along this streamline determine an expression for (a) the rate of change of the v component of velocity with respect to y and (b) the acceleration of the particle.

Ans. (a) $\dfrac{\partial v}{\partial y} = -2x,$ (b) $a = 2x^3 i$

11.6. The velocity components in the vicinity of the corner in Fig. P11.6 are given by the expressions $u = 2x, v = -2y, w = 0$. Assuming that the fluid is incompressible, (a) show that the continuity equation is satisfied, (b) determine an expression for the convective acceleration along the x axis, and (c) determine an expression for the local acceleration along the x axis.

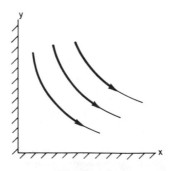

Fig. P11.6

11.7. Determine the third velocity component which, when combined with the two components below, will satisfy the continuity equation for an incompressible fluid.

$$u = x^2 + y^2 + z^2, \qquad v = xy + yz + z^2, \qquad w = ?$$

Ans. $w = -3xz - z^2/2 + f(x, y)$

11.8. Water under a pressure of 10 psi flows with a velocity of 10 fps through a 60° horizontal bend having a uniform diameter of 12 in. Neglecting the pressure drop in the bend, determine the magnitude and direction of the resultant fluid force on the bend.

11.9. The horizontal nozzle of Fig. P11.9 having a tip diameter of 1 in. is attached to the end of a 3-in. diameter water line. A pressure gage located near the end of the pipe indicates a pressure of 50 psi. The jet velocity just outside the nozzle tip is 83.2 fps, and the area at this section is 0.9 of the tip area. Determine the resultant force the nozzle exerts on the pipe.

Ans.   294 lb →

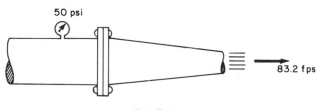

50 psi

83.2 fps

Fig. P11.9

11.10. An incompressible fluid flows steadily through the 180° horizontal bend in Fig. P11.10. At the entrance to the bend both the velocity and the pressure are uniform over the cross section. At the outlet section the pressure is uniform, but the velocity distribution is parabolic, i.e., $u = 2U[1 - (r/R)^2]$ where U is the average velocity and R is the pipe radius. Determine the force the fluid exerts on the bend in the x direction. Express your answer in terms of p_1, p_2, U, R, and ρ.

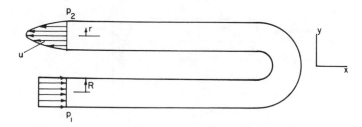

p_2

r

u

y

x

R

p_1

Fig. P11.10

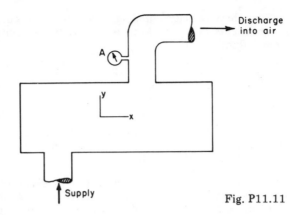

Fig. P11.11

11.11. Water is supplied to the closed tank of Fig. P11.11 through a horizontal pipe at the rate of 6.28 cfs. The fluid is discharged through a 6-in. pipe and has its direction changed through a horizontal 90° bend. The pressure gage at point A indicates a pressure of 8 psi. Determine (a) the magnitude and sense of the force in the x direction required to hold the system in equilibrium and (b) the force exerted by the fluid on the pipe bend in the y direction.

Ans. (a) 390 lb →, (b) 616 lb↑

11.12. Determine the vertical component of the force the jet exerts on the vane in Example 11.4.

11.13. A horizontal jet of water moving with a velocity of 30 m/s impinges on a vertical plate (Fig. P11.13). If the plate moves with a constant velocity of 5 m/s, determine the horizontal force the jet exerts on the plate.

Ans. 4910 N →

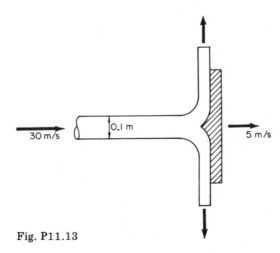

Fig. P11.13

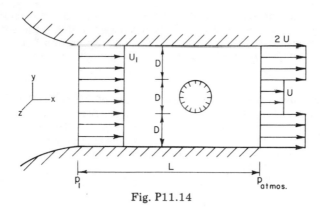

Fig. P11.14

11.14. An incompressible fluid flows steadily through a horizontal two-dimensional channel past a fixed circular cylinder and discharges into the atmosphere. At the entrance to the channel the velocity distribution is uniform and the gage pressure is p_1. At the exit the velocity distribution can be approximated as shown in Fig. P11.14. The average shearing stress acting along the channel wall is τ_0. Determine an expression for the drag on the cylinder in terms of p_1, U_1, D, L, τ_0, and the fluid density ρ.

11.15. For the flow system shown in Fig. P11.14 and described in Problem 11.14 determine an expression for the rate at which heat must flow through the channel walls (over the length L), assuming that (a) the cylinder is perfectly insulated, (b) the internal energy of the fluid remains constant, (c) there is no heat flow through the fluid at the entrance or exit of the channel, and (d) there is no heat flow in the z direction. Express your answer in terms of p_1, U_1, D, and ρ.

Ans. $\dot{Q}_h = 0.336\rho DU_1^3 - 3DU_1p_1$

11.16. A viscous Newtonian fluid is contained between two wide, horizontal, parallel plates (Fig. P11.16). The bottom plate is fixed and the upper plate moves with a constant velocity U_0. There is no pressure variation in the direction of flow, and the velocity in the x direction varies linearly with y. The other velocity components (v, w) are zero. If the internal energy of the fluid remains constant, determine an expression for the rate at which heat is transferred through the walls of the channel in a length L (per unit width). There is no heat flow in the x or z directions.

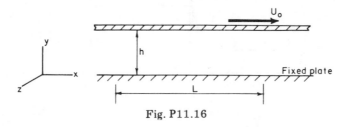

Fig. P11.16

11.17. Determine the efficiency of a pump motor unit that discharges 4 cfs of water from a free surface 30 ft below the ground into the bottom of an open tank with a fluid surface 30 ft above the ground. The head loss in the system is 5 ft of water and 40 hp is supplied to the pump motor unit.

Ans. 73.8%

11.18. Water flows through a turbine at the rate of 0.8 m^3/s (Fig. P11.18) and discharges through a draft tube into a large reservoir. The pressure in the 0.5-m diameter pipe preceding the turbine is 35×10^4 N/m^2. Determine the output power (in watts) of the turbine assuming no losses occur and the kinetic energy at the exit of the draft tube is negligible.

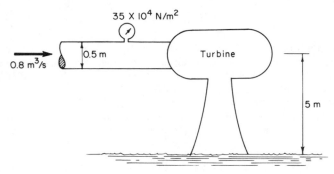

Fig. P11.18

11.19. Consider an abrupt horizontal expansion in a pipe (Fig. P11.19). It has been determined experimentally that the pressure p_1 over the entire section 1 is constant and equal to the pressure in the smaller pipe. Further downstream at section 2 the pressure is p_2, and the steady mean velocity is V_2. The flow in the smaller pipe and at section 2 may be assumed to be one dimensional. Make use of the integral momentum and energy equations to determine an expression for the energy loss in the expansion per unit of mass flowing. Neglect wall shearing stresses. Express your answer in terms of V_2 and the diameter ratio D_2/D_1.

Ans. $(V_2^2/2)[(D_2/D_1)^2 - 1]^2$ loss per unit mass

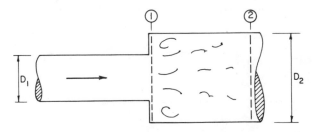

Fig. P11.19

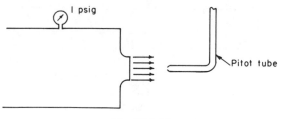

Fig. P11.20

11.20. Air (density = 0.0024 slugs/ft³) flows out of a large tank in which the pressure is 1 psig through a well-rounded opening (Fig. P11.20). Assume the air is incompressible and nonviscous. (a) Determine the velocity of the jet leaving the tank. (b) If a Pitot tube were placed in the jet (Fig. P11.20), what would be the pressure at its tip?

11.21. A nonviscous incompressible fluid ($\gamma = 50$ lb/ft³) flows through the horizontal converging circular tube of Fig. P11.21. A mercury U-tube manometer is connected across the tube as shown. If the discharge through the tube is 0.5 cfs, what is the differential reading Δh on the manometer?

Ans. 0.0947 ft

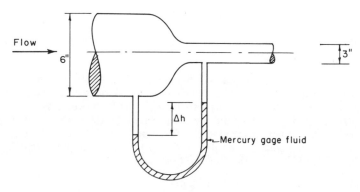

Fig. P11.21

11.22. During the calibration of a $6'' \times 4''$ (inlet and throat diameters respectively) Venturi meter, 6000 lb of oil (sp. gr. = 0.85) was discharged in 32 s. The mercury U-tube manometer connected to the Venturi indicated a differential reading of 18 in. Determine the meter coefficient.

11.23. The differential reading on a U-tube mercury manometer connected between inlet and throat of a Venturi meter is 6 in. If the flowing fluid is water and the meter coefficient is 0.98, determine the discharge through the meter.

Ans. 0.445 cfs

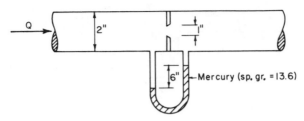

Fig. P11.24

11.24. Water at 60 F flows through a flat-plate orifice (Fig. P11.24). The differential reading on the U-tube mercury manometer is 6 in. If the meter coefficient is 0.65, what is the discharge through the pipe?

11.25. In a certain steady two-dimensional flow field the fluid may be assumed to be incompressible and nonviscous and the weight of the fluid (sp. wt. = 50 lb/ft^3) is the only body force. The x component of velocity is known to be $u = 6x$, which gives the velocity in fps when x is measured in feet. The y axis is vertical. (a) Determine the y component of velocity so that the continuity equation is satisfied. At $x = y = 0$ the velocity is zero. (b) Can the difference in pressures between the points $x = 1$, $y = 1$, and $x = 1$, $y = 4$ be determined from the Bernoulli equation? If so, determine the value in psi.

Ans. (a) $-6y$, (b) yes; 3.95 psi

11.26. Which of the following two-dimensional flow fields are irrotational? (a) $u = 2x, v = -2y$; (b) $u = 1 - y^2, v = 0$; (c) $u = 3(x^2 - y^2), v = -6xy$.

11.27. A viscous fluid is contained between two infinite horizontal parallel plates spaced 0.5 in. apart. The bottom plate is fixed and the upper plate moves with a constant velocity U_0. (a) If the velocity of the upper plate is 2 fps and the fluid has a viscosity of 0.03 lb-s/ft^2 and a specific weight of 70 pcf, determine the force per square foot of the upper plate required to maintain the velocity U_0. (b) What is the pressure differential in the fluid between the top and bottom plate?

Ans. (a) 1.440 psf, (b) 2.92 psf

11.28. A thin film of viscous liquid of constant thickness (no velocity perpendicular to plate) flows steadily down an infinite inclined plane (angle of inclination = α). Determine the relationship between the film thickness h and the discharge per unit width Q. Assume that air resistance is negligible so that the shearing stress at the free surface is zero.

11.29. Oil, having a viscosity of 10^{-4} lb-s/ft^2 and a specific weight of 50 lb/ft^3, flows steadily through a 2-in. diameter horizontal smooth pipe (Fig. P11.29). The average velocity in the pipe is 0.5 fps. Determine the differential reading Δh on the inclined-tube manometer.

Ans. 1.202×10^{-3} ft

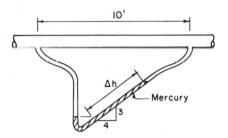

Fig. P11.29

11.30. A liquid having a specific weight of 64.4 lb/ft^3 flows steadily through a horizontal converging section of pipe (Fig. P11.30). A liquid having a viscosity of 10^{-3} slugs/ft-s and a density of 2 slugs/ft^3 is sucked into the throat of the converging section through a 0.1-in. horizontal pipe. The fluid in the large pipe can be assumed to be nonviscous; but due to the small size, viscous effects are important in the attached pipe. If the pressure at the inlet to the small pipe is 2 psig, what will be the average velocity of the fluid flowing through the small pipe when the velocity in the larger pipe is 2 fps and the pressure is 2 psig?

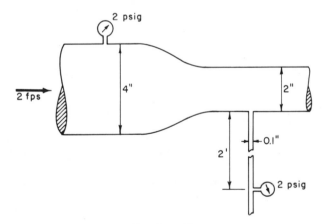

Fig. P11.30

11.31. Water at 32 F is pumped through a horizontal 0.05-m diameter smooth steel pipe at the rate of 3×10^{-3} m^3/s. Two pressure gages located 170 m apart indicate a pressure increase of 28×10^3 N/m^2. If the pump is located between the gages, what power is the pump adding to the fluid?

Ans. 346 W

11.32. In a laboratory experiment to determine the friction coefficient f in the loss equation, the following data were recorded for one run: pressure gage readings at the ends of a 50.0-ft horizontal run of 1.50-in. (inside diameter) pipe were 12 and 4 psi; the differential reading on a manometer connected to inlet and throat of a $1.50'' \times 1.00''$ Venturi meter (meter coefficient = 0.98) was 6.00 in. of mercury. The fluid in the pipe was oil (sp. gr. = 0.80). Determine the friction coefficient for this run.

11.33. Oil having a specific gravity of 0.90 and a viscosity of 0.002 slugs/ft-s is pumped at the rate of 0.8 cfs from one large open tank into another through 600 ft of 6.00-in. new steel pipe. The ends of the pipe are submerged. The surface of the oil in the receiving tank is 50 ft above that in the supply tank. Determine the power input to the pump if it is 80% efficient.

Ans. 6.24 hp

11.34. Fuel oil (sp. wt. = 53.0 pcf and absolute viscosity = 124×10^{-6} slugs/ft-s) flows from a pressurized reservoir through 2000 ft of new 6-in. steel pipe to discharge into the air at a point 40 ft above the level of the oil in the reservoir. Determine the discharge if the pressure above the oil in the reservoir is held constant at 80 psig.

chapter 12

Dimensional Analysis and Model Theory

12.1 Introduction

Many practical problems cannot be solved analytically, and in these commonly occurring instances use must be made of experimentally obtained information. In the typical problem it is desirable to know how some particular characteristic of the system is influenced by others. For example, in the study of fluid flow through a pipe one may want to know how the pressure drop is affected by the velocity of the fluid, pipe size, fluid viscosity, etc. To determine the relationship between pressure drop and velocity, tests can be run in which the velocity is varied and the corresponding pressure drop measured, while all other characteristics or variables are held constant. This process can then be repeated for each variable under consideration. Although in principle this is a satisfactory procedure, it will be difficult and time consuming to use in practice; e.g., to vary the viscosity, different fluids or at least fluids at different temperatures must be used; to vary the pipe size, different pipes must be used.

Also, for experimental data to be of the most value they must be obtained and presented in such a manner as to be applicable to systems other than the one used to acquire the data. If we think of the laboratory system being used to obtain experimental results as a model, then we wish to use these results to predict the behavior of another system (the one of actual interest), which we call the *prototype*. Thus the concept of modeling, or similitude, is intimately associated with the planning and execution of experimental programs.

In many cases the prototype is of a different size, utilizes different materials, and operates under different conditions than the model; and it is difficult to deduce prototype behavior from model data obtained by the method previously mentioned, i.e., by determining the effect of each characteristic on an individual basis. Fortunately, experimental programs can be greatly enhanced through the grouping of variables so that it is not necessary to work with each individually. This idea of working with selected groups of variables is based on the fact that each variable or characteristic of a given system can be qualitatively described by its basic dimensions (see Section 1.7) and that we must be able to group variables so that each term appearing in an equation relating the variables must have the same basic dimensions. An analysis of a given problem from this viewpoint is called a *dimensional analysis*, and the mathematical basis for this approach is described in the following section.

12.2 Dimensional Analysis

When dealing with physical phenomena, we describe them in terms of various quantities such as velocity, acceleration, density, area, etc. These so-called *secondary quantities* in turn are described and measured in terms of another set of quantities considered to be *primary quantities*. In mechanics the primary quantities or basic dimensions are normally taken to be length L; time T; and mass M. It can be shown that any secondary quantity s_i is expressible in terms of the primary quantities in the form $s_i = L^a T^b M^c X^d Y^e$ where, for example, X and Y are other basic dimensions such as temperature and electrical charge that may be required to describe the secondary quantity.[1] Common examples of secondary quantities and their basic dimensions include:

area	$A = L^2$	density	$\rho = ML^{-3}$
volume	$V = L^3$	stress	$\sigma = MT^{-2}L^{-1}$
velocity	$v = LT^{-1}$		

To obtain the basic dimensions of stress, use is made of the fact that stress is a force F divided by an area, but force and mass are related through Newton's second law of motion; i.e., $F = MLT^{-2}$. It is thus apparent that an equivalent set of basic dimensions to be used in mechanics problems is L, T, and F.

In a given problem, several variables (secondary quantities) u_1, $u_2, \ldots, u_k$ usually are required to describe the phenomenon of interest. A number of dimensionless products of these variables can be formed by combining the variables in the form $u_1^{x_1} u_2^{x_2} \cdots u_k^{x_k}$

---

1. P. W. Bridgman, *Dimensional Analysis* (New Haven: Yale Univ. Press, 1931), Ch. 2.

where the exponents $x_1, x_2, \ldots, x_k$ are selected so that the resulting product is dimensionless. Thus if we let any one of the variables, say u_i, have the basic dimensions

$$u_i = L^{a_i} T^{b_i} M^{c_i} X^{d_i} Y^{e_i}$$

we can express the product as

$$(L^{a_1} T^{b_1} M^{c_1} X^{d_1} Y^{e_1})^{x_1} (L^{a_2} T^{b_2} M^{c_2} X^{d_2} Y^{e_2})^{x_2}$$

$$\cdots (L^{a_k} T^{b_k} M^{c_k} X^{d_k} Y^{e_k})^{x_k}$$

For the product to be dimensionless, the exponents of the various basic dimensions must combine to give a zero value for each basic dimension. Thus

$$a_1 x_1 + a_2 x_2 + \cdots + a_k x_k = 0$$

$$b_1 x_1 + b_2 x_2 + \cdots + b_k x_k = 0$$

$$c_1 x_1 + c_2 x_2 + \cdots + c_k x_k = 0 \tag{12.1}$$

$$d_1 x_1 + d_2 x_2 + \cdots + d_k x_k = 0$$

$$e_1 x_1 + e_2 x_2 + \cdots + e_k x_k = 0$$

There will be as many equations as basic dimensions, say m in number, and k unknown x's where k is equal to the number of original variables in the problem. From the theory of equations it is known there are $k - r$ linearly independent solutions to Eqs. (12.1) where r is the rank of the matrix of coefficients

$$\begin{bmatrix} a_1 & a_2 & \cdots & a_k \\ b_1 & b_2 & \cdots & b_k \\ c_1 & b_2 & \cdots & c_k \\ d_1 & d_2 & \cdots & d_k \\ e_1 & e_2 & \cdots & e_k \end{bmatrix}$$

This matrix is commonly called the *dimensional matrix*. Since the rank of a matrix is the highest order nonzero determinant contained in it the rank cannot exceed the number of equations, but may be smaller. Thus the number of independent dimensionless products that can be formed is equal to the number of original variables k minus the rank of the coefficient matrix. Such a set of dimensionless products is called a *complete set*. Once a complete set of dimensionless products is found, all other possible dimensionless combinations can be formed as multiples of powers of the products contained in the complete set.

An essential postulate of dimensional analysis is that the form of any functional relationship between a given set of variables does not depend on the system of units used; i.e., the functional relationship is dimensionally homogeneous. If this condition of homogeneity is utilized, it can be proved that a functional relationship between a given set of variables can be reduced to a relationship among a complete set of dimensionless products of these variables.[2] This result is known as the *Buckingham pi theorem* and can be stated concisely as follows: if an equation involving k variables is dimensionally homogeneous, it can be reduced to a relationship among $k - r$ independent dimensionless products, where r is the rank of the dimensional matrix.

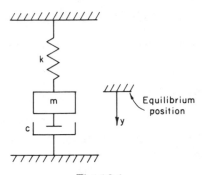

Fig. 12.1

To illustrate the use of the Buckingham pi theorem, we will apply it to a simple vibration problem involving the spring-mass dashpot system of Fig. 12.1. The problem is to determine a functional relationship between the displacement y and the time t by means of dimensional analysis.

The first step in the analysis is to list the variables and their dimensions as follows:

y = displacement L 　　　　　　　　　v_0 = initial velocity LT^{-1}

m = mass M 　　　　　　　　　　　　y_0 = initial displacement L

　c = damping coefficient MT^{-1} 　　　　t = time T

　k = spring constant MT^{-2}

We must list all variables that will influence y. We now form the product $y^{x_1} m^{x_2} c^{x_3} k^{x_4} v_0^{x_5} y_0^{x_6} t^{x_7}$, and with the substitution of the basic dimensions for each variable we obtain

2. H. L. Langhaar, *Dimensional Analysis and Theory of Models* (New Haven: Wiley, 1951), Ch. 3.

$$(L)^{x_1} (M)^{x_2} (MT^{-1})^{x_3} (MT^{-2})^{x_4} (LT^{-1})^{x_5} (L)^{x_6} (T)^{x_7}$$

with the corresponding set of equations

$$L:\ x_1 + 0 + 0 + 0 + x_5 + x_6 + 0 = 0$$

$$T:\ 0 + 0 - x_3 - 2x_4 - x_5 + 0 + x_7 = 0 \qquad (12.2)$$

$$M:\ 0 + x_2 + x_3 + x_4 + 0 + 0 + 0 = 0$$

The dimensional matrix is

	y	m	c	k	v_0	y_0	t
L	1	0	0	0	1	1	0
T	0	0	-1	-2	-1	0	1
M	0	1	1	1	0	0	0

Consider now the determinant on the left side of the matrix:

$$\begin{vmatrix} 1 & 0 & 0 \\ 0 & 0 & -1 \\ 0 & 1 & 1 \end{vmatrix} = 1$$

Since this is nonzero, it follows that the rank of the dimensional matrix is three; thus there are four dimensionless products required to describe this problem.

To find a suitable set of dimensionless products, commonly called *pi terms*, we assign values to four of the x's in Eq. (12.2) and solve for the remaining three. For example, let $x_4 = 1$, $x_5 = 0$, $x_6 = 0$, and $x_7 = 0$. The only restriction here is that the determinant of the remaining coefficients must be nonzero so that we can solve for the remaining x's. In this case we have previously shown that this determinant is nonzero. With $x_4 = 1$, $x_5 = 0$, $x_6 = 0$, and $x_7 = 0$, it follows that Eqs. (12.2) are satisfied if $x_1 = 0$, $x_2 = 1$, and $x_3 = -2$. Thus one dimensionless product is

$$\pi_1 = y^0 m^1 c^{-2} k^1 v_0^0 y_0^0 t^0 = km/c^2$$

Now let $x_4 = 0$, $x_5 = 1$, $x_6 = 0$, and $x_7 = 0$, and we find that $\pi_2 = v_0 m/yc$. This process can be continued with $x_4 = 0$, $x_5 = 0$, $x_6 = 1$, and $x_7 = 0$ and with $x_4 = 0$, $x_5 = 0$, $x_6 = 0$, and $x_7 = 1$ to give $\pi_3 = y_0/y$ and $\pi_4 = ct/m$.

The specific form of the pi terms depends on which of the x's are assigned values and on the values themselves. However, once an independent set is determined, all other possible independent sets can be formed as products of powers of the original set. Using this procedure, we can form various combinations to arrive at what we consider to be the most useful set. In the present example the obvious disadvantage of the pi terms developed is that the displacement appears in two of the pi terms. It is usually convenient to have the variable of primary interest appearing in only one pi term. We can form a new set by letting

$$\pi'_1 = (\pi_3)^{-1} = y/y_0, \qquad \pi'_2 = (\pi_1)^{-1/2} = c/\sqrt{km}$$

$$\pi'_3 = (\pi_1)^{1/2}\pi_4 = t\sqrt{k/m}, \qquad \pi'_4 = \pi_2/[(\pi_1)^{1/2}\pi_3] = (v_0/y_0)\sqrt{m/k}$$

Thus we conclude that a functional relationship between y and t can be expressed in the form

$$y/y_0 = f\,[c/\sqrt{km},\, t\sqrt{k/m}, (v_0/y_0)\sqrt{m/k}] \qquad (12.3)$$

Dimensional analysis does not give us the complete answer to the problem, since the form of the function is unknown and generally cannot be obtained from a dimensional analysis alone, but it always simplifies the problem by reducing the number of variables to be considered.

In essence, dimensional analysis allows us to study a problem described by the functional relationship

$$u_1 = \phi(u_2, u_3, u_4, \ldots, u_k) \qquad (12.4)$$

in terms of a set of dimensionless quantities

$$\pi_1 = f(\pi_2, \pi_3, \ldots, \pi_{k-r}) \qquad (12.5)$$

An obvious advantage is the reduction in the number of variables (from k to $k - r$) to be controlled in an experiment. In addition, it is usually much easier to control the dimensionless products than the original variables.

In most problems the dimensionless products can be obtained by inspection since the required number can easily be determined and the variables can simply be combined into dimensionless groups. An independent set can be assured if each dimensionless product contains one variable not contained in any other product.

EXAMPLE 12.1

The velocity v of small-amplitude waves on the surface of a liquid is assumed to be a function of g, the acceleration of gravity; h, the depth of fluid; ρ, the fluid density; and σ, the surface tension FL^{-1}. Use the formal procedure (not by inspection) to determine a suitable set of pi terms, one of which should be $v/\sqrt{gh}$.

SOLUTION

The first step in performing a dimensional analysis is to list the variables and their basic dimensions. For this example we have:

$$v = LT^{-1} \qquad \rho = ML^{-3} = FT^2 L^{-4}$$

$$g = LT^{-2} \qquad \sigma = FL^{-1}$$

$$h = L$$

We are using F, L, and T as basic dimensions, and the density ρ has been expressed in terms of these dimensions, rather than ML^{-3}, through Newton's second law. We could also work with M, L, and T if we so desired.

We next form the product $v^{x_1} g^{x_2} h^{x_3} \rho^{x_4} \sigma^{x_5}$, and in terms of the basic dimensions we can write

$$(LT^{-1})^{x_1} (LT^{-2})^{x_2} (L)^{x_3} (FT^2 L^{-4})^{x_4} (FL^{-1})^{x_5}$$

For this to be a dimensionless product, the following equations must be satisfied:

$$L: \; x_1 + x_2 + x_3 - 4x_4 - x_5 = 0$$

$$T: \; -x_1 - 2x_2 + 0 + 2x_4 + 0 = 0$$

$$F: \; 0 + 0 + 0 + x_4 + x_5 = 0$$

The dimensional matrix is

	v	g	h	ρ	σ
L	1	1	1	-4	-1
T	-1	-2	0	2	0
F	0	0	0	1	1

Consider the determinant on the right side of the matrix; i.e.,

$$\begin{vmatrix} 1 & -4 & -1 \\ 0 & 2 & 0 \\ 0 & 1 & 1 \end{vmatrix} = 2$$

Since this is nonzero, the rank of the matrix is 3 and the pi theorem tells us that the five variables can be expressed in terms of two pi terms. To find these, we arbitrarily assign values to two of the x's, say $x_1 = 1$ and $x_4 = 0$, and solve for x_2, x_3, x_5. For this case

$$x_2 = -1/2, \qquad x_3 = -1/2, \qquad x_5 = 0$$

and one pi term is $\pi_1 = v/\sqrt{gh}$. To find the second pi term we set $x_1 = 0$ and

$x_4 = 1$ and obtain $\pi_2 = gh^2 \rho/\sigma$. Thus this type of wave phenomenon can be studied in terms of the relationship

$$v/\sqrt{gh} = f(gh^2 \rho/\sigma) \qquad\qquad Ans.$$

EXAMPLE 12.2

The buoyant force acting on a body submerged in a fluid is a function of the specific weight of the fluid and the volume of the body. Prove by means of dimensional analysis that the buoyant force is directly proportional to the specific weight.

SOLUTION

The variables are:

B = buoyant force F, γ = specific weight FL^{-3}, V = volume L^3

and we note there are three variables and two basic dimensions so that only one pi term is required. This can be readily obtained as $\pi_1 = B/\gamma V$. The only way we can form a functional relationship with one pi term is to set it equal to a constant, thus $B/\gamma V = K$ or

$$B = K\gamma V \qquad\qquad Ans.$$

and we see that the buoyant force is directly proportional to the specific weight.

In those special instances when the phenomena can be described in terms of a single pi term, the explicit form of the relationship among the variables can be deduced from dimensional analysis alone. However, the value of the constant must be obtained by experiment.

EXAMPLE 12.3

An expression for the drag on a smooth rigid sphere held in a stream of viscous incompressible fluid is to be developed by dimensional analysis (Fig. E12.3a). The fluid is considered to be unbounded and moves with a constant velocity U at a great distance from the sphere.

SOLUTION

The variables assumed to apply are as follows:

R = drag F ρ = mass density of fluid, $FT^2 L^{-4}$

d = sphere diameter L μ = viscosity of fluid, FTL^{-2}

U = velocity of fluid LT^{-1}

Since there are five variables and three independent basic dimensions, two pi terms are required. We can form these by inspection since the only requirements are: (a) there must be two pi terms, (b) they must be dimensionless, and

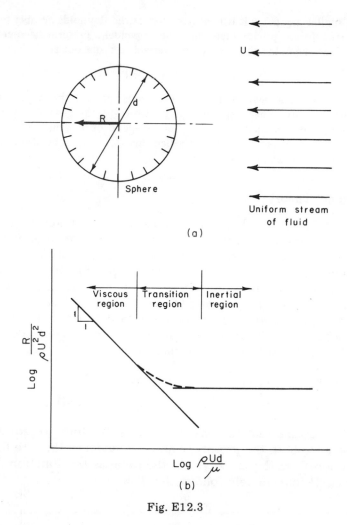

Fig. E12.3

(c) they must be independent. We let $\pi_1 = R/\rho U^2 d^2$ and $\pi_2 = \rho U d/\mu$. It can readily be checked that these are dimensionless and independent since π_1 contains R and π_2 does not; i.e., you cannot form π_2 from π_1. The general expression for the drag can therefore be written as

$$R/\rho U^2 d^2 = f(\rho U d/\mu) \qquad\qquad Ans.$$

The pi term in parenthesis is a form of Reynolds number. As discussed in Section 12.5, the physical significance of this parameter is that it is an index of the ratio of inertia forces in the fluid to the viscous forces. For a highly viscous fluid moving slowly past the sphere, it is expected that inertial forces will be small in comparison with viscous forces; suggesting that for this case $R/\rho U^2 d^2 = $ constant $\times (\rho U d/\mu)^{-1}$ i.e., the variable ρ drops out of the problem.

By a similar argument it is expected that as the Reynolds number becomes large, inertial forces predominate and the dependent pi term becomes independent of the Reynolds number; i.e., the variable μ drops out and

$$R/\rho U^2 d^2 = \text{constant}$$

The two constants appearing in these equations are not the same and their values must be determined experimentally. However, the general shape of the curve relating the dependent pi term to the independent pi term can be deduced as in Fig. E12.3b. Experiments have verified that this is the general shape of the $R/\rho U^2 d^2$ versus Reynolds number curve.

12.3 Model Theory

As mentioned in Section 12.1, measurements taken on one system are commonly utilized to predict the behavior of another; the former is called the model and the latter the prototype. To make predictions between the model and prototype, it is first necessary to establish a set of relationships between the two systems. These relationships are commonly referred to as *modeling laws*, *scaling laws*, or *similarity requirements* and can be developed from Eq. (12.5) in the following manner. Assume we have two systems, the prototype and the model, each described by the equations

$$\pi_1 = f(\pi_2, \pi_3, \ldots, \pi_{k-r}) \quad \text{(prototype)}$$

$$\pi_{1m} = f_m(\pi_{2m}, \pi_{3m}, \ldots, \pi_{(k-r)m}) \quad \text{(model)}$$

We further assume that the phenomenon with which we are dealing is the same for both the prototype and the model so that the form of the function f for the prototype is the same as the function f_m for the model. It immediately follows that if we let

$$\pi_2 = \pi_{2m}$$

$$\pi_3 = \pi_{3m}$$

$$\begin{matrix} \cdot \\ \cdot \\ \cdot \end{matrix} \qquad\qquad (12.6)$$

$$\pi_{k-r} = \pi_{(k-r)m}$$

then

$$\pi_1 = \pi_{1m} \qquad\qquad (12.7)$$

and thus these equations provide us with the necessary similarity requirements. Equations (12.6) specify the required relationships

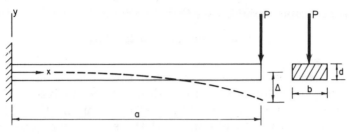

Fig. 12.2

between prototype and model so that we predict π_1 from a measured π_{1m} taken on the model. These equations represent the *model design conditions* and Eq. (12.7) the *prediction equation*. To illustrate the method more clearly, the following simple example is given.

Let it be required to establish the similarity relationships for predicting the end deflection of a cantilever beam of rectangular cross section (Fig. 12.2) due to a concentrated load P. It is assumed that the deformation is small, the material behaves elastically, shearing deflections are negligible, and the beam is loaded in a plane of symmetry so there is no twist. With these conditions the following variables are applicable:

Δ = end deflection L d = depth of beam L

a = length of beam L P = load F

b = width of beam L E = modulus of elasticity FL^{-2}

Application of the pi theorem reveals that since there are six variables expressible in terms of two basic dimensions, four dimensionless parameters are required to describe this problem. One possible set is

$$\Delta/a = f(b/a,\ d/a,\ P/Ea^2)$$ (12.8)

It now follows that if

$$b/a = b_m/a_m,\qquad d/a = d_m/a_m,\qquad P/Ea^2 = P_m/E_m a_m^2$$ (12.9)

for two systems then

$$\Delta/a = \Delta_m/a_m$$ (12.10)

Equation (12.9) yields the model design conditions for this problem, and Eq. (12.10) is the prediction equation between the model and prototype. Since there are two basic dimensions in this problem, two *scales* can be arbitrarily selected; e.g., let $a/a_m = n_a$, where n_a is the *length scale*, and $P/P_m = n_P$, where n_P is the *force scale*. The scales for all other variables are then fixed; i.e.,

$$\Delta/\Delta_m = b/b_m = d/d_m = n_a$$

and $E/E_m = n_P n_a^{-2}$.

The foregoing example reveals that three basic steps are used in establishing modeling laws from a dimensional analysis. These are (1) the selection of the variables, (2) the application of the pi theorem, and (3) the development of the similarity requirements by equating pi terms.

Although in principle this procedure is straightforward and relatively simple, two major difficulties are frequently encountered. The first is in the selection of the pertinent variables. It is clear that a good understanding of the problem must be achieved to ascertain the pertinent variables. The selection of variables is usually based on the experience of the investigator and a knowledge of the fundamental equations that govern the phenomenon. This does not imply that a detailed mathematical model of the system must first be established, but simply that certain fundamental laws such as Newton's laws of motion are known to be applicable to the system. Common errors encountered at this stage are the inclusion of nonindependent variables and the omission of pertinent variables or parameters (such as the acceleration of gravity) on the basis that they are constant. The inclusion of a group of variables that are not independent (such as the beam cross-sectional area, width, and depth in the cantilever beam example) is clearly unnecessary. In addition, all pertinent independent variables or parameters must be included, and the fact that they may or may not be constant is of no consequence at this stage in the analysis. Omissions or the listing of unnecessary variables will normally not be detected without the aid of experiments.

The second difficulty that frequently arises is in the control of the pi terms. Each pi term yields a requirement between the model and prototype system. In certain instances it is difficult if not impossible to satisfy one or more of these requirements. The classical example of this difficulty is in fluid flow problems in which both the Reynolds number and the Froude number (defined below) are important. Similarity requirements arising from these two dimensionless parameters are

$$\rho U \lambda / \mu = \rho_m U_m \lambda_m / \mu_m \ \ldots \text{(Reynolds number)}$$

and

$$U^2 / g\lambda = U_m^2 / g_m \lambda_m \ \ldots \text{(Froude number)}$$

where U is a velocity, λ is a length, μ is the viscosity of the fluid, and

g the acceleration of gravity. If the same fluid is used in both model and prototype and both systems operate in the same gravitational field, it follows that

$$U/U_m = \lambda_m /\lambda = n_\lambda^{-1} \qquad (12.11)$$

from the Reynolds number condition, and

$$U/U_m = \sqrt{\lambda/\lambda_m} = \sqrt{n_\lambda} \qquad (12.12)$$

from the Froude number condition.

Equations (12.11) and (12.12) give different values for the velocity scale. A conflict therefore exists, and one of these design conditions cannot be satisfied. A model for which at least one of the similarity requirements is not satisfied is said to be *distorted*. Numerous problems could be cited to show that distorted models are not rare exceptions but may frequently occur. A further discussion of distorted models can be found in the books by Murphy[3] and Langhaar.[4]

EXAMPLE 12.4

A simply supported timber beam ($E = 1.8 \times 10^6$ psi) carries a distributed static load over half its span. The load varies linearly from zero at the left support to k (lb/ft) at the center (Fig. E12.4). The beam is 6 in. wide, 3 in. deep, and 10 ft long. A model is to be constructed from a 0.5-in. wide aluminum strip ($E = 10.3 \times 10^6$ psi). Strips of various depths are available. Determine the design conditions and the equation for predicting the deflection at midspan. Indicate the size of the model and give a description of the model loading. The weight of the beam and shearing deflections may be neglected.

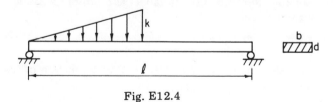

Fig. E12.4

SOLUTION

The deflection y_0 of the beam at midspan can be expressed as

$$y_0 = \phi(\ell, b, d, E, k)$$

---

3. G. Murphy, *Similitude in Engineering* (New York: Ronald Press, 1950).
4. H. L. Langhaar, *Dimensional Analysis and Theory of Models* (New York: Wiley, 1951).

and a dimensional analysis yields

$$y_0/\ell = f(b/\ell, d/\ell, k/E\ell)$$

The design conditions are therefore

$$b/\ell = b_m/\ell_m, \quad d/\ell = d_m/\ell_m, \quad k/E\ell = k_m/E_m\ell_m \qquad Ans.$$

and the prediction equation is

$$y_0/\ell = y_{0m}/\ell_m \qquad Ans.$$

Since $b = 6$ in. and $b_m = 0.5$ in., the length scale $\ell/\ell_m = 12$, and it immediately follows that

$$d_m = d/12 = 0.25 \text{ in.} \qquad Ans.$$

The loading condition is obtained from the last design condition; i.e.,

$$k_m = \frac{E_m}{E}\frac{\ell_m}{\ell}k = \frac{(10.3 \times 10^6)}{(1.8 \times 10^6)}\frac{1}{12}k = 0.476k \qquad Ans.$$

Thus the model loading must be a distributed load over half the span of the beam, varying linearly from zero to $0.476k$ (lb/ft) at midspan.

EXAMPLE 12.5

A solid sphere having a specific weight γ_s is immersed in a liquid of specific weight $\gamma_f(\gamma_f > \gamma_s)$ and then released. A model system having a length scale of 4 is to be used to determine the maximum height above the liquid surface that the sphere will rise upon release from a given depth. The significant properties of the liquid are known to be density, specific weight, and viscosity. Establish the design conditions and prediction equation for the model system and determine whether the same liquid can be used in model and prototype systems. Neglect the effect of air resistance.

SOLUTION

The height h that the sphere will attain will be a function of the following variables:

h = height L $\qquad\qquad\qquad$ ρ_f = density of fluid ML^{-3}

d = sphere diameter L $\qquad\qquad$ g = acceleration of gravity LT^{-2}

a = initial depth of submergence L $\quad$ μ = fluid viscosity $ML^{-1}T^{-1}$

ρ_s = density of sphere ML^{-3}

The inclusion of the three variables ρ_s, ρ_f, and g implies that both mass densities and weights are important since the specific weight γ of a material is related to its density through the relationships $\gamma = \rho g$.

In functional form we can express h as

$$h = \phi(d, a, \rho_s, \rho_f, g, \mu)$$

and a dimensional analysis yields

$$h/d = f(a/d, \rho_s/\rho_f, \mu/d^{3/2} \rho_f \sqrt{g})$$

The design conditions are

$$a/d = a_m/d_m, \qquad \rho_s/\rho_f = \rho_{sm}/\rho_{fm}, \qquad \mu/d^{3/2} \rho_f \sqrt{g} = \mu_m/d_m^{3/2} \rho_{fm} \sqrt{g_m} \qquad Ans.$$

and the prediction equation is

$$h/d = h_m/d_m \qquad\qquad Ans.$$

We note from the last design conditions that if the same fluid is used for both model and prototype ($\mu = \mu_m$, $\rho_f = \rho_{fm}$) and the acceleration of gravity is the same in both systems, then $d/d_m = 1$. However, since it is specified in the problem that the length scale is 4, it follows that the same fluid cannot be used in the model and prototype. To satisfy the last design condition, a model fluid must be selected such that $\mu_m/\rho_{fm} = (4)^{-3/2} (\mu/\rho_f)$.

12.4 Structural Modeling

The use of structural models for predicting not only deflections or displacements but also stresses and strains is widespread. In this section we will consider from a general viewpoint the prediction equation and the model design conditions for static elastic problems. For these problems we assume that the material obeys Hooke's law and can be described by Young's modulus E and Poisson's ratio ν. In addition, any stress component σ at some point x_i will be a function of the geometry of the system as characterized by some length ℓ and other required lengths λ_i. The subscript i will be used to designate a set of variables. Thus λ_i is equivalent to a set of lengths $\lambda_1, \lambda_2, \lambda_3, \ldots$.

The loading may be specified with the loads P and P_i and any prescribed boundary displacements by η_i. The stress can therefore be expressed in the functional form

$$\sigma = \phi(x_i, \ell, \lambda_i, P, P_i, \eta_i, E, \nu) \qquad (12.13)$$

We now apply dimensional analysis to this set of variables to obtain

$$\sigma \ell^2/P = f(x_i/\ell, \lambda_i/\ell, \eta_i/\ell, P/E\ell^2, P_i/P, \nu) \qquad (12.14)$$

Similarity requirements are obtained by making the pi terms on the right side of Eq. (12.14) equal between model and prototype. Equality of the first three pi terms, x_i/ℓ, λ_i/ℓ, and η_i/ℓ, means that we must maintain geometric similarity between model and prototype not only with regard to shape but also with respect to prescribed displacements and coordinates.

The loading scale is established from the relationship

$$P/E\ell^2 = P_m/E_m \ell_m^2$$

or

$$P/P_m = (E/E_m)(\ell^2/\ell_m^2) \qquad (12.15)$$

Note that the model and prototype materials need not be the same, but the elastic moduli scale E/E_m and the length scale ℓ/ℓ_m fix the loading scale. All additional loads P_i must be in the same ratio; i.e.,

$$P_i/P_{im} = P/P_m \qquad (12.16)$$

The last pi term in Eq. (12.14) imposes the rather stringent similarity requirement that Poisson's ratio must be equal for model and prototype materials. If the prototype and model are constructed of the same material, this condition is satisfied. For plane-strain or plane-stress problems involving simply connected bodies (for which the body forces are zero, constant, or vary linearly with position) the stress distribution is known to be independent of Poisson's ratio.[5] Similar problems involving multiply connected bodies containing holes can also be modeled without regard to Poisson's ratio if the resultant force acting on the boundary of the hole is zero. However, if these conditions are not met and different materials are used, the Poisson's ratio condition will not generally be satisfied, and for this case a judgment must be made with respect to the significance of Poisson's ratio for the specific problem under consideration.

If all the aforementioned design conditions are satisfied, it follows that

$$\sigma\ell^2/P = \sigma_m \ell_m^2/P_m$$

or

$$\sigma/\sigma_m = (P/P_m)(\ell_m^2/\ell^2) = E/E_m \qquad (12.17)$$

Since any displacement component u or strain component ϵ will be a function of the same variables given in Eq. (12.14), the same model-design conditions are required for displacements and strains as for stresses. The corresponding prediction equations become

$$u/\ell = u_m/\ell_m$$

5. J. W. Dally and W. F. Riley, *Experimental Stress Analysis* (New York: McGraw-Hill, 1965), p. 247.

or

$$u/u_m = \ell/\ell_m \qquad (12.18)$$

and

$$\epsilon = \epsilon_m \qquad (12.19)$$

i.e., the displacements scale as the length scale, whereas the strains are equal in model and prototype. These scaling laws for elastic structures are valid for both small and large deformations as long as the material in both model and prototype obeys Hooke's law. Other types of loads, e.g., line, surface, and volume loads, can be readily incorporated into the analyses.

12.5 Fluid-Flow Modeling

Due to the general complexity of fluid-flow phenomena, experiments and models are frequently used to solve specific problems. In Table 12.1 the variables most often found in the study of flow phenomena are listed along with the several dimensionless groups that can be formed from these variables. These dimensionless groups appear so often that special names have been assigned to them (Table 12.1). We encountered the Reynolds number in the discussion of energy losses and turbulence in Chapter 11. In any problem in which velocity, geometry, density, and viscosity are important, the

Table 12.1. Common fluid flow variables and dimensionless groups

Variables	Dimensionless groups	Name	Significance
Pressure difference Δp Velocity U	$\Delta p/\rho U^2$	Euler number	$\dfrac{\text{pressure force}}{\text{inertia force}}$
Length ℓ Acceleration of gravity g	$U^2/g\ell$	Froude number	$\dfrac{\text{inertia force}}{\text{gravitational force}}$
Viscosity μ	$\rho\ell U/\mu$	Reynolds number	$\dfrac{\text{inertia force}}{\text{viscous force}}$
Density ρ Surface tension σ	$\rho\ell U^2/\sigma$	Weber number	$\dfrac{\text{inertia force}}{\text{surface tension force}}$
Speed of sound c	U/c	Mach number	$\dfrac{\text{inertia force}}{\text{compression force}}$

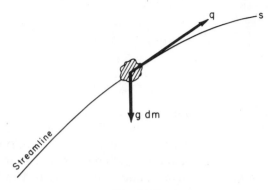

Fig. 12.3

Reynolds number can be utilized as an appropriate dimensionless group.

As indicated in the last column of Table 12.1, each of the dimensionless terms can be interpreted as an index of the ratio of two forces. For example, the Froude number is an index of the ratio of the force due to the acceleration of a fluid particle to the force due to gravity (weight). Consider a fluid particle moving along a streamline (Fig. 12.3). The magnitude of the component of inertia force F_I along the streamline can be expressed as $F_I = a_s \, dm$, where a_s is the magnitude of the acceleration along the streamline of the particle having a mass dm. From our study of particle motion along a curved path we know that

$$a_s = \frac{dq}{dt} = q \, \frac{dq}{ds}$$

where s is measured along the streamline.

If we write the velocity and length s in dimensionless form; i.e.,

$$q* = q/U, \qquad s* = s/\ell$$

where U and ℓ represent some characteristic velocity and length respectively, then

$$a_s = \frac{U^2}{\ell} \, q* \, \frac{dq*}{ds*}$$

and

$$F_I = \frac{U^2}{\ell} \, q* \, \frac{dq*}{ds*} \, dm$$

The magnitude of the weight of the particle F_g is $F_g = g \, dm$, so the ratio of the inertia to the gravitational force is

$$\frac{F_I}{F_g} = \frac{U^2}{g\ell} \, q* \, \frac{dq*}{ds*} \qquad (12.20)$$

Thus the force ratio F_I/F_g is proportional to $U^2/g\ell$, which is the Froude number. In a similar manner the other force ratios can be developed. The Weber number is important in problems involving the interface between two fluids where surface tension may become a significant force. If fluid compressibility is important, the Mach number becomes a significant factor. For example, in high-speed gas flows large changes in density may occur and the force associated with this compression must be taken into account. To illustrate how some of these numbers arise naturally in the analyses of fluid-flow phenomena, three specific problems will be considered.

Drag on Bodies—In Example 12.3 an expression for the drag on a sphere immersed in a fluid was developed. We can generalize this result for bodies of any shape. Assume that the geometry of the body can be described by a series of lengths λ_i in addition to some characteristic length ℓ. For incompressible steady flow of a Newtonian fluid the drag R is expressible as

$$R = \phi(U, \ell, \lambda_i, \rho, \mu) \qquad (12.21)$$

where U is the approach velocity and ρ and μ are fluid density and viscosity respectively. Application of the pi theorem yields

$$R/\rho U^2 \ell^2 = f(\lambda_i/\ell, \rho\ell U/\mu) \qquad (12.22)$$

Note that for a sphere the only geometric variable required was the diameter, and in this case Eq. (12.22) reduces to the result obtained in Example 12.3.

It is common practice to express drag in the form

$$R = C_D A (\rho U^2/2) \qquad (12.23)$$

where C_D is called the *drag coefficient* and A is some appropriate cross-sectional area. A comparison of Eq. (12.23) with (12.22) shows that the drag coefficient is a function of the geometry and the Reynolds number. In most instances the drag coefficient must be obtained experimentally, and some typical data are shown in Fig. 12.4. In this particular problem the Froude, Weber, or Mach numbers do not appear since neither gravity nor surface tension is important (no fluid interfaces), and we have assumed the fluid to be incompressible. The parameter $R/\rho U^2 \ell^2$ can be thought of as a form of Euler number since the drag could be expressed as some average pressure differen-

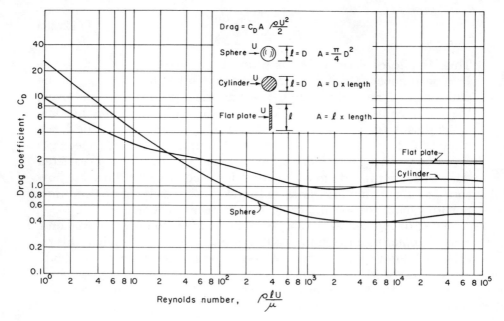

Fig. 12.4

tial times an area; i.e.,

$$\frac{R}{\rho U^2 \ell^2} = \frac{(\Delta p_a)\ell^2}{\rho U^2 \ell^2} = \frac{\Delta p_a}{\rho U^2}$$

For geometrically similar bodies

$$(\lambda_i/\ell)_{\text{body 1}} = (\lambda_i/\ell)_{\text{body 2}}$$

The drag coeffcient will be the same if the Reynolds number is the same for the two bodies; i.e.,

$$(\rho \ell U/\mu)_{\text{body 1}} = (\rho \ell U/\mu)_{\text{body 2}}$$

Pressure Drop in a Pipe—Consider an incompressible fluid flowing steadily through a straight horizontal circular pipe. Assume that the pressure drop per unit length $\Delta p/\ell$ is a function of the mean velocity U, the pipe diameter D, the fluid viscosity μ, the fluid density ρ, and some average height r which is an index of the pipe-wall roughness. Thus in functional form

$$\frac{\Delta p}{\ell} = \phi(U, D, \mu, \rho, r) \tag{12.24}$$

From a dimensional analysis we can obtain

$$\frac{D\,\Delta p}{\ell\rho U^2} = F\left(\frac{r}{D},\ \frac{\rho DU}{\mu}\right)$$ (12.25)

or

$$\frac{\Delta p}{\rho U^2} = \frac{\ell}{D}\ F\left(\frac{r}{D},\ \frac{\rho DU}{\mu}\right)$$ (12.26)

Thus the pressure drop, expressed as the Euler number, is a function only of the Reynolds number once the geometric factors are fixed. Equation (12.26) can also be written in the form

$$\frac{\Delta p}{\rho g} = f\left(\frac{r}{D},\ \frac{\rho DU}{\mu}\right)\frac{\ell}{D}\frac{U^2}{2g}$$ (12.27)

and we recognize Eq. (12.27) as the head-loss equation given in Section 11.8 where $f = 2F$ and is called the friction factor. The friction factor is a function of the pipe roughness; the Reynolds number and typical experimental data are given in Fig. 11.17.

Open Channel Flow—As a final example of fluid-flow phenomena we consider the steady flow of a liquid through a rectangular channel that is open to the atmosphere. A common method for measuring the rate of flow through an open channel is to place a flat, sharp-edged plate called a *weir* across it (Fig. 12.5). We wish to develop a suitable

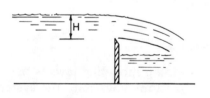
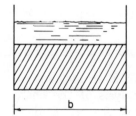

Fig. 12.5

expression for the average velocity U of the fluid flowing over the weir crest. Since the flow in an open channel is due to gravity, g will certainly be an important variable. Also, the fluid density ρ, viscosity μ, and surface tension σ will possibly be important. Thus we can write

$$U = \phi(g, H, \mu, \rho, \sigma)$$ (12.28)

or

$$U^2/gH = f(\rho HU/\mu,\ \rho HU^2/\sigma)$$ (12.29)

Thus we have the Froude number as a function of both the Reynolds number and the Weber number. In general it would be difficult to model this phenomena since if the same fluid is used in both model and prototype and $g = g_m$, different velocity scales are predicted from the Froude, Reynolds, or Weber number similarity requirements. See Eqs. (12.11) and (12.12). This is a common difficulty encountered in open-channel modeling.

For this particular example we might assume that the influence of viscous and surface tension forces is small so that

$$U^2/gH = C \quad \text{(constant)} \tag{12.30}$$

We can define U in terms of the discharge Q through the equation $Q = UbH$ so Eq. (12.30) can be written as

$$Q = \sqrt{Cg}bH^{3/2} \tag{12.31}$$

Experiments have shown that an equation of the form

$$Q = 3.33bH^{3/2} \tag{12.32}$$

(with b and H in feet) predicts satisfactorily the flow (in cfs) over a sharp-crested rectangular weir where the coefficient 3.33 is empirically determined. A measurement of the head H can therefore be used to predict the discharge flowing through the open channel. If the weir does not extend completely across the channel or if the head is very small so that viscous and surface tension forces become important, Eq. 12.32 is no longer valid.

12.6 Characteristic Equation Method

An alternate method for determining modeling laws is based on an analysis of the characteristic equations governing the system of interest. In performing a dimensional analysis, only the variables influencing the phenomenon are required; the specific form of the equation(s) relating the variables need not be specified. However, if the equations describing the problem (usually differential equations based on a mathematical model) are known, use can be made of this information to develop the necessary similarity relationships. The method based on the characteristic equations essentially consists of rewriting the equations in dimensionless form and determining from the transformed equations the conditions under which the behavior of two systems will be similar. This method is illustrated with the following simple example of the spring-mass system previously studied by means of dimensional analysis.

Consider the spring-mass-dashpot system of Fig. 12.1. As before, the problem is to determine the displacement y as a function of time

t by means of a model study. It is well known that the displacement of the mass is described by the differential equation

$$m \frac{d^2 y}{dt^2} + c \frac{dy}{dt} + ky = 0 \qquad (12.33)$$

along with the initial conditions

$$y = y_0 \text{ and } \frac{dy}{dt} = v_0$$

at $t = 0$. If we now introduce two dimensionless parameters $y^* = y/y_0$ and $t^* = t/\tau$ where $\tau = \sqrt{m/k}$, Equation (12.33) can be written as

$$\frac{d^2 y^*}{dt^{*2}} + \frac{c}{\sqrt{mk}} \frac{dy^*}{dt^*} + y^* = 0 \qquad (12.34)$$

The initial conditions become $y^* = 1$, and

$$\frac{dy^*}{dt^*} = \frac{v_0}{y_0} \sqrt{m/k}$$

at $t^* = 0$. From a consideration of Eq. (12.34) it is seen that for any two systems governed by an equation of this form, the solution for y^* will be the same; i.e.,

$$y^* = y_m^* \qquad (12.35)$$

if

$$c/\sqrt{mk} = c_m/\sqrt{m_m k_m}$$
$$(v_0/y_0)\sqrt{m/k} = (v_{0m}/y_{0m})\sqrt{m_m/k_m} \qquad (12.36)$$
$$t\sqrt{k/m} = t_m\sqrt{k_m/m_m}$$

where the subscript m refers to the model. The last condition specifies the time scale for the problem. The similarity requirements are the same as would be obtained from Eq. (12.3), which is based on the method of dimensional analysis.

If the characteristic equation(s) are known for the system, the procedure described in this example can be followed to establish the necessary relationships between the prototype and model. However, in many problems the characteristic equations are more numerous and complicated than in the example given or may even be unknown, and this method cannot be readily used. For these more complicated problems the method based on dimensional analysis appears to be more suitable.

PROBLEMS

12.1. (a) Distinguish between units and dimensions. (b) Indicate four important applications of dimensional analysis. (c) Is it ever possible to make quantitative predictions solely on the basis of dimensional analysis? Explain.

12.2. (a) Can erroneous variables be detected through dimensional analysis? Explain. (b) If in a given problem it is found that the number of variables is equal to or less than the number of basic dimensions, what conclusion can be drawn? (c) Is it correct to refer to the "scale" of a model? Explain.

12.3. The natural frequency of vibration of an elastic wire stretched between two supports is a function of the distance between supports ℓ, the wire diameter d, the modulus of elasticity E, and the density of the wire ρ. Determine a suitable set of pi terms, using the formal procedure (not by inspection) for investigating the effect of these variables on the frequency.

Ans. $\omega d \sqrt{\rho/E} = f(\ell/d)$ where ω is the frequency

12.4. A solid sphere of mass M and diameter d is projected vertically upward with an initial velocity V_0 in air. It is desired to predict the maximum height h to which the sphere will rise. Assume that the only significant characteristic of the air is the air density ρ. Using the formal procedure (not by inspection), develop a suitable set of pi terms for predicting h.

12.5. The velocity of sound in a gas is dependent upon the pressure and the density. With the aid of dimensional analysis, prove that the speed of sound is proportional to the square root of the pressure.

12.6. The force P due to fluid drag acting on a solid sphere settling with a constant velocity V in a cylindrical tank is to be determined experimentally. It is assumed that the force P is a function of the velocity V, the fluid viscosity $\mu (FTL^{-2})$, fluid density ρ, sphere diameter d, and cylinder diameter D. Prove by means of dimensional analysis that the expression for P can be written as $P = C_D \rho d^2 V^2$ where the drag coefficient C_D is expressed as $C_D = f(\rho V d/\mu, d/D)$.

12.7. A simply supported beam carries a uniformly distributed static load k (lb/ft) over its entire length. Develop a set of pi terms that could be used for obtaining a general equation for predicting the maximum deflection of the beam. Assume that the weight of the beam is negligible in comparison with the load, that the pertinent cross-sectional variable is the moment of inertia, and that the beam is operated in the elastic range. Use the formal procedure for developing the pi terms.

Ans. $y_0/\ell = f(I/\ell^4, E\ell/k)$ where y_0 = deflection, I = moment of inertia, ℓ = beam length, E = modulus of elasticity

12.8. It has been observed that as a fluid such as air flows past a wire, a distinct tone is emitted. An investigator desiring to study the frequency of this tone considered the following variables: frequency ω, wire diameter d, and fluid velocity v. In addition he felt that some fluid property was needed, but he was unable to decide between absolute viscosity FTL^{-2} or kinematic viscosity $L^2 T^{-1}$. (a) Which viscosity would you choose as the pertinent variable? Why? (b) Develop an equation for ω. (c) If the phenomenon were actually independent of fluid properties, what type of curve would be obtained from the experimental data if it were obtained in the most efficient manner?

12.9.   The following data were obtained experimentally in a study of the phenomenon described in Example 12.1.

v	(fps)	1.27	1.79	2.20	2.54
h	(ft)	0.05	0.10	0.15	0.20

For all tests $\rho = 1.94$ slugs/ft^3, $\sigma = 0.00498$ lb/ft, $g = 32.2$ fps^2. Determine the general equation for the velocity v.

Ans. $v = 1.00 \sqrt{gh}$

12.10. It is known that for a thin-walled cylindrical pressure vessel the stress acting in the direction of the longitudinal axis is a function of the cylinder radius r, the pressure p, and the wall thickness t. Experiments performed with a series of cylinders having radii which vary from 10 in. to 50 in., but all with a constant wall thickness of 1/2 in., indicate that the stress in psi when plotted versus the radius r in in. follows the equation $\sigma = 300r$. The tests were all run at an internal pressure of 300 psi. With the aid of dimensional analysis, predict the longitudinal stress for a cylinder having a radius of 60 in., a wall thickness of 1 in., and operated under a pressure of 400 psi.

12.11. The pressure drop between two points in a horizontal constant-diameter pipe containing a slowly moving viscous fluid is a function of the following variables:

Δp = pressure drop μ = viscosity (0.024 lb-s/ft)
d = pipe diameter (1 in.) v = mean velocity (1 fps)
ℓ = distance between pressure taps

Experimentally a plot of Δp versus ℓ was obtained holding the other variables constant at the values indicated. The data plotted as a straight line on log paper with a positive slope of 1 and for $\ell = 1$ ft, $\Delta p = 111$ psf. The length ℓ was varied from 0.5 ft to 50 ft. Predict the pressure drop in a 2-in. diameter pipe over a length of 10 ft for a fluid of the same viscosity but flowing at the rate of 3 fps.

Ans.   832 psf

12.12. A test of a true model of the spring-mass system of Fig. P12.12 indicates that the natural frequency of vibration of the model is 5 cps. The scales between the model and prototype are $\ell/\ell_m = 10$, $k/k_m = 9$, and

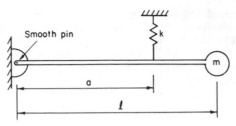

Fig. P12.12

$m/m_m = 4$. Predict the natural frequency of the prototype system. The mass of the supporting bar is negligible.

12.13. A horizontal simply supported timber ($E = 10.0 \times 10^9 \, N/m^2$) beam 3 m long with a square cross section 5 cm wide and 5 cm deep carries a vertical concentrated load. The flexural stresses in the beam are to be studied by means of a steel ($E = 20.0 \times 10^{10} \, N/m^2$) model. A length scale of 10 is to be used. Determine the relationship between stresses in model and prototype.

Ans. $\sigma = 0.05\sigma_m$

12.14. A simply supported beam supports a concentrated mass at its midpoint. The mass is connected to the beam by means of a coil spring ($k = 20$ lb/in.). (a) Develop the design conditions and prediction equation for a model to predict the deflection of the beam from its equilibrium position when the system is set in motion by displacing the concentrated mass from its equilibrium position and then releasing it. Note that the weight of the system need not be considered. (b) If the model beam is to be constructed of the same material as that of the prototype and the time scale is to be 2, determine the length scale and the required spring constant for the model.

12.15. A general equation for predicting the discharge of liquid from a large reservoir over a triangular, sharp-edged weir (see Fig. P12.15) is to be determined. It can be assumed that both surface tension and viscous effects are negligible and the head remains constant. Data from previous tests indicate that the discharge Q, in cfs, is given by the equation $Q = 2.54 \tan \theta$, $30° \leqslant \theta \leqslant 60°$, when the head H is 1 ft. Can a general equation for the discharge be developed from this information? If so,

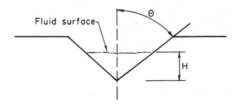

Fig. P12.15

determine the general equation and indicate any restrictions or limitations. If not, indicate any additional tests that must be performed.

Ans. $Q/\sqrt{gH^5} = 0.447 \tan \theta$,

where g is the acceleration of gravity, $30° \leqslant \theta \leqslant 60°$

12.16. If the stresses in an elastic structure are due entirely to its weight, show that for all geometrically similar structures of the same material the stresses in the largest will be the greatest.

12.17. A large motor is mounted on four rubber mounts, each having an effective spring constant k. Due to unbalance in the shaft of the motor, a vertical shaking force $F = F_0 \cos \omega t$ is developed, where F_0 is the magnitude of this force, ω is the angular frequency at which the motor is running, and t is time. (a) Establish a suitable set of design conditions and the prediction equation for predicting the vertical displacement of the motor at any instant. (b) If the time scale between the model and prototype is to be unity, at what speed should the model be operated?

Ans. (b) $\omega = \omega_m$

12.18. A model of an open channel is to be designed on the basis of the Froude criteria. If the model is not distorted geometrically, determine (a) the discharge scale and the time scale and (b) the kinematic surface tension σ/ρ scale if surface tension is assumed to be a significant variable.

12.19. It is desired to predict the drag on a high-speed airplane by testing a scale model (length scale = 50) in a variable density wind tunnel. The prototype operates in standard air ($\rho = 0.002378$ slugs/ft^3; $t = 59$ F) and the temperature of the air in the wind tunnel remains constant at 59 F. The model is to be tested at the same Reynolds number and Mach number as the prototype. Determine (a) the required air density in the tunnel and (b) the drag on the prototype corresponding to a 1-lb force on the model. Assume that the viscosity and speed of sound are functions only of temperature.

Ans. (a) 0.1189 slugs/ft^3,   (b) 50 lb

12.20. An important consideration in the design of certain farm implements is the value of the force P required to move a cutting tool at a given angle of inclination through a soil at a constant velocity V (Fig. P12.20). It is desired to study this problem by means of a model. (a) Establish a suitable set of design conditions and the prediction equation for predicting

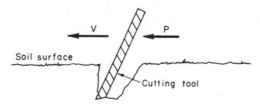

Fig. P12.20

the force P. Assume that both gravitational and inertial effects are important and, in addition, that the pertinent soil characteristic is its shearing strength FL^{-2}. (b) Can the same soil be used in both model and prototype systems? Explain.

12.21. An incompressible fluid flows through a horizontal pipe of constant diameter. A perforated plate is placed across the pipe and it is found experimentally that in a geometrically similar model of this system the pressure drop across the plate is 3×10^3 N/m^2 for a velocity of 3 m/s in the model. If the same fluid is used in model and prototype and the length scale is 5, determine (a) the required corresponding velocity in the prototype and (b) the corresponding pressure drop.

Ans.   (a) 0.600 m/s,   (b) 120.0 N/m^2

12.22. A viscous fluid flows slowly over a rectangular weir. For the flow conditions of interest viscous, gravitational, and surface tension forces are all important. A model having a length scale of 4 is to be constructed to predict the mean velocity of the fluid at the weir crest. A special device is to be constructed so that the effective gravitational field of the model can be changed. The model fluid has the same density and viscosity as the prototype fluid, but its surface tension can be varied (by means of a chemical additive) between the values of σ and 4σ, where σ is the prototype surface tension. If possible the model is to be undistorted. Establish the required value of the model acceleration of gravity g_m and surface tension σ_m.

APPENDIX

APPENDIX

Conversion Tables for Units

Table A1. Length

Multiply number of → by ↘ to get ↓	Inches (in.)	Feet (ft)	Centimetres (cm)	Metres (m)
Inches (in.)	1	12	3.937×10^{-1}	3.937×10
Feet (ft)	8.333×10^{-2}	1	3.281×10^{-2}	3.281
Centimetres (cm)	2.540	3.048×10	1	10^2
Metres (m)	2.540×10^{-2}	3.048×10^{-1}	10^{-2}	1

Table A2. Area

Multiply number of → by ↘ to get ↓	Sq. inches $(in.^2)$	Sq. feet (ft^2)	Sq. centimetres (cm^2)	Sq. metres (m^2)
Sq. inches $(in.^2)$	1	1.440×10^2	1.550×10^{-1}	1.550×10^3
Sq. feet (ft^2)	6.944×10^{-3}	1	1.076×10^{-3}	1.076×10
Sq. centimetres (cm^2)	6.452	9.290×10^2	1	10^4
Sq. metres (m^2)	6.452×10^{-4}	9.290×10^{-2}	10^{-4}	1

Table A3. Volume

Multiply number of → by ↘ to get ↓	Cu. inches $(in.^3)$	Cu. feet (ft^3)	Cu. centimetres (cm^3)	Cu. metres (m^3)
Cu. inches $(in.^3)$	1	1.728×10^3	6.102×10^{-2}	6.102×10^4
Cu. feet (ft^3)	5.787×10^{-4}	1	3.531×10^{-5}	3.531×10
Cu. centimetres (cm^3)	1.639×10	2.832×10^4	1	10^6
Cu. metres (m^3)	1.639×10^{-5}	2.832×10^{-2}	10^{-6}	1

Table A4. Force

Multiply number of ⟶ by ↘ to get ↓	Pounds (lb)	Dynes	Newtons (N)
Pounds (lb)	1	2.248×10^{-6}	2.248×10^{-1}
Dynes	4.448×10^{5}	1	10^{5}
Newtons (N)	4.448	10^{-5}	1

Table A5. Mass

Multiply number of ⟶ by ↘ to get ↓	Slugs	Grams (gm)	Kilograms (kg)
Slugs	1	6.852×10^{-5}	6.852×10^{-2}
Grams (gm)	1.459×10^{4}	1	10^{3}
Kilograms (kg)	1.459×10	10^{-3}	1

Index